Building Materials
Product Emission and
Combustion Health Hazards

Building Materials
Product Emission and Combustion Health Hazards

Kathleen Hess-Kosa

CRC Press
Taylor & Francis Group
Boca Raton London New York

CRC Press is an imprint of the
Taylor & Francis Group, an **informa** business

CRC Press
Taylor & Francis Group
6000 Broken Sound Parkway NW, Suite 300
Boca Raton, FL 33487-2742

First issued in paperback 2019

ISBN-13: 978-0-4987-1493-8 (hbk)
ISBN-13: 978-0-367-87248-9 (pbk)

Library of Congress Cataloging-in-Publication Data

Names: Hess-Kosa, Kathleen.
Title: Building materials : product emission and combustion health hazards/ Kathleen Hess-Kosa.
Description: Boca Raton : Taylor & Francis, CRC Press, 2017. | Includes bibliographical references.
Identifiers: LCCN 2016036689| ISBN 9781498714938 (hardback : alk. paper) | ISBN 9781315371269 (ebook)
Subjects: LCSH: Buildings--Health aspects. | Materials--Toxicology. | Hazardous substances. | Flammable materials.
Classification: LCC RA566.6 .H47 2017 | DDC 363.17--dc23
LC record available at https://lccn.loc.gov/2016036689

Visit the Taylor & Francis Web site at
http://www.taylorandfrancis.com

and the CRC Press Web site at
http://www.crcpress.com

Contents

Section II   Polymers in Construction

Section III   Building Materials by Function

Preface

The twenty-first century presents old challenges to managing old building materials and new challenges to managing a world gone wild with chemistry. The more we endeavor to go forward, the further behind we seem to find ourselves.

The purpose of this book is to provide a broad spectrum of building material health hazard information—not easily attained—to environmental professionals, construction management firms, architects, and first responders. Building material composition is in a constant state of flux. An effort has been made herein to present generic information, not detailed information regarding each and every building product. The intent is to provide direction and guidance.

Section I, "All Things Considered" is an introduction—the rationale and background—for all that follows within the other sections. Building materials are introduced—both natural and synthetic. Health hazards associated with older buildings being renovated and newer buildings under construction are briefly discussed. Product emissions are historically the cause of indoor air quality health concerns. Chemistry has gone wild! Safety data sheets can't keep up. The green movement responds. The efficacy of these different approaches appears to be costly and/or impractical.

Section II, "Polymers in Construction" is a discussion of the twentieth century world of plastics—millions of formulations with new formulations being added every day. Polymers commonly used in building materials are introduced and discussed at length as are plasticizers and additives which are not chemically bound within the polymers. Many of the plasticizers and additives are not regulated by OSHA, yet they may cause eye, skin, and respiratory irritation. As most modern building products are comprised of plastics, an understanding of polymers is extremely important when discussing building materials.

Section III, "Building Materials by Function" is a discussion of building materials as they relate to function and their product emissions/combustion products. Function and trade may be considered synonymous as the chapters progress from foundation to finish-out. This approach has been chosen because some similar materials are used in different applications and present different exposure concerns, and when planning and choosing building materials, each trade bears its own responsibility for assessing building materials. Not only are the product emissions/combustion products discussed, but some of the material and design defects are discussed as are some of the materials that could leach out into the environment and/or drinking water. In Chapter 10, there is an extensive discussion of thermal decomposition and combustion products as they relate to fire first responders.

The information contained within this book has been the result of considerable research. Yet, the challenge was exciting. There are many surprises! Hopefully, you too will find it captivating.

Acknowledgments

I wish to dedicate this book to my husband, Mike Kosa, who suffered through my late night and weekend trials-and-tribulations in researching and writing this publication. Without the benefit of other similar books and the constant change in technical information, I had turned into a reclusive troll. It is a wonder that those dearest to me are still standing—unscathed by my ramblings. And I would also like to thank Irma Shagla Britton, Senior Editor, Environmental and Engineering, CRC Press/Taylor & Francis Group. She has been a delight to work with and an angel of patience during all the turmoil I have thrown her way.

Author

Kathleen Hess-Kosa, MS, CIH, is an industrial hygiene/environmental consultant with Omega Environmental Consulting in Canyon Lake, Texas. In 1972, she earned her Bachelor of Science in microbiology with a minor in chemistry from Oklahoma State University. In 1979, she earned a Master of Science in industrial hygiene from the College of Engineering at Texas A&M University. After her travails through the education system, she finally entered the real world and became challenged with all problems great and small within an insurance company.

While working for Firemen's Fund Insurance Companies, she had the opportunity to become involved in a variety of unique industrial hygiene assessments including her first indoor air quality study which involved a newly constructed 800-occupant high rise office building. This was back in 1982, and the 90% complaint rate dropped to 10%—after throwing everything but the kitchen sink at the problem. Her success sparked that which has become her passion for indoor air quality. Subsequently, she has pursued multiple avenues to effectively respond to the new challenges.

Upon starting her own consulting firm in 1986, Ms. Hess-Kosa has continued to pursue her passion, added to her list of curiosities mold/moisture damage, sewage contamination, and all things out of the ordinary. She has taken several courses in construction techniques at the local community college. She has designed and been involved in construction projects involving two of her own homes. One of her hobbies it that of building massive southwest wood doors. She has also been a consultant to several construction firms.

Beyond this book, Ms. Hess-Kosa is credited with several other technical books. They include *Environmental Sampling for Unknowns, Indoor Air Quality: Sampling Methodologies* (2nd edition), *Environmental Site Assessment Phase I: A Basic Guide* (2nd edition), *An Environmental Health and Safety Auditing Made Easy, A Checklist Approach for Industry, Indoor Air Quality: The Latest Sampling and Analytical Methods.*

Ms. Hess-Kosa is an active outdoor enthusiast and a world traveler, photographer, and avid reader of adventure novels. Life is full of challenges!

Section I

All Things Considered

1

Building Material Components: Old and New, Natural and Synthetic

In 2013, Denis Hayes, an environmental activist commented: "That new building smell—that's the smell of poisons!" The statement reflects the "dangers of the vast and largely undisclosed stew of chemicals hiding in the building materials of modern homes, schools, hospitals and offices—indoor spaces in which Americans now spend an estimated 90 percent of their time" (Peeples 2014). Many of today's building materials are a cauldron of surprises! Yet, those of the past are not without their demons as well.

We will first revisit past toxic building materials in order to manage the toxic dilemma as a whole—not only because they were used in the past, but they are still used today. Although they do not emit gases, their impact on society has been and will continue to be too great to ignore.

Asbestos was first used in building materials in the late 1800s. Subsequently, stories of asbestos exposures, suffering and loss, financial ruin, and long-drawn-out court cases abound even to this day. Remediation of old asbestos-containing material is ongoing. It seems to be an endless journey into the abyss. Yet, just as we think we may gain some modicum of control, we falter and reflect. Thought by many to be a building material of the past, asbestos-containing building materials are still being manufactured worldwide.

> Despite irrefutable scientific evidence calling out the dangers of asbestos, 2 million tons of asbestos are exported every year to the developing world, where it's often handled with little to no regulation (Vice 2016).

Introduced during the rise of the Roman Empire, lead was used too extensively to line their aqueducts and for potable water pipes. It has since been used in a multiple building materials—the most notorious of which are lead-based paint and lead water supply lines.

> The state of Maryland tested and found more than 65,000 children in the city with dangerously high blood-lead levels from 1993 to 2013. Across the United States, more than half a million kids are poisoned by lead each year, and the majority come from cities like Baltimore: rust belt towns built up during the first half of the 20th century when leaded paint was dominant. As populations and employment opportunities shrank in recent decades, poverty and neglect combined with older housing allowed lead paint poisoning to plague the city (Barry-Jester 2015).

> While a harsh national spotlight focuses on the drinking water crisis in Flint, Michigan, a USA Today Network investigation has identified almost 2,000 additional water systems spanning all 50 states where testing has shown excessive levels of lead contamination over the past four years. The water systems, which reported lead levels exceeding EPA standards, collectively supply water to 6 million people. About 350 of those systems provide drinking water to schools or day care centers. The investigation also found at least 180 of the water systems failed to notify consumers about the high lead levels to the community (Young and Nichols 2016).

Into the twentieth century, as mankind makes greater strides in polymer technology, formaldehyde polymer resins, particularly urea formaldehyde, has been used with greater frequency in building materials. Formaldehyde emissions resulted in irritation of the eyes, nose, and throat—symptoms that are typically associated with indoor air quality. Thus, formaldehyde has become a "modern day toxin."

> In the 2006, the Federal Emergency Management Agency (FEMA) provided travel trailers, recreational park trailers and manufactured homes for habitation by displaced victims of Hurricane Katrina and Rita. Some of the people who moved into the FEMA trailers complained of breathing difficulties, nosebleeds, and persistent headaches. Formaldehyde emissions from building materials were blamed.

To be discussed herein are the following:

- Natural building products—asbestos and lead
- Plastics' notorious formaldehyde
- Modern building materials—plastics, organic solvents, inorganic chemicals, wood preservatives, and more

---

Natural Building Products

Prior to the twentieth century, buildings were constructed by the ingenious manipulation of Earth's natural resources. Materials were extracted from surrounding resources such as the soil (e.g., clay and mud), surface, quarried, and mined stone (e.g., granite and limestone), metals extracted from rock (e.g., copper, iron, and lead), plants (e.g., bamboo, paints, and dyes), trees (e.g., lumber), animals (e.g., hides), and seepage from the bowels of the Earth (e.g., asphalt). Nature serves up her share of toxins that man has used in building materials from ancient times until today. However, many of the more toxic natural materials are not likely to contribute to poor indoor air quality due to emissions. Yet, natural toxins in building materials when disturbed can serve up a nasty stew. Exposures to workers and building occupants are preventable. Take heed. Awareness is the charge of the day!

Some of nature's more toxic substances used in building materials are not emission products. They are, however, worthy of discussion. Some of the more noteworthy of these include asbestos, lead, and crystalline silica.

Asbestos

Asbestos has been used throughout history as a fireproof material and ultimately a thermal insulator until the EPA partial ban on the use of friable asbestos-containing building materials in 1978. The first recorded use of asbestos was in the first century by the Roman Emperor Charlemagne. He was said to have had a tablecloth which had supernatural powers. When he hosted dinner parties, with people eating and drinking off the tablecloth. Then, at the end of the evening, when the table was cleared, he would pick up the cloth and throw it into the fire. When it failed to burn his guests were amazed.

Later, the miracle fiber was used in clothing and jewelry. Then, in the later part of the nineteenth century, asbestos was used in roofing materials and sprayed-on acoustical insulation. Henceforth, its use in building materials exploded. Into the twentieth century, there was an asbestos boom. Some of the materials in which asbestos was used were hot-water-pipe insulation cement board, floor tiles, caulk, and asphalt—a whole host of building materials.

Asbestos-containing building materials can be a double-edged sword. The benefits of asbestos products can be lifesaving while asbestos fiber releases can be devastating. One case in point is the "Fall of the Twin Towers."

It has been argued that asbestos fireproofing on steel girders and beams of the World Trade Centers would have saved lives (Milloy 2001).

> It is estimated that the World Trade Center contained approximately 400 tons of asbestos fiber at the time of the attacks—much, if not all of which, was also released into the ambient air with the smoke plume, collapsing towers, and subsequent fires.
>
> In 1971, New York City banned the use of asbestos in spray fireproofing. At that time, asbestos insulating material had only been sprayed up to the 64th floor of the World Trade Center towers.
>
> In September 2001, two hijacked airliners crashed into Floors 96 to 103 of One World Trade Center and Floors 87 to 93 of Two World Trade Center—floors that had not been sprayed with asbestos. Instead lasting up to four hours before melting as would have been the case if the steel had been sprayed with asbestos, the steel girders of lasted 1 hour and 40 minutes in Tower One and 56 minutes in Tower Two before collapsing.

On the other hand, when the Twin Towers collapsed, a massive cloud of smoke, dust, and debris released these hazardous asbestos fibers and other toxic substances into the air. This may have contributed to health effects of volunteers and workers at Ground Zero.

> The World Trade Center Health Registry estimates about 410,000 people were exposed to a host of toxins including asbestos during the rescue, recovery and clean-up efforts that followed 9/11 … . Nearly 70 percent of recovery personnel have suffered from lung problems, including a condition that was later coined "World Trade Center Cough" (Asbestos 2015).

When disturbed, asbestos fibers are released. Fiber release and occupant exposures to airborne asbestos can be financially devastating to a building owner and/or contractor. Public opinion and reputation are at stake as well. Therefore, during the planning stages of renovation and demolition projects, an asbestos assessment and remediation must be performed. Most suggest an assessment be performed on all buildings built prior to 1978, but asbestos-containing building materials were still manufactured and used beyond this year. See Table 1.1.

TABLE 1.1

Asbestos-Containing Materials Usage Ban

Year	Asbestos-Containing Material Usage Ban
1973	EPA banned spray-applied surfacing asbestos-containing material for fireproofing/insulating purposes.[a]
1975	EPA banned installation of asbestos pipe insulation and asbestos block insulation on facility components, such as boilers and hot-water tanks, if the materials are either preformed (molded) and friable or wet applied and friable after drying.[a]
1978	EPA banned spray-applied surfacing materials for purposes not already banned.[a]
1977	The CPSC banned the use of asbestos in artificial fireplace embers and wall-patching compounds.[b]
1989	EPA issued a final rule banning most asbestos-containing products.[c,d]
1991	The EPA 1989 rule was vacated and remanded by the Fifth Circuit Court of Appeals. As a result, most of the original ban on the manufacture, importation, processing, or distribution in commerce.
1999	For the majority of the asbestos-containing products originally covered in the 1989 final rule was overturned.[d]

[a] See National Emission Standards for Hazardous Air Pollutants (NESHAP) at 40 CFR Part 61, Subpart M.
[b] See 16 CFR Part 1305 and 16 CFR 1304.
[c] See 40 CFR 763, Subpart I.
[d] EPA. 2015. U.S. Federal bans on asbestos. *EPA*. December 24. Accessed March 24, 2016. https://www.epa.gov/asbestos/us-federal-bans-asbestos.

TABLE 1.2

Asbestos-Containing Materials Not Banned

Cement-corrugated sheet
Cement flat sheet
Clothing
Pipeline wrap
Roofing felt
Vinyl floor tile
Cement shingle
Millboard
Cement pipe
Automatic transmission components
Clutch facings
Friction materials
Disk brake pads
Drum brake linings
Brake blocks
Gaskets
Non-roofing coatings
Roof coatings

Source: Adapted from EPA. 2015. U.S. Federal bans on asbestos. *EPA.* December 24. Accessed March 24, 2016. https://www.epa.gov/asbestos/us-federal-bans-asbestos.

It should furthermore be noted that there was no ban, either implied or stated, on a whole host of asbestos-containing products. See Table 1.2. Although most suppliers and consumers voluntarily chose avoidance, asbestos-containing materials might still be unknowingly sold and purchased. Therefore, in new construction, it is advisable that the purchaser seek asbestos-free affirmation regarding building materials.

Lead

Lead, first used in ancient Roman water delivery systems, is soft, malleable, durable, and waterproof. Prior to the twentieth century, lead was used in plumbing, roofing, roof flashing, and vent pipe covers. Lead roofs have actually withstood the test of time. The roof of St. Paul's Cathedral in London, built in the seventeenth century, after the Great Fire of London, was made of lead, and it remains to this day, a testament to durability of lead. Lead roof flashing is still used to this day. You might ask, "What about lead-based paint?"

Prior to the twentieth century, corrosion-resistant, durable lead-based paint was in great demand in the United States and internationally, and it became more readily available through mass production in 1880. Shortly after its wide usage became common practice, in 1887, U.S. medical authorities began to observe lead poisoning in children.* At the time, they were, however, unable to identify the source. In 1904, childhood lead poisoning was linked to lead-based paints. In 1909, several European countries banned the use of interior lead-based paints, and the League of Nations, not inclusive of the United States, banned the paint in 1922. In 1943, a report concluded that children eating lead paint chips could suffer from neurological disorders. In 1955, public health officials encouraged a voluntary national standard to prohibit the use of lead pigments in interior residential paints. Finally, with the passage of the Lead-Based Paint Poisoning Prevention Act of 1971, interior usage of lead-based house paint began to be phased out in the United States. In 1978, the U.S. Consumer Product Safety Commission (CPSC) banned the use of interior and exterior lead-based paint in residential housing.

Homes built prior to 1978 may have lead-based paint either inside or outside, and homes and apartments built prior to 1950 are very likely to have lead-based paint both inside and outside. The paint is most commonly found on exterior surfaces, interior wood, doors, and windows.

Any active disturbances of lead-based paint in a residence and other child-occupied facility—built prior to 1978—are subject to the EPA Lead-Based Paint Renovation, Repair, and Painting Rule.† Active disturbance of lead-based painted materials includes remodeling, repair, maintenance, electrical work, painting, plumbing, carpentry, window replacement, and other renovation projects as well as demolition. By extension, a "child-occupied facility" is generally inclusive of office, commercial, and institutional buildings. All projects involving the disturbance of lead-based painted building materials should be properly assessed and remediated in accordance with federal, state, and local requirements before proceeding.

Unless lead-based paint is spray applied to surfaces, lead and lead-containing building materials are not an emission concern. Lead and lead-containing building materials may, however, pose an environmental problem. Examples are lead roof rainwater runoff collection systems and lead water pipes contributing to lead in the drinking water!

Crystalline Silica

Crystalline silica, the crystalline form of silica (i.e., SiO_2), is one of the most abundant minerals in the Earth's crust. The most common formation is quartz which is a component of sand. Sand was and still is used in concrete,

---

* Lead poisoning causes a host of problems, many worse in children, including decreased intelligence, impaired neurobehavioral development, stunted physical growth, hearing impairment, and kidney problems.

† EPA's Lead Renovation, Repair, and Painting Rule requires that firms performing renovation, repair, and painting projects that disturb lead-based paint in homes, child care facilities, and preschools built before 1978 have their firm certified by EPA (or an EPA-authorized state), use certified renovators who are trained by EPA-approved training providers, and follow lead-safe work practices.

brick, cement, and mortar. Today, fine sand is used as filler for paints, plastics, and rubber.

Fine crystalline silica dust is hazardous when inhaled. The dust particles can penetrate deep into the lungs and cause disabling and sometimes fatal lung diseases, including silicosis and lung cancer, as well as kidney disease. Occupational exposures to respirable crystalline silica occur when cutting, sawing, drilling, crushing, and grinding of concrete, brick, ceramic tiles, rock, and stone products (e.g., quartz countertops).

Prior to the twenty-first century, exposure concerns in construction were minimal. Yet, according to a recent OSHA study, about 2-million construction workers are exposed to respirable crystalline silica in over 600,000 workplaces. OSHA estimates that about 800,000 of the 2,000,000 workers are exposed to excessive levels of silica as proposed in the Crystalline Silica Rule. The rule was finalized to be effective on June 23, 2016. In the Construction Standard (which is different from the General Industry and Maritime Standards), compliance is required within a year of the effective date—June 23, 2017.

Although not an emissions concern, fine quartz and sand dust particles do pose a worker an exposure problem. This is a friendly wake-up call. As not all building materials that contain crystalline silica are known or understood, the presence of silica in a building material is mentioned by a product.

Non-Emission Toxins

Although not within the scope of this book, non-emission toxic building materials may impact the environment during renovation and demolition projects. Beyond that, they may pose an airborne or waterborne exposure threat to workers and/or building occupants. Forewarned is forearmed!

Additionally, asbestos and lead-based paint may be encountered in new building materials. Not all asbestos-containing building materials have been banned, and asbestos products are still being made and sold worldwide. Lead-based paint has not been banned for non-child-related buildings. Buyer beware!

Crystalline silica dust and welding fumes from torch cutting and welding on lead-based paint (typically white or red) or lead–chromate paint (typically chrome yellow) that has been applied to beams and structural steel pose an OSHA-regulated occupation exposure risk. Crystalline silica is in multiple building materials—old and new. Painted structural steel should be tested prior to welding.

Plastics' Most Notorious

The twentieth century begot the Age of Plastics. Innovation and technology brought in synthetic building materials that could mimic, replace, and preserve many of our valuable natural resources. It does not however come without a price!

While natural products are generally not emission products, plastics and other petrochemical products serve up a medley of irritant/toxic emissions. Yes, plastics are possible only through the world of chemistry. Yet, there is a trade-off.

Plastics are the least-known, least-understood building material. In 2016, it was reported that over 297.5 million tons of plastics were sold worldwide in 2015 (PRWEB 2016). Construction products own a large piece of the plastic pie—how much is unclear. And as plastics mimic nature, differentiating the real from the unreal can be a daunting task. Manmade miracle polymers are full of surprises—particularly those that result in formaldehyde emissions!

Formaldehyde

Owing to its ubiquitous nature and extensive use in building materials and furniture, formaldehyde was the main target health hazard in indoor air-quality investigations from the late 1970s onward. Known health effects due to low-level exposures typically found in indoor air quality include irritation of the eyes, nose, and throat. Symptoms may include watery eyes, burning eyes and nose, and coughing. People chronically exposed to low levels may also experience asthma, chronic bronchitis, severe headaches, sleep disorders, chronic fatigue, and nausea. In more severe cases, there may be lower respiratory tract irritation and bronchospasm. Formaldehyde has also been classified by International Agency for Research on Cancer (IARC) as "carcinogenic to humans."

Commonly known building materials that off-gas formaldehyde are form-aldehyde-bonded resin products such as urea–formaldehyde foam insulation (UFFI) and composite wood products (e.g., plywood, particleboard, pressboard, and MDF). Most people in the United States believe that UFFI has been banned. This is a topic that needs to be revisited.

In March 1982, the CPSC called for a ban on UFFI in residences and schools. The ban came in the wake of numerous complaints that UFFI was associated with respiratory problems, dizziness, nausea, and eye and throat irritation. However, on August 24, 1983, the ban was lifted after the Solicitor General's office decided not to appeal a Fifth Circuit Court of Appeals' ruling that overturned the ban. At this time, there are no UFFI bans in place within the United States or Europe—other than a few isolated states (e.g., Connecticut, Massachusetts, New Hampshire, New Jersey, and New York) (CBIS 2016). California has set limits on the sale and application of UFFI. Under the Hazardous Products Act, Canada banned UFFI in 1980—prior to the U.S. ban and retraction (Aaron 2009).

Although today, composite wood is the main source of formaldehyde emissions, the products that are composed of composite wood are not clearly understood. Composite lumber and sheathing is used in framing. Veneered and laminated composite wood is used in cabinetry, doors, windows, walls, ceilings, and flooring—the appearance of fine wood without the expense of single-species solid wood. Although many manufacturers claim that formaldehyde emissions from their products are limited, health complaints due to formaldehyde emissions from composite products still persist.

On July 23, 2014, Global Community Monitor, an environmental advocacy group, filed a lawsuit on behalf of California consumers, claiming that Lumber Liquidators was selling formaldehyde-emitting laminate flooring that exceeded California limits. They tested over 50 samples using a variety of different testing methods and sample batches. The results showed average initial formaldehyde emissions greater than 100 times the amount allowed under California's Proposition 65 (Global Community Monitor 2014).

In February 2015, 60 Minutes aired an investigative report, "Lumber Liquidators Selling Toxic Chinese-Made Laminate Flooring in U.S" (Business Spotlight 2015).

> Anderson Cooper investigated laminate flooring that's in thousands of American homes. Chinese-made laminate flooring sold in Lumber Liquidators outlets in the U.S. contained toxic formaldehyde which is a chemical known to cause cancer.

Currently, Lumber Liquidators faces lawsuits across the United States. The claims allege that the company's Chinese flooring products are toxic, emitting high levels of formaldehyde. Lumber Liquidators defective flooring class action lawsuits were filed as far back as 2013 but more lawsuits have since been filed in various states across the United States after the 60 Minutes investigative report. The plaintiffs claim that the products are defective, and their homes had lost their value (Lawyers and Settlements 2016).

However, for now, formaldehyde emissions from UFFI and laminate flooring may be the tip of the iceberg. Several other building materials likely to emit formaldehyde are fiberglass insulation, doors, windows, countertops, and carpeting. New products are being developed daily, and manufacturer disclosures are vague. Anticipate another tsunami of surprises!

Modern Building Materials

It wasn't until the early 1970s with advances in technology and tighter, more energy-efficient structures that product emissions gave rise to "sick building syndrome." Modern technology brought with it tougher, stronger, more durable building materials and quality, and affordable buildings. The wonderful world of plastics exploded!

Plastics took on a whole new life of their own. Competition gave rise to plasticizers and additives that are for the most part unregulated toxins and irritants. As they seek to protect their technology, manufacturers today either do not disclose unregulated toxic additives, or they declare proprietary secret.

Plastics are used in most building materials, many of which have plasticizers, most of which have additives. A partial list of plastic building materials follows:

- Composite wood products (e.g., cabinets)
- Thermal/acoustical insulation (e.g., fiberglass batting)
- Windows and doors (e.g., vinyl-clad windows and molded doors)
- Faux-building components (e.g., synthetic stone walls)
- Engineered countertops (e.g., Corian®)
- Surface applications (e.g., paints)
- Adhesives (e.g., construction glues)
- Sheeting (e.g., vinyl roofing and house wrap)
- Glazing (e.g., bulletproof glass)
- Coatings (e.g., electrical wire covering)
- Flooring (e.g., vinyl tiles)

Organic solvents and inorganic chemicals have been used extensively in processing plastics and to a lesser extent in some wood treatments as well as in formulating paints, adhesives, and varnishes. Most paints, adhesives, and varnishes are also composed of polymerized plastics.

Wood-preservative chemicals are used to treat timber and lumber. Toxic metal salts and metalloids are used in paint pigments, plastic color additives, and fire retardants. Toxic, corrosive gases may be emitted from gypsum board. Asphalt fumes may be emitted during roof-surfacing materials.

Summary

The stew of toxic building materials—old and new, regulated and unregulated, and known and unknown—is coming to a boil. The old and the new are potentially all components of today's building materials.

The older, regulated, and known building materials are also new, unregulated products in different formats. They are as follows:

- Asbestos and lead are environmentally regulated toxins with exemptions such as asbestos in vinyl floor tiles and lead roofing.
- Crystalline silica is an occupationally regulated toxin.

The newer building materials are composed of regulated and unregulated, known, and unknown components of building materials, most of which emit irritant and/or toxic gases. They are as follows:

- Formaldehyde, a known regulated toxin, is emitted from products composed of formaldehyde-based resins such as veneered particle-board cabinets.
- Plastic building material are partially known and mostly misunderstood; some are regulated, and most are not regulated. See Section II, Polymers in Construction.
- Volatile organic solvents are generally known and regulated.
- Inorganic chemicals are generally known and regulated.
- Toxic wood preservatives are generally known and regulated.
- Toxic metal salts and metalloids are generally known and regulated as occupational and environmental exposures.
- Potentially corrosive emissions from gypsum are poorly understood, and the corrosive gases are not identified.
- Asphalt fumes are known, and components are regulated.

Today's known and unknown, irritant, and toxic building material emissions are just the beginning.

In the next chapter, learn the foibles for obtaining manufacturer information, green building guidelines, and the growing voids in information. The age of information and advancements in technology are riddled with holes.

References

Aaron, Bob. 2009. "Perhaps Health Canada Should Review UFFI Ban." *The Star.* March 21. Accessed March 28, 2016. http://www.thestar.com/life/homes/2009/03/21/perhaps_health_canada_should_review_uffi_ban.html.

Asbestos. 2015. "Asbestos and The World Trade Center." *Asbestos.* July 23. Accessed March 27, 2016. http://www.asbestos.com/world-trade-center.

Barry-Jester, Anna Maria. 2015. "Baltimore's Toxic Legacy of Lead Paint." *FiveThirtyEight.* May 7. Accessed March 27, 2016. http://fivethirtyeight.com/features/baltimores-toxic-legacy-of-lead-paint.

Business Spotlight. 2015. "Lumber Liquidators Selling Toxic Chinese-Made Laminate Flooring in U.S." *2Paragraphs.* March 1. Accessed March 28, 2016. http://2paragraphs.com/2015/03/lumber-liquidator-selling-toxic-chinese-made-laminate-flooring-in-us.

CBIS. 2016. "Urea Formaldehyde Notice." *CBIS.* Accessed March 28, 2016. http://www.cbisinc.com/urea.html.

EPA. 2015. "U.S. Federal Bans on Asbestos." *EPA*. December 24. Accessed March 24, 2016. https://www.epa.gov/asbestos/us-federal-bans-asbestos.

Global Community Monitor. 2014. Tests show flooring from Lumber Liquidators contains hazardous levels of formaldehyde. *Global Community Monitor*. July 23. Accessed March 28, 2016. http://www.gcmonitor.org/llprop65pr.

Lawyers and Settlements. 2016. "Lumber Liquidators." *Lawyers and Settlements*. February 24. Accessed March 28, 2016. https://www.lawyersandsettlements.com/lawsuit/lumber-liquidators-toxic-flooring.html.

Milloy, Steve. 2001. "Asbestos Could Have Saved WTC Lives." *Fox News*. September 14. Accessed March 27, 2016. http://www.foxnews.com/story/2001/09/14/asbestos-could-have-saved-wtc-lives.html.

Peeples, Lynne. 2014. "Are Toxic Chemicals in Building Materials Making Us Sick?" *Huffington Post Green*. January 23. Accessed March 30, 2016. http://www.huffingtonpost.com/2013/12/12/building-materials-asthma_n_4427243.html.

PRWEB. 2016. "Global Plastics Consumption to Reach 297.5 Million Tons by 2015, According to New Report by Global Industry Analysts, Inc." *prweb*. February 19. Accessed February 19, 2016. http://www.prweb.com/releases/plastics_bioplastics/engineered_plastics/prweb9194821.htm.

Vice. 2016. "Why the Deadly Asbestos Industry Is Still Alive and Well." *Vice*. Accessed March 27, 2016. http://www.vice.com/en_uk/video/why-the-deadly-asbestos-industry-is-still-alive-and-well.

Young, Alison and Mark Nichols. 2016. "Beyond Flint: Excessive Lead Levels Found in Systems. *WCSH*. March 17. Accessed March 27, 2016. http://www.wcsh6.com/news/health/beyond-flint-excessive-lead-levels-found-in-maine-water-systems/86451908.

2

Product Emissions and Evolution of Indoor Air Pollution

According to the U.S. EPA, indoor air pollution is one of the country's top five most urgent environmental risks to public health. As we spend 90% of our time indoors, poor air quality places a huge burden on public health. Studies were performed, and the results were startling!

- Our indoor environment is 2–5 times more toxic than our outdoor environment, and in some cases, the air measurements indoors have been found to be 100 times more polluted.
- Up to 75-million people risk getting sick because of the buildings within which they work (Griffin 1993).
- An estimated 1.34-million office buildings experienced some form of building-related health problems, and approximately 30% of all office employees were affected (Hess-Kosa 2011).
- Studies show that one-half of our nation's 115,000 schools have problems linked to indoor air quality (U.S. EPA 2015).
- Nearly 55-million children and approximately 6-million adults spend a significant portion of their days in more than 132,000 public and private school buildings in the United States (U.S. EPA 2011).

As building-related health complaints came in by the drove, "sick building syndrome" became a common mantra. Building occupants grew increasingly alarmed. Media gloomed onto sensationalized story narratives. Lawyers smelled red meat! Indoor air-quality investigations were on the rise. Many a culprit had been identified, and many had not. Is it a particle or a chemical? What is the source? Is it product emissions—consumer products, industrial pollution, furnishings, personal activities such as tobacco smoke, or building materials? Who is responsible? Is it the manufacturer, retailer, building owner, custodial/maintenance personnel, or building contractor? Regulatory agencies, scientific guidelines, and environmental consultants became indispensable.

Today, however, rumblings about sick building syndrome are not as prevalent as it was in the 1900s. But has the source been properly diagnosed? A doctor not able to diagnose that which causes an illness will treat the symptoms. This is the case with health complaints related to indoor air quality. If the source of poor indoor air quality is not properly diagnosed, the solution may resolve the symptoms.

The present-day solution to building-related health concerns is to increase the HVAC fresh air intake which increases the energy requirements to condition a building and to restrict building material emissions. Yet, increasing the energy costs is a poor option, and identifying toxic building material components is a daunting task.

As product emissions have been on the rise, there has been a rise in public awareness. Product emissions have been subject to discovery, escalating as new products are introduced. And the green movement proposed energy and resource conservation building methods and building material product emissions procedures. The search for solutions to indoor air pollution and product emissions is discussed herein.

Evolution of Public Awareness

Since the worldwide energy crisis in 1973, advances in energy efficiency building construction have not been without a downside. In an effort to conserve fuel in commercial and residential buildings, builders started constructing airtight structures with inoperable, airtight windows, sealed openings on the exterior walls (e.g., electrical outlets). The new buildings were "weatherized" for energy efficiency.

In a well-weatherized building, the air exchange rate is 0.2–0.3 air changes per hour—a considerable decrease from 2 changes per hour in older non-weatherized buildings. Thus, while the older buildings tend to dilute and exchange indoor air contaminants, the newer weatherized buildings retain contaminants.

Adding further insult to injury, HVAC fresh air intake was minimized in office buildings in an effort to reduce the energy usage required to condition the outdoor air. Some building owners only accommodated fresh air intake when the temperature difference between the inside air and outdoor air was minimal. Energy cost was the driving force for limiting fresh air taken into the building(s). Subsequently, in the 1970s, "tight building syndrome" began to raise its ugly head. This was the dawn of awareness, the rise of indoor air-quality complaints in office buildings.

By 1986, the news media began to sensationalize the condition and coined the term "sick building syndrome." Sick building syndrome is a condition whereby the occupants of a building experience health and comfort problems that are perceived to be building related, and the cause is unknown.

Originally, formaldehyde off-gassing from furnishings inside buildings and from particleboard in mobile homes was targeted as the single-most investigated culprit. One article, published in 1987, refers to formaldehyde as a "deadly sin." New media touted, "It Could Be Your Office That Is Sick," "Tight Homes, Bad Air," and "The Enemy Within." Sensational! Insurance claims were on the rise, and insurance companies began to exclude "claims arising directly or indirectly out of formaldehyde whether or not the formaldehyde is airborne as a fiber or particle, contained in a product, carried or transmitted on clothing contained in or a part of any building, building material, insulation product or any component part of any building."

Since 1970, sick building syndrome was epidemic. Developers, owners, and scientists struggled to find solutions to the illusive building-related health problems. While contractors, building owners, and scientists struggled to find solutions, to pinpoint the source of complaints, lawsuits were on the rise. Some examples follow:

- In Missouri, a family was awarded $16.2 million after alleging their home was contaminated by formaldehyde "seeping from the particle board in the floor" (Kerch 1990).
- In Texas, a $4.5 billion lawsuit was filed—alleging contamination from toxic substances used to construct an elementary school (Kerch 1990).
- In Washington, employees in one office building sued their employer for negligently failing to provide a smoke-free environment (Kerch 1990).
- According to a nationwide study, there were somewhere between 800,000- and 1.2-million buildings in the United States with poor indoor air quality. One of the buildings was in Wheaton, Illinois. The new Du Page County Courthouse costs $53 million to build and will likely cost millions more to make it healthy (Becker and Gregory 1992).

With the passage of time, it became clear that the problem was not a simple one, and looking for unknowns was not a simple process. As industrial hygienists and environmental consultants scrambled to identify other possibilities, office building investigations became research projects. The cost was staggering—into the thousands of dollars.

The original hit list evolved to include not only formaldehyde but carbon monoxide, carbon dioxide (fresh air/indicator gas), and total organics. Industrial hygienists further attempted to identify volatile organic compounds (VOCs). Many began looking at carpet emissions (e.g., 4-phenylcyclohexene), tobacco smoke, and airborne/surface allergens.

With the passage of time and increased public concerns, product emissions were addressed with greater frequency and expanded contaminants. By 1986, the list of potential product emissions included formaldehyde, carbon monoxide, carbon dioxide, allergens, electromagnetic radiation, radon, and a medley of products based on content (e.g., cleaning products containing VOCs). Still, problems persisted.

In 2000, mold became that which was later to be referred to as "The Mold Rush." Media headlines heralded, "The Diss on Hotel Air," "Moldy Attitudes on Indoor Air Need a Good Scrubbing," "The Good, The Bad and The Moldy," and "Fungal Sleuths." The shift was phenomenal. In the public's eye, mold had become the single-most cause of indoor air-quality complaints. This was a little shortsighted, yet it was the perceived reality. Requests to perform indoor air-quality studies were equivalent to a "mold study" even when it wasn't visibly apparent. Many in the public had turned a blind eye to other possibilities.

In many cases, when the elusive mold blame game failed and complaints persisted, investigators were forced to revert back to the basics and carpet emissions. Yet, even today, very few investigators go the extra mile of spending the extra money to identify VOC components and to consider other possibilities (e.g., forensic dust and fine particles). Other considerations became emissions from copy machines, sewer gases, ozone, and outdoor air.

Now, in the twenty-first century, the term "green building" has become the solution of the day. Although energy and resource conservation are the primary focus, healthy buildings are part of the mix. Subsequently, green buildings are generally perceived to be synonymous with healthy buildings. Product emissions testing and green certification of products has raced to the forefront, beginning to parallel green building concerns, and indoor air-quality standards for high-performance buildings were recommended in 2009 in the ANSI/ASHRAE Standard 189.1.

After all the endeavors to resolve the building-related health complaints, indoor air-quality complaints still persist. Not only do they persist, but there are ongoing discoveries of product emissions—some of which are tainted Chinese drywall and laminate flooring.

Tainted Chinese Drywall

In January 2009, Chinese drywall became the suspect of causing unpleasant odors and possibly electrical problems in Florida homes. Residents were complaining of health problems and declared their homes unlivable. Builders were accused of shoddy construction.

> The problematic drywall, much of it imported from China, emitted foul odors and frequently caused mysterious failures of new appliances and electronics. Worse yet, some residents complained of serious respiratory problems, bloody noses, and migraines. (Steiger 2011)

By April 2009, The Associated Press indicated that "imports of potentially tainted Chinese building materials exceeded 500 million pounds during a four-year period of soaring home prices. The drywall may have been used in more than 100,000 homes (built or) rebuilt after Hurricane Katrina in August 2005." Some speculate that the problem may go back as far as 2001, but the purchase and installation of Chinese drywall on a massively large scale did not occur until the devastation that was wrought by the 2004 hurricane in Florida (Hess-Kosa 2011, 305).

In the meantime, in March, another battleground was forthcoming in Louisiana. There were complaints, once again, of the drywall emitting a rotten egg smell, causing respiratory problems, and corroding electrical equipment. A couple in a suburb of New Orleans filed a lawsuit, this time against drywall manufacturers. Knauf Plasterboard Co. Ltd. of China, a German-owned, drywall manufacturer, was identified as the biggest supplier of Chinese drywall sent to the United States shortly after Hurricane Katrina in 2006. In 2011, public advocates/journalists—ProPublica's Joaquin Sapien and the Sarasota Herald-Tribune's Aaron Kessler, reported:

> Despite an investigation by the U.S. Consumer Product Safety Commission (CPSC), most of the primary issues remained unresolved. Some builders and suppliers knew early-on that some Chinese-made drywall was problematic but continued using it anyway. Health and structural complaints from people who lived in homes built by Habitat for Humanity in the post-Katrina "Musicians' Village" are virtually ignored. A family-owned German company is closely involved in the operations of a Chinese subsidiary that produced some of the tainted drywall. A proposed settlement for customers who bought bad drywall from Lowe's offered small payouts to victims and big fees to attorneys. And bad drywall might be related to 12 infant deaths at an Army base in North Carolina. They discovered that almost 7,000 homeowners said they had been affected by tainted drywall. This was almost double the number listed on the CPSC's website. (Steiger 2011)

A Houston, Texas consortium of Chinese drywall inspectors commented (Chinese Drywall Inspectors 2011):

> Many more US homes, particularly in Florida, Southeast Texas, Alabama, Mississippi and Louisiana, have toxic Chinese drywall than previous calculations. Many homes in the US have toxic Chinese drywall intermixed with US made drywall. If these homes were built or remodeled after 2001, they believed a "small amount of imported Chinese drywall is enough to make an entire house toxic."
>
> By 2011, the U.S. had 3,756 cases. The estimated number of affected homes [was estimated to be as high as] 500,000. New homes, in the 2001 to 2009 time frame, are not the only affected structures—remodeled and storm damage restoration homes are also impacted. Homeowner's insurance companies have so far been silent but are expected to try to dis-include this from their coverage. Foreclosure banks and lenders have heard about this and are requiring "hold harmless" disclaimers from the purchasers.

As of 2015, German-owned Knauf Plasterboard Tianjin Co., and four companies it supplied agreed in 2010 to pay for home repairs. That settlement is expected to total $1.1 billion, attorneys have said. Taishan Gypsum Company, Ltd. was also sued. They refused to respond. In 2012, they were ordered to pay $2.7 million to seven Virginia homeowners and their lawyers. Taishan finally paid up the $2.7 million plus $500,000 in interest. This is likely the tip of the iceberg. Lawsuits are ongoing as legal dealings with Chinese businesses in the U.S. financial securities sector are currently coined—"China's Great Legal Firewall" (Mcconnaughey 2015).

Laminate Flooring

On March 1, 2015, "60 Minutes" Anderson Cooper investigated formaldehyde-emitting laminate flooring—"Lumber Liquidators Linked to Health and Safety Violations" (Nagel 2015). Lumber Liquidators, the largest and fastest-growing retailer of hardwood flooring in North America, was selling laminate flooring that emitted toxic levels of formaldehyde. The investigation also found that of the 31 boxes of Chinese laminated flooring tested, only one was compliant with California formaldehyde emission standards, and flooring made in Chinese factories was falsely labeled as CARB 2 compliant (the standard for formaldehyde emissions in wood flooring) (CBS News 2015). The media had a tremendous impact on public opinion and sales.

Prior to the bad publicity, Lumber Liquidators sold more than 100-million square feet of the laminate annually in America—most of which was manufactured in China (CBS News 2015). Despite their allegations that there were no problems associated with Chinese laminate flooring, Lumber Liquidators suspended sales of all its Chinese-made laminate flooring, and replaced it with European and American-made laminate.

As of October 2015, a $10 million Department of Justice settlement was composed of fines, forfeited earnings, and community-service contributions to organizations such as the Rhinoceros and Tiger Conservation Fund and the National Fish and Wildlife Foundation—for the purchase and use of exotic wood species, not for the laminate flooring. This settlement resolved criminal charges—not civil charges (Nagel 2015).

Currently, 138 civil lawsuits have been filed, consolidated into a class action lawsuit filed in the U.S. District Court for the Eastern District of Virginia (Nagel 2015). The charge is against Lumber Liquidators for selling products containing excessive levels of formaldehyde. Additionally, they had been notified by regulators in California that some samples of its flooring exceeded state formaldehyde limits in preliminary testing. The California Air Resources Board said it was investigation retailers. A week preceding California's announcement and after an accusatory blogger alleged other retailers may have formaldehyde issues in some of their Chinese laminate flooring Lowe's ceased to sale Chinese flooring, Lowe's halted all sales of the product and ordered independent testing—"out of an abundance of caution."

Product emissions can clearly have a divesting impact on the construction industry, building owner's, insurance companies, retailers, and many more. Building material emissions are certainly being viewed with a jaundiced eye!

The Wonderland of Product Emissions

Indoor exposures to pollutants are typically 2 to 5 times higher than those found outdoors. This is primarily attributed to emissions from construction materials, furnishings, office supplies/equipment, fixtures, and maintenance/cleaning products. Other contributing sources include individual use products (e.g., perfumes) and outdoor air pollutants. Indoor air pollutants are truly a wonderland of surprises!

Furthermore, indoor industrial exposures to toxic substances are generally 10–100 times those found in nonindustrial environments. This extreme difference raises a skeptical eyebrow whenever environmental professionals respond to office building complaints. Exposures are much lower in nonindustrial environments while complaints are greater. Why are the health complaints in office buildings disproportionate to those in industry?

Some feel the discrepancies lie in the complex array of chemical exposures in offices and other nonindustrial environments. Furthermore, nonindustrial environments are generally exposed to low levels of more than 300 chemicals—an amalgam that may exceed high industrial exposure levels to a limited number of chemicals. The combination of overwhelming numbers and the synergistic irritant/health effects of indoor air contaminants set the stage for occupant health complaints. This scenario is exacerbated by the type of containment(s) and level of buildup.

Due to increased energy efficiency requirements and reduced makeup air, indoor air is ripe for the buildup of airborne contaminants. While in industrial environments chemicals are removed by local exhaust and dilution ventilation, energy-efficient commercial and residential buildings retain and/or recycle most air contaminants. Trapped chemicals can and often does result in a mixture of multiple irritant/toxic substances—an occupant health hazard in the making!

What are the sources? In 1989, the National Institute for Occupational Safety and Health (NIOSH) completed an investigation of approximately 500 sick buildings. They concluded that the primary source of indoor air-quality problems were inadequate ventilation (58%), contaminants from inside the building (16%), contaminants outside the building (10%), microbial contaminants (5%), building materials (4%), and unknown sources (13%). Inadequate ventilation is not a source—the source being unknown, the resolution being improvement of the air ventilation. Contaminants from inside the building may include interior building materials (e.g., particle board kitchen cabinets). And unknown sources are open ended. There was a great deal of uncertainty.

Add up all the unknowns, and the number is staggering! Approximately 71% of all buildings investigated had unknown sources of product emissions resulting in health complaints, an additional 4% of the sick buildings associated with building materials, and a portion of the 16% interior building materials that may have been due to interior building materials.

Fifteen years following the NIOSH investigation, the source focus became building materials. Building material emissions are getting a lot of attention as building criteria and standards are coming to the forefront.

The Green Movement

Into the twenty-first century, there has been a drastic shift from an all-out assault on all product emissions to building material emissions. These were advanced in the form of the Leadership in Energy and Environmental Design (LEED) Certification Program for newly constructed buildings and the ANSI/ASHRAE 189.1 "Standard for the Design of High-Performance, Green Buildings—Except Low-Rise Residential Buildings" (ASHRAE Standard).

The U.S. Green Building Council LEED Program

In 1998, the U.S. Green Building Council developed a technical criteria for certifying "green building" construction. The program comes under the heading of LEED. It is a voluntary, market-driven approach to encourage the construction of energy and resource-efficient buildings that are healthy to live in. Some state and local governments have adopted the LEED Certification Program for public owned and/or funded buildings, and the U.S. General Services Administration (GSA) initially required all new construction and large renovation projects of nonmilitary federal buildings to meet minimum standards. In 2010, the GSA mandated higher green building standards for their buildings—in order to help deliver on President Obama's push for sustainability (GSA 2010).

The LEED Certification Program is point driven, and there are different point qualifiers for (1) homes and multifamily low rise, (2) multifamily midrise, and (3) high-performance high-rise buildings. The latest LEEDv4, launched in the fall of 2013, expanded program inclusion to data centers, warehouses, and distribution centers (Hardcastle 2013).

Although the point system varies by building type, many of the categories are similar, particularly as they relate to Materials and Resources and to Indoor Environmental Quality Management. Required under the "Materials and Resources" category is the storage and collection of recyclables and a recycling management plan for construction and demolition waste—neither of which are likely to impact air quality in the building. Optional items that may, however, have an indirect impact on air quality are (1) the use of at least 20 permanently installed building materials, from at least five different manufacturers, that have published reports showing the chemical content of the material (e.g., Safety Data Sheets); and (2) the use of at least 20 permanently installed building materials, from at least five different manufacturers, that have publicly released a report showing their extraction locations, methods, and a commitment to sustainable extraction practices.

As for the "Indoor Environmental Quality Management," minimum requirements are strict fresh air intake measures and tobacco smoke prohibitions. Optional items likely to have a more implicit effect on occupant health are (1) the use of low VOC-containing building materials (e.g., interior paints and coatings, adhesives, flooring, composite wood products, and insulation); (2) develop and implement an indoor air-quality management plan to include, but not be limited to, compliance with SMACNA IAQ Guidelines for Occupied Buildings Under Construction, protection of absorptive materials that are installed or stored on-site from moisture, the use of temporary HVAC equipment or use of a minimum of MERV 8 filters on-site, and prohibitions on smoking; and (3) compliance with LEED's Maximum Concentration of Air Pollutants following indoor air-quality monitoring which may be performed after construction has been completed. See Table 2.1. Other optional items are more comfort related such as temperature, lighting, sound, and quality views. It should be noted that the main concern regarding building material emissions is VOCs, no reference to formaldehyde which is, if they should so choose, one of the chemicals targeted during the "optional" air sampling.

Several unheralded stories from industrial hygienist and others abound—even LEED-certified buildings have been subject to building material emissions and building-related health complaints. Frequently, the findings are formaldehyde emissions from the building materials. Sometimes, the cause

TABLE 2.1

LEED v4 Updated

Contaminant	Maximum Concentration LEED v4
Particulates	
PM 10	50 μg/m^3
PM 2.5[a]	15 μg/m^3
Carbon monoxide	9 ppm or no greater than
	2 ppm above outdoor levels
Formaldehyde	27 ppb
4-PCH[b]	6.5 μg/m^3
Ozone[a]	0.075 ppm
Total VOC	500 μg/m^3
Speciated VOCs, except formaldehyde[a]	California DPH Table 4-1

[a] Not listed in previous LEED 2009.
[b] LEED 2009 listed only, NA thereafter. This test was only required under LEED 2009 if carpets and fabrics with SBR latex backing material are installed as part of the base building systems.

and/or source of poor air quality is unknown. After considerable expense and effort to attain LEED status, it would truly be a tragedy to conclude a construction project on an unhealthy note!

The Standard for the Design of High-Performance, Green Buildings: Except Low-Rise Residential Buildings

In 2009, ANSI and ASHRAE proposed the "Standard for the Design of High-Performance Green Building—Except Low-Rise Residential Buildings" (ANSI/ASHRAE Standard 189.1). The Standard may be adopted by enforcement authorities to provide a minimum acceptable level of design criteria for construction and/or renovation of high-performance green buildings. It does not apply to single-family residential structures, multifamily structures less than three-stories-high, manufactured homes, and buildings that don't have utilities. The purpose of the Standard is to provide minimum requirements for the "siting, design, construction, and operation while supporting indoor environments that support the activities of building occupants in high-performance buildings."

A slight reflection of the voluntary LEED Certification Program, the Standard has teeth. The categories are similar but with different headings, and the Standard is mandatory while the LEED is voluntary. Components of the Standard that directly and/or indirectly impact air quality all fall under the heading of Indoor Environmental Quality and include the following:

- Minimum outdoor airflow rates (in accordance with ANSI/ASHRAE/ASHE Standard 170).
- Particulate matter filtration minimums in health care facilities.
- Ozone-cleaning devices in areas designated "non-attainment" with the National Ambient Air Quality Standards.
- Sealing gaps between air filters.
- Restrictions on tobacco smoking.
- Emissions chamber testing, Green Seal labeling, and/or VOC emission limits for the following building materials:
  - Adhesive and sealants
  - Paints and coatings
  - Floor coverings (carpet and hard surface flooring)
  - Composite wood, wood structural panels, and agrifiber products—no "added" urea formaldehyde*

---

* Structural panel components such as plywood, particle board, wafer board, and oriented strand board identified as "Exposure 1," "Exterior," or "HUD-approved" are considered acceptable for interior use.

- Ceiling and wall systems (wall insulation, acoustical ceiling panels, tackable wall panels, gypsum wallboard and panels, and wall coverings)

In accordance with the Standard, the HVAC system(s) shall remain covered, clean, and non-operational during construction. Upon construction completion, prior to occupancy, a building flush-out is mandated with optional baseline air monitoring. An extensive list of Maximum Concentration of Air Pollutants far exceeds that of the LEED list—30 organic compounds (including formaldehyde and 4-PCH), total VOCs, and four non-VOCs. See Table 2.2. Although involved and expensive, optional air monitoring is a proactive approach to assess building material emissions post occupancy and allows for post occupancy remedy.

Green building programs are primarily focused on energy and resource conservation, and indoor air quality and building material emissions are secondary, if not an afterthought. The additional expense for implementing one of these programs is beyond the reach of most homeowners, and the expense for implementing the Standard is beyond the reach of most building owners.

Summary

Since the worldwide energy crisis in 1973, tighter, greater energy efficiency requirements and less fresh air infiltration into newly constructed buildings entrapped and contained toxic and irritant gases emitted from the new high-technology building materials. The newer high-tech plastic building materials began to dominate, or replace the more natural building products. Furthermore, the world of chemistry was on fire. New untested chemicals were added to the mix. These were the ingredients for a perfect storm. Tight building syndrome became the watchword of the day!

By 1986, formaldehyde exposures due to urea–formaldehyde-impregnated particleboard became Public Enemy #1. Then, the hit list was extended to carbon monoxide (a toxic combustion gas), carbon dioxide (an indicator of inadequate fresh air), VOCs, carpet emissions (e.g., 4-PCH), tobacco smoke, and airborne allergens (e.g., dust mites and mold). Environmental consultants scrambled to address rising complaints from building occupants.

Then, in 2009, in the wake of Hurricane Katrina, building occupants in southern coastal regions (e.g., Louisiana and Florida) alleged "tainted Chinese drywall" was destroying their homes and damaging their health. In 2015, formaldehyde was revisited in "Chinese laminate flooring."

The "green movement" claimed to address the new construction product emissions concerns within the LEED sustainable building certification program with very little focus in indoor air pollution. The ANSI/ASHRAE

TABLE 2.2

ANSI/ASHRAE Standard 189.1-2014

Compound Name	Maximum Concentration, μg/m³ (unless otherwise noted)
Inorganic Contaminants	
Carbon monoxide (CO)	9 ppm and no greater than 2 ppm above outdoor levels
Ozone	0.075 ppm (8-hours)
Particulates (PM 2.5)	35 (24-hours)
Particulates (PM 10)	150 (24-hours)
Volatile Organic Compounds	
Acetaldehyde	140
Acrylonitrile	5
Benzene	3
1,3-Butadiene	20
t-Butyl methyl ether (methyl-4-butyl ether)	8000
Carbon disulfide	800
Caprolactam[a]	100
Carbon tetrachloride	40
Chlorobenzene	1000
Chloroform	300
Dichlorobenzene (1,4-)	800
Dichloromethane (methyl chloride)	400
Dioxane (1,4)	3000
Ethylbenzene	2000
Ethylene glycol	400
Formaldehyde	33 (~26 ppb)
2-Ethylhexanoic acid[a]	25
n-Hexane	7000
1-Methyl-2-pyrrolidinone[a]	160
Naphthalene	9
Nonanal[a]	13
Octanal[a]	7.2
Phenol	200
4-Phenylcyclohexene (4-PCH)[a]	2.5
2-Propanol (isopropanol)	7000
Styrene	900
Tetrachloroethene (tetrachloroethylene, perchloroethylene)	35
Toluene	300
1,1,1-Trichloroethane (methyl chloroform)	1000
Trichloroethane (trichloroethylene)	600
Xylene isomers	700
Total volatile organic compounds (TVOC)	500[b]

[a] This test is only required if carpets and fabrics with an SBR latex-backing material are installed as part of the base-building system.

[b] TVOC reporting shall be in accordance with CDPH/EHLB/Standard Method v1.1 and shall be in conjunction with the individual VOCs listed above.

Standard 189.1 set forth criteria for sustainable high-rise buildings with a more aggressive approach to new product emissions. They set building material VOC emission limits and established a preoccupancy flush-out. They also developed a long laundry list of chemicals that should be evaluated. Compliance with the Standard is voluntary.

Building material emissions are a major contributor to indoor air pollution. The problem deserves a thorough investigation and response that will minimize future occupant exposures to irritating and/or toxic building materials. See the next chapter for product information gathering, green guidance programs, and anticipated complications.

References

ASHRAE and U.S. Green Building Council. 2014. *Standard for the Design of High-Performance Green Buildings Except Low-Rise Residential Buildings.* International Green Construction Code, Atlanta: ASHRAE, p. 45.

Becker, Robert and Ted Gregory. 1992. "Sick Buildings Are Difficult to Diagnose, Often Costly to Cure." *Chicago Tribune.* September 6. Accessed January 21, 2016. http://articles.chicagotribune.com/1992-09-06/news/9203210145_1_sick-building-syndrome-du-page-county-courthouse-million-courthouse.

CBS News. 2015. "Lumber Liquidators Linked to Health and Safety Violations." *CBS News.* March 1. Accessed January 25, 2016. http://www.cbsnews.com/news/lumber-liquidators-linked-to-health-and-safety-violations.

Chinese Drywall Inspectors. 2011. "Chinese Tainted Drywall." *Defective Chinese Drywall.* Accessed January 22, 2016. http://defectivechinesedrywall.com/what_is_it.htm.

Griffin, Katherine. 1993. "Sick Buildings Indoor Air Pollution Is One of the Country's Five Most Urgent Environmental Issues, Epa Says." *philly.com.* March 2. Accessed January 21, 2016. http://articles.philly.com/1993-03-02/living/25952729_1_indoor-air-pollution-pam-connolly-ventilation.

GSA. 2010. "GSA Moves to LEED Gold for All New Federal Buildings and Major Renovations." *GSA.* October 28. Accessed January 27, 2016. http://www.gsa.gov/portal/content/197325.

Hardcastle, Jessica Lyons. 2013. "LEED v4 Approved, Launches This Fall." *Environmental Leader News.* July 3. Accessed January 27, 2016. http://www.environmentalleader.com/2013/07/03/leed-v4-approved-launches-this-fall.

Hess-Kosa, Kathleen. 2011. *Indoor Air Quality: The Latest Sampling and Analytical Methods.* Boca Raton, FL: CRC Press.

Kerch, Steve. 1990. "Buildings Can Make You Sick." *Chicago Tribune.* July 27. Accessed January 21, 2016. http://articles.chicagotribune.com/1990-07-22/business/9003010879_1_sick-building-syndrome-indoor-air-tenants.

Mcconnaughey, Janet. 2015. "Chinese Drywall Suit and Chinese Government Agencies." *Business Insider-Associated Press.* September 7. Accessed January 25, 2016. http://www.businessinsider.com/ap-chinese-drywall-suit-and-chinese-government-agencies-2015-9.

Nagel, Tara. 2015. "Lumber Liquidators Update: What's Going On?" *Class Action*. October 12. Accessed January 25, 2016. http://www.classaction.org/blog/lumber-liquidators-update.

Steiger, Paul. 2011. "Story So Far: 'Tainted Drywall'". *ProPublica*. April 4. Accessed January 22, 2016. http://www.propublica.org/article/story-so-far-tainted-drywall.

U.S. EPA. 2011. "EPA, Schools to Tackle Indoor Air Quality." *EPA*. January 14. Accessed January 22, 2016. http://yosemite.epa.gov/opa/admpress.nsf/d0cf6618525a9efb85257359003fb69d/4c9742d4a1796670852578005cf1cd!OpenDocument.

———. 2015. "Schools: Indoor Air Quality." *EPA*. November 2. Accessed January 22, 2016. http://www.epa.gov/schools-air-water-quality/schools-indoor-air-quality.

3

Safety Data Sheets, Green Movement, and More

Owing to technological advances and greater energy efficiency requirements, indoor air has become a fertile ground for the buildup of a medley of indoor air pollutants—chemicals with no means of escape. Are they regulated toxins? Are they unregulated toxins? Are they regulated irritants? Are they unregulated irritants? Are they carcinogens? Are they reproductive hazards?

According to the Chemical Abstracts Service (CAS), there are as of January 2016 an excess of 106 million registered unique organic and inorganic substances worldwide, and the numbers continue to rise at breakneck speed (CAS 2016).

- On September 2, 2010, there were 54,973,018 registered substances
- On January 21, 2016, there were 106,014,352 registered substances, an increase of 100,521, double that of 2010
- On January 28, 2016 at 2:45 PM, there were 106,439,490 registered substances, an increase of 425,138 in a week
- On January 28, 2016 at 3:00 PM, there were 106,439,639 registered substances, an increase of 149 in 15 minutes

The increase in newly registered substances has been staggering. Are they introduced into the latest and greatest products as untested, unregulated toxins? Keeping track of innovation and change in the world of chemistry is becoming an insurmountable information challenge.

In the early 1980s, the National Academy of Sciences' National Research Council completed a 4-year study and found that 78% of the chemicals in the highest-volume commercial use had not even had minimal toxicity testing. Thirteen years later, it is doubtful there has been any significant improvement, if any at all. So, while advances in formulations are improving our options, we are faced with a dilemma of the possible introduction of newly minted toxins, chemicals yet to be studied. While the market is facing a massive rush toward progress, research foundations and regulating agencies cannot possibly keep up with the introduction of the newer, cutting-edge chemical formulations.

As we seek to maintain some semblance of control over product emissions, product composition and toxicity information is coveted. Indoor environmental professionals typically rely on safety data sheets (SDSs) as their go-to source of information. Other pertinent references include "green" certification products, VOC emissions from building materials, and formaldehyde emissions from composite wood. Herein, you will find the readily available resources of information and limitations thereof.

Safety Data Sheets

SDSs, previously referred to as material safety data sheets (MSDSs), are typically the "benchmark" for builders and environmental consultants to seek building material information. Until its 2016 implementation in the United States, SDSs were referred to as MSDSs. The rationale for the change was to provide a globally harmonized system—an internationally accepted hazard information format. See Table 3.1.

In the United States, the Occupational Safety and Health Administration (OSHA) Hazard Communication Standard requires that the product manufacturer, distributor, and/or importer provide SDSs to end users in order to communicate hazardous components that may impact the health and safety of workers. Most SDSs also contain information regarding product environmental impact, disposal considerations, transportation information, and regulatory information regarding specific U.S. states (e.g., California), countries (e.g., Canada), and politico-economic unions (e.g., European Union). That said, content information as required by the various state and international regulatory agencies vary considerably. The list of regulated chemicals varies. The toxic chemical exposure limits varies. And the management of hazardous chemicals varies. In other words, SDS information is not internationally standardized. Only the format is "harmonized."

Although the focus is worker exposures in occupational settings, SDSs have become the consumer's go-to place for product information—particularly as they relate to building materials. Public awareness is on the rise!

According to OSHA (i.e., U.S. SDSs), a "health hazard" is a chemical for which there is scientifically valid evidence, based on at least one study, conducted in accordance with established scientific principles, that acute (e.g., short term, immediate impact) or chronic (e.g., long term, slow progression) health effects may occur in exposed employees.

> Health hazards are chemicals/substances which are carcinogens and toxic or highly toxic agents. These include, but are not limited to, strong and mild irritants, reproductive toxins, corrosives, sensitizers, liver and kidney toxins, central nervous system toxins, lung damage toxins, agents that act on the formation and development of blood cells, and/or agents that affect the skin, eyes, or mucous membranes.

TABLE 3.1

SDS Information[a]

Section 1	Identification
Section 2	Hazard(s) Identification
	Summary of hazards as presented within the SDS
Section 3	Composition/Information on Ingredients
	Substances, mixtures, and trade secrets defined
Section 4	First Aid Measures
Section 5	Fire Fighting Measures
	Hazards that develop during a fire such as any hazardous combustion products created when the chemical burns
Section 6	Accidental Release Measures
	Response actions for spills, leaks, and releases
Section 7	Handling and Storage
Section 8	Exposure Controls/Personal Protection
	Exposure limits, engineering controls, and recommended personal protective equipment
Section 9	Physical and Chemical Properties
	Appearance, flammability, odor, odor threshold, evaporation rate, and many more
Section 10	Stability and Reactivity
Section 11	Toxicological Information
	Toxicological [routes of exposure, health effects, relative toxicity (e.g., LD_{50}), symptoms, and carcinogenicity]
Section 12	Ecological Information (nonmandatory)—environmental impact information
Section 13	Disposal Considerations (nonmandatory)
Section 14	Transport Information (nonmandatory)
Section 15	Regulatory Information (nonmandatory)
	Safety, health and environmental regulations specific to the product, not indicated anywhere else in the SDS
Section 16	Other Information (nonmandatory)

Source: OSHA. 2012. Hazard Communication Standard: Safety Data Sheets. *OSHA Brief.* Accessed September 10, 2016. https://www.osha.gov/Publications/OSHA3514.html.

[a] The SDSs must contain Sections 12 through 15, to be consistent with the UN Globally Harmonized System of Classification and Labeling of Chemicals (GHS), but OSHA will not enforce the content of these sections because they concern matters handled by other agencies.

On the U.S. SDS, hazardous substances are designated by reference to the OSHA Toxic and Hazardous Substances (29 CFR 1910, Subpart Z), the American Conference of Government Industrial Hygienists (ACGIH) Threshold Limit Values for Chemical Substances and Physical Agents, and carcinogens or potential carcinogens are designated according to three sources (i.e., 29 CFR 1910, Subpart Z, National Toxicology Program Annual Report on Carcinogens, and International Agency for Research on Cancer Monographs). The toxic/irritant substance references are limited to substances predominantly used in manufacturing, and not all substances used in manufacturing. Toxicological studies are limited to substances predominantly used in manufacturing, and not all substances used in manufacturing and certainly not inclusive of the latest-and-greatest, and the studies are losing ground as research and development accelerates.

The global community of scientists are continually adding new substances to the list of unknowns—registering new chemicals to the CAS registry, ticking off second by second numbers similar to the U.S. National Debt Clock. Look it up on the official CAS website—"Organic and Inorganic Substances to Date"! (American Chemical Society 2016)

SDS require the identity of substances and mixtures that are designated by the Hazard Communication Standard as hazardous. A substance is a single chemical or component which has a unique CAS number and may be inclusive of low level, high risk chemicals/components such as the carcinogen asbestos. Mixtures are two or more substances that are classified as health hazards. The concentration of each ingredient must be specified (i.e., exact percentage by weight) except where (1) there are batch-to-batch variations; (2) the same SDS is used for a group of similar mixtures; and (3) a component is a trade secret. Furthermore, if not listed by OSHA or the ACGIH as a toxic substance, a substance or component of a mixture does not have to be listed. For example, caprolactam, an irritant and a mild toxin, is widely used in manufacturing synthetic nylon, but it is not listed as a toxin by either OSHA or the ACGIH. Caprolactam has, however, been identified by NIOSH as a dust health hazard with a recommended exposure limit (REL) of 1 mg/m³ and a vapor health hazard of 0.22 ppm. NIOSH RELs do not have to be listed!

OSHA has published exposure limits for approximately 500 substances, and the ACGIH has published limits for about 700 substances. Time weighted average (TWA) exposure limits are established for 8-hour, 40-hour workweek exposures. However, the public is exposed to product emissions 24 hours a day, 7 days a week. Although residential levels may be considerably less, exposure durations are considerably more. In occupational environments, healthy workers are exposed to limited, known toxic substances. However, in residential and office environments, people of all ages and health are exposed to multiple unknowns. For this reason, building product emissions of chemicals listed under the purview of OSHA and the ACGIH are worthy of note. SDSs and/or MSDSs have thus been the backbone for assessing building material emissions and, in some cases, combustion products for emergency responders.

The SDS and MSDS information provide similar information, in different formats. Sometimes, when one requests an MSDS on the Internet, the SDS format is automatically provided. In 2016, however, an MSDS request may get an old, outdated MSDS.

Beyond the SDSs for worker protection, the California Office of Environmental Health Hazard Assessment (OEHHA) set forth health-based building material specifications. California standards are reputed to be the go-to place for state-of-the-art information.

California Green Buildings Emissions Testing for VOCs

In 1999, the California Sustainable Building Task Force was formed to direct State "green building" specifications out of which sprang the OEHHA's chronic reference exposure limits (CREL) that set material testing criteria for VOC emissions and expanded to exposure limits for indoor air pollutants in state buildings. The state building specifications were rewritten into the Construction Specification Institute format that came to be referred to as "Section 01350." In 2004, the specifications were expanded. It became the standard practice for the *Testing of Volatile Organic Emissions from Various Sources Using Small-Scale Environmental Chambers* (VOC Testing Standard).

In 2007, the *Standard Practice* was incorporated into the State "purchasing criteria." It was later adopted or adapted by the following entities:

- State of Minnesota
- LEED *Indoor Air Quality Criteria*
- National Green Building Standard
- Business and Institutional Furniture Sustainability Standard
- Carpet and Rug Institute
- Green Label Plus
- Scientific Certification Systems
- Indoor Advantage Gold
- Resilient Floor Covering Institute Floor Score
- GREENGUARD
- ANSI/ASHRAE Standard 189.1 *Maximum Concentration of Air Pollutants*

According to the VOC Testing Standard, environmental small-scale chamber testing must be or have been performed on all building materials. The measured VOC emission rates are converted by a modeling approach into an estimated airborne concentration—based on several factors such as average room volume, fresh air exchange (e.g., HVAC makeup air and/or openings to outdoors), and occupancy.

The calculated modeling findings should not exceed half of the latest CREL limits with the exception of formaldehyde. Formaldehyde is assigned a stand-alone limit. Chemical substances (on the CREL or other lists) that are not VOCs (e.g., metals, acids, and pesticides) are not required to be analyzed according to the published standard. Target VOCs are presently posted within the VOC Testing Standard—"target CREL VOCs and their maximum allowable concentrations" as published within the standard. See Table 3.2.

Target CREL VOCs were excerpted from the following publications:

- Cal/EPA OEHHA List of reference exposure limits (RELs)—organic chemicals identified as chronic inhalation hazards and are likely to result in serious adverse systemic effects (exclusive of cancer).
- Cal/EPA OEHHA Safe Drinking Water and Toxic Enforcement Act of 1986 (Proposition 65)—organic chemicals that are known or probable human carcinogens and reproductive/developmental toxins.
- Cal/EPA Air Resources Board list of Toxic Air Contaminants (TACs)—organic chemicals that are on the Cal/EPA list of Hazardous Air Pollutants exclusive of, but are not limited to, dibenzo-p-dioxins and dibenzofurans (chlorinated in the 2, 4, 7, and 8 positions and containing 4 to 7 chlorine atoms), perchloroethylene, and environmental tobacco smoke.

Within the REL list—which is also published within the VOC Testing Standard—all "organic and inorganic" are included, not only the organic chemicals. The all-inclusive REL list is identified in the publication as the "Acute, 8-hour, and Chronic RELs." See Table 3.3. And it is more than a list! It is an added source of information that is inclusive of all California designated likely indoor air pollutants—toxic substances. Component of the REL list are as follows:

- Exposure types—acute, 8-hour, and chronic exposures
- Inhalation and/or oral exposures anticipated
- Hazard index of target organs
- Primary animal species involved in principle toxicological study

The products that are designated health hazards which are likely to contribute to indoor air pollution, and product emissions are neither indicated within the document nor clearly connected to consumer products and/or building materials. Some of the listed substances may indirectly impact indoor air quality such as welding fumes (e.g., hexavalent chromium); some of those on the list are outdoor sources (e.g., diesel exhaust); some are combustion by-products (e.g., hydrogen cyanide); and some are used in the manufacture of synthetic polymers (e.g., butadiene). The rational for the California EPA's assignment of listed RELs is unclear.

TABLE 3.2

Target CREL VOCs for Emissions Testing: Their Maximum Allowable Concentrations (California DPH 2010, p. 37) and Chronic Inhalation Exposure Limits

	Compound Name	CAS Number	Allowable Concentration[a] ($\mu g/m^3$)	Inhalation REL ($\mu g/m^3$)
1	Acetaldehyde	75-07-0	70	140
2	Benzene	71-43-2	1.5	3
3	Carbon disulfide	75-15-0	400	800
4	Carbon tetrachloride	56-23-5	20	40
5	Chlorobenzene	108-90-7	500	1000
6	Chloroform	67-66-3	150	300
7	Dichlorobenzene (1,4-)	106-46-7	400	800
8	Dichloroethylene (1,1)	75-35-4	35	70
9	Dimethylformamide (N,N-)	68-12-2	40	80
10	Dioxane (1,4)	123-91-1	1500	3000
11	Epichlorohydrin	106-89-8	1.5	3
12	Ethylbenzene	100-41-4	1000	2000
13	Ethylene glycol	107-21-1	200	400
14	Ethylene glycol monoethyl ether	110-80-5	35	70
15	Ethylene glycol monoethyl ether acetate	111-15-9	150	300
16	Ethylene glycol monomethyl ether	109-86-4	30	60
17	Ethylene glycol monomethyl ether acetate	110-49-6	45	90
18	Formaldehyde	50-00-0	16.5	9
19	Hexane (n-)	110-54-3	3500	7000
20	Isophorone	78-59-1	1000	2000
21	Isopropanol	67-63-0	3500	7000
22	Methyl chloroform	71-55-6	500	1000
23	Methylene chloride	75-09-2	200	400
24	Methyl t-butyl ether	1634-04-4	4000	8000
25	Naphthalene	91-20-3	4.5	9
26	Phenol	108-95-2	100	200
27	Propylene glycol monomethyl ether	107-98-2	3500	7000
28	Styrene	100-42-5	450	900
29	Tetrachloroethylene (Perchloroethylene)	127-18-4	17.5	35
30	Toluene	108-88-3	150	300

(Continued)

TABLE 3.2 (*Continued*)

Target CREL VOCs for Emissions Testing: Their Maximum Allowable Concentrations
(California DPH 2010, p. 37) and Chronic Inhalation Exposure Limits

	Compound Name	CAS Number	Allowable Concentration[a] ($\mu g/m^3$)	Inhalation REL ($\mu g/m^3$)
31	Trichloroethylene	79-01-6	300	600
32	Vinyl acetate	108-05-4	100	200
33–35	Xylenes, technical mixture	108-38-3	350	700

[a] Refer to http://www.oehha.ca.gov/air/chronic_rels/AllChrels.html. All maximum allowable concentrations are one-half the corresponding CREL adopted by Cal/EPA OEHHA
with the exception of formaldehyde. For any future changes in the CREL list by OEHHA,
values in Table 4.1 shall continue to apply until these changes are published in the Standard
Method.

Green Product Emissions Testing and Certification Programs

As early as 1978, the Germans introduced the "Blue Angel." This was, and
still is, a certification program for environmentally "friendly" products and
services. Other countries began to mimic the Germans and began develop-
ing their own. Out of this arose the Nordic Swan, Canadian Environmental
Choice, U.S. Green Seal, and others—all eventually becoming part of the
Global Ecolabelling Network (GEN).

In 1994, GEN was formed as an international cooperative, putting
a global face on product emissions testing and labeling. It is marketed
as a nonprofit association dedicated to "improve, promote, and develop
ecolabelling" of products and services (Global Ecolabelling Network
2016a,b).

In layperson terms, ecolabeling is a world recognized certification label
that identifies environmental performance of a product or service based on
"life cycle considerations." In contrast to green symbols and claims made by
manufacturers and service providers, an ecolabel is awarded by an impar-
tial third party. In other words, the manufacturers must pay for testing and
certification of their products—the right to display the GEN label on their
products.

Since inception, the network has grown, embracing national organizations (i.e., members) around the globe, stretching from North America to
the European Union to Hong Kong to New Zealand. As of February 2016,
there were 26 GEN members, some of which had been added since 2007
(e.g., The Standards Institute of Israel), some of which had withdrawn from
the GEN (e.g., India's Central Pollution Control Board) (Hess-Kosa 2011).
See Table 3.4.

TABLE 3.3

Sample Acute, 8-Hour, and Chronic REL Information

Substance (CAS)	REL Type[a]	Inhalation REL (ug/m³)	Oral REL (ug/kg BW-day)	Hazard Index Target Organs	Species[b]
Acetaldehyde (75-07-0)	A	470[c]		Eyes; respiratory system (sensory irritation)	H
	8-hour	300[c]		Respiratory system	R
	C	140[c]		Respiratory system	R
Acrylic acid (79-10-7)	A	6,000		Respiratory system; eyes	R
Ammonia (7664-41-7)	A	3200[c]		Respiratory system; eyes	H
	C	200		Respiratory system	H
Caprolactam (105-60-2)	A	50		Eyes (sensory irritation)	H
	8-hour	7		Respiratory system	R
	C	2.2		Respiratory system	R
Chromium (hexavalent) (18540-29-9) and soluble hexavalent chromium compounds (except chromic trioxide)	C	0.2	20	Inhalation: Respiratory system Oral: Hematologic system	R
Hydrogen cyanide (74-90-8)	A	340		Nervous system	H
	C	9		Nervous system; endocrine system; cardiovascular system	H
Phenol (108-95-2)	A	5,800		Respiratory system; eyes	H
	C	200		Alimentary system; cardiovascular system; kidney; nervous system	R
Toluene	A	37,000		Respiratory, nervous systems; eyes;development	H
	C	300		Respiratory, nervous systems; eyes;development	R
Toluene diisocyanates (2,4- and 2,6-)	A	2		Respiratory system	H
	8-hour	0.15		Respiratory system	H
	C	0.008		Respiratory system	H

Source: California OEHHA. 2014. "Acute, 8-hour and Chronic Reference Exposure Level (REL) Summary." California Office of Environmental Health Hazard Assessment. June 28. Assessed August 2016. http://oehha.ca.gov/air/general-info/oehha-acute-8-hour-and-chronic-reference-exposure-level-rel-summary.

[a] REL types: A = acute, 8-hour = 8-hour, C = chronic. Exposure averaging time for acute RELs is 1 hour. For 8-hour RELs, the exposure averaging time is 8 hours, which may be repeated. Chronic RELs are designed to address continuous exposures for up to a lifetime: the exposure metric used is the annual average exposure.

[b] Species used in key study for REL development: D = dog; Gb = gerbil; GP = guinea pig; H = human; Ha = hamster; M = mouse; Mk = monkey; R = rat; Rb = rabbit.

[c] REL based on benchmark dose (BMC) approach.

TABLE 3.4

GEN Members

Australia	Good Environmental Choice Australia Ltd.
Brazil	Associacae Brasileira de Normas Tecnicas (ABNT)
China	China Environmental United Certification Center and China Quality Center
China	China Quality Center
EU	European Commission-DG Environment
Germany	Federal Environmental Agency (FEA): The Blue Angel
Hong Kong (GC)	Green Council
Hong Kong (HKFEP)	Hong Kong Federation of Environmental Protection (HKFEP) Limited
Indonesia	Ministry of Environment
Israel	The Standards Institution of Israel
Japan	Japan Environment Association (JEA)
Korea	Korea Environmental Industry & Technology Institute (KEITI)
Malaysia	SIRIM QAS International Sdn Bhd
New Zealand	Environmental Choice New Zealand
Nordic Five Countries	Nordic Ecolabelling Board: The Nordic Swan
North America	Green Seal
North America	ECOLOGO
Philippines	Philippine Center for Environmental Protection and Sustainable Development, Inc. (PCEPSDI)
Russia	Saint-Petersburg Ecological Union
Chinese Taipei	Environment and Development Foundation (EDF)
Singapore	Singapore Environment Council
Sweden (SSNC)	Swedish Society for Nature Conservation (SSNC): Good Environmental Choice
Sweden (TCO)	TCO Development
Thailand	Thailand Environment Institute (TEI)
Ukraine	Living Planet

Source: Adapted from Global Ecolabelling Network. 2016a. List of members and genics status. *Global Ecolabelling Network.* Accessed February 5, 2016. http://www.globalecolabelling. net/members_associates/map/index.htm.; Global Ecolabelling Network. 2016b. What is ecolabelling? *Global Ecolabelling Network.* Accessed February 5, 2016. http://www. globalecolabelling.net/what_is_ecolabelling/index.htm.

Green Seal

Green Seal is a U.S.-based member of GEN. The organization provides testing and certification of product emissions and performance from cradle to grave. Most Green Seal emissions testing and certification is focused on the U.S. EPA environmental emissions restrictions on environmental pollution of sunlight reactive volatile organic compounds and on the California Air Resources Board criteria for indoor air pollution of volatile organic compounds. Building materials tested are (1) paints, coatings, and sealants; and (2) adhesives for commercial use. The generally accepted threshold in paint is 0.01% VOCs below which further testing is generally not required, and there are certain chemicals prohibited in most, if not all, certified products. As of February 2016, substances prohibited in Green Seal products include methyl methacrylate, MDI, phthalates, bisphenol A, and crystalline silica.

Green Label Plus

Although not in GEN, Green Label is an industry-monitored, third-party certification organization for carpets and carpet pads and is managed by the Carpet and Rug Institute (CRI). The Green Label and Green Label Plus programs certify the "lowest emitting" carpet backing, adhesives, and cushions on the market. The Green Label program tests for volatile organic compounds in cushions (e.g., pads) used under carpet whereas the Green Label Plus test is far more all-inclusive of carpet backing, adhesives, and cushions. See Tables 3.5 and 3.6.

GREENGUARD and GREENGUARD Gold

TABLE 3.5

Green Label Cushion Criteria

TVOCs:	1000 μg/m² h
Butylated hydroxytoluene (BHT)	300 μg/m² h
Formaldehyde	50 μg/m² h
4-PCH	50 μg/m² h

In 2001, Air Quality Services, a United States based product emissions testing laboratory, established the GREENGUARD Environmental Institute ("GREENGUARD"). Its purpose was to test and certify indoor air emissions from furniture, building products, and interior finishes. At the time, Air Quality Services was the only internationally accredited laboratory in the United States. Subsequently, in 2011, Air Quality Services was acquired by Underwriters Laboratories (UL), and today, the certification program is under the UL Environment division. They perform third-party dynamic environmental chamber testing and certification of building products and interior finishes, furniture and mattresses, individual office furniture products, and office furniture seating. See Table 3.7 for the UL Environment Certification Criteria applicable to building materials and interior finishes. Beyond the certification criteria, UL Environment also performs exposure modeling as per the ANSI/ASHRAE Standard 62.1-2007, *Ventilation for Acceptable Indoor Air Quality* or the U.S. EPA's recommended exposure factors for residential applications.

So, let us digress and take a quick look at Blue Angel certification of construction products as it stands today. Why? Because emission considerations meet and exceed that of most other certification organizations. For paints and beyond, the German certification not only address VOCs and formaldehyde but plasticizers, additives, and preservatives—product emissions often overlooked by most testing and certifying services. There is a world beyond VOCs and formaldehyde!

TABLE 3.6

Green Label Plus Components

- Initial testing evaluates carpet against the 35 compounds listed on the California 01350 version 1.1.
- Carpet products are tested annually for emission levels for seven chemicals as required by Section 01350, plus six additional chemicals.
- Annual and interim testing of certified carpet is based on 24-hour chamber testing for targeted chemicals and the total level of volatile organic compounds (TVOC).
- Adhesive products are tested for emission levels for 10 chemicals as required by Section 01350, plus five additional chemicals.
- Subsequent annual and semi-annual testing of certified adhesive products is based on 24-hour chamber testing for targeted chemicals and the TVOC.

TABLE 3.7

UL Environment Certification Criteria for Building Materials and Interior Finishes

Criteria	Maximum Allowable Predicted Concentrations GREENGUARD Tier Compliance Criteria	
	Certified	Gold
TVOC[a]	500 µg/m³	220 µg/m³
Formaldehyde	61.3 µg/m³ (50 ppb)	9 µg/m³ (7.3 ppb)
Total aldehydes[b]	100 ppb	43 ppb
Individual "Gold" VOCs (33 VOCs listed)	1/10th the ACGIH TLVs	1/100th the ACGIH TLVs
4-Phenylcyclohexene (4-PCH)	6.5 µg/m³	6.5 µg/m³
Particle matter (<10 µm in size)[d]	50 µg/m³	20 µg/m³

Source: Adapted from UL Environmental. 2016. GREENGUARD Certification criteria for building products. *UL*. Accessed July 28, 2016. http://industries.ul.com/wp-content/uploads/sites/2/2014/09/GG_VOC_tables.pdf.

Note: ACGIH TLVs are subject to annual changes.

[a] Defined to be the total response of measured VOCs falling within C_6 to C_{16} with responses calibrated to a toluene surrogate.

[b] The sum of all measured normal aldehydes from formaldehyde through nonanal, plus benzaldehyde, individually calibrated to a compound specific standard. Heptanal through nonanal are measured via TD/GC/MS analysis and the remaining aldehydes are measured using HPLC/UV analysis.

[c] Thirty-three VOCs list (July 2016), subject to frequent additions and changes, may be found on the ULE website. Any VOC not listed must produce an air concentration level no greater than the acceptable fraction of the Threshold Limit Value (TLV) individual work place standard (Reference: ACGIH in Cincinnati, Ohio).

[d] Particle emission requirement only applicable to HVAC duct products with exposed surface area in air streams (a forced air test with specific test method) and for wood finishing (e.g., sanding) systems.

As public concerns escalate, manufacturers seek to deliver certified low-emission product labels. However, there is a downside—the cost for certifying. It is speculated that the cost for going "green" increases product costs 20% or more. The cost difference may change as the market adapts, but, for now, product labeling is limited—represented by a microcosm of all building materials.

U.S. EPA Toxic Substances Control Listings and Formaldehyde Standards for Composite Wood Products

In 1976, the EPA formed the Toxic Substances Control Act (TSCA). Its purpose was to "protect the public" from unreasonable risk of injury to health or the environment by regulating the use and/or volume of toxic substances by banning them entirely. The original list of 62,000 chemicals that were manufactured and sold within the United States at the time was "grandfathered in"—exempted from the newly minted regulatory requirements. The untested "existing chemicals" were listed in the TSCA Inventory without the benefit of toxicological studies. In a 1997 study by the Environmental Defense Fund, "nearly three quarters (71%) of the sampled high-priority chemicals did not meet the minimum data requirements for health hazard screening … in accordance with an internationally accepted definition of a minimum screening information data set" (EDF 1997, p. 15).

As of 2015, only 250 of the 84,000 chemicals registered for commercial use had been tested, 3000 of which are "high production volume"* chemicals. Only nine have been banned or restricted (e.g., PCBs, chlorofluorocarbons, lead-based paint, asbestos in building materials, and formaldehyde) (Center for Effective Government 2015).

As of 2016, the toxic substances restricted and/or defined by the TSCA have become inclusive of the followings:

- Asbestos Hazard Emergency Response of 1986
- Indoor Radon Abatement of 1988
- Lead Exposure Reduction of 1992
- High Performance Schools of 2007
- Formaldehyde Standards for Composite Wood Products of 2010

In regard to indoor air pollution, formaldehyde is front-and-center.

The "Formaldehyde Emissions Standards for Composite Wood Products" (15 USC 53 Subchapter VI) defines product emission limits for manufacturers of composite wood products. Within the standard, product emissions of formaldehyde are defined for manufacturer claims of "no emissions" or "low emissions." Many manufacturers seek to attain these limits in their composite wood products. See Table 3.8.

Information and Interpretation Limitations

In a 2015 publication, "Volatile Emissions from Common Consumer Products," University of Melbourne Professor Anne Steinemann reported that "of the volatile ingredients emitted, fewer than 3% were disclosed on any product label or material safety data sheet." Her research involved air

---

* High production volume chemicals are those that are produced and/or imported in excess of 1,000,000 pounds or 600 tons per year.

TABLE 3.8

Formaldehyde Emission Standards for Manufacturer Product Claims

Formaldehyde Emissions Designation	Material Type[a]	Testing 3 Months[b]	Testing 6 Months[b]	Ceiling[c]
No-added formaldehyde-based resins	1	<0.04 ppm	–	<0.05 ppm
(i.e., soy, polyvinyl acetate, and methylene diisocyanate)	2,3,4	<0.04 ppm	–	<0.06 ppm
Ultra low-emitting formaldehyde-based resin[d]	1	–	<0.05 ppm	<0.09 ppm
(i.e., melamine–urea–	2	–	<0.05 ppm	<0.08 ppm
formaldehyde, phenol	3	–	<0.06 ppm	<0.09 ppm
formaldehyde, and resorcinol formaldehyde)	4	–	<0.08 ppm	<0.11 ppm

[a] Material types
   (1) Hardwood plywood.
   (2) Particleboard.
   (3) Medium-density fiberboard.
   (4) Thin medium-density fiberboard.
[b] Routine quality control tests over a period of months.
[c] A single test or part of routine quality control tests.
[d] These "ultra low-emitting" designations are stricter than the mandated California "Airborne Toxic Control Measures to Reduce Formaldehyde Emissions from Composite Wood Product Regulation."

fresheners, laundry products, cleaners, and personal care products—31 fragranced, 6 fragrance-free. Of the fragranced products, 15 were regular and 15 were "green" products. Of the fragrance-free products, 2 were regular and 4 were "green" products. The top chemicals found in 50% of the products were ethanol and acetone. With the exception of "green" products, acetaldehyde—a volatile regulated substance[*]—was found in 45% of all the other products. However, camphor—a waxy, regulated substance—was found in 41% of the "green" products, not in any of the others. The conclusion was stunning (Steinemann 2015):

- Over 97% of the ingredients were not disclosed in the MSDSs.
- Over 94% of the potentially hazardous ingredients were not disclosed in the MSDSs.

Additionally, chemical changes can and do occur when terpenes combine with ozone in the air. Terpenes—not regulated substances—are the primary constituents of essential oils, impart a pleasant aroma to the product (e.g.,

---

[*] Within this book, all references to "regulated substance(s)" denote a substance or substances for which there is an occupational exposure—as mandated by federal OSHA and/or recommended by ACGIH. The percentage of each regulated component and the regulation exposure limit is required information in all SDSs.

TABLE 3.9

Summary of Research Findings of Effects of Total VOC
Mixtures

TVOC (mg/m³)	Health Effects/Irritancy Response
<0.20	No response
0.20–3.0	Irritation and discomfort
3.0–25	Increased discomfort (probable headache)
>25	Neurotoxic/health effects (e.g., toxic exposures)

perfumes and cleaners). Yet, terpenes react with ozone to form formaldehyde, acetaldehyde, and ultrafine particles. Thus, unbeknownst to consumers, fragranced products which contain terpenes—pinene and limonene—could potentially be converted in the air to regulated substances.

Building materials are complex and emit a multitude of VOCs. In combination multiple VOCs can have synergistic health effects which exceed the effects of each individual organic components in indoor air. In 1990, Lars Molhave, a Denmark researcher who has received considerable attention internationally, used chamber challenge testing to develop a dose response relationship to multiple volatile organic compounds (VOCs). The study concluded that subjects experienced irritation and discomfort at airborne levels between 0.20 mg/m³ and 3.0 mg/m³. See Table 3.9.

Chemically induced irritants—eye, skin, and respiratory irritation—are a moving target. While most professionals have focused on organic chemicals, emissions from building materials, consumer products, and outdoor environmental pollutants, inorganics are rarely discussed. Whereas we know mixtures of organic chemicals can cause irritation and other health effects, inorganic substances (e.g., inorganic acids, ozone, hydrogen sulfide, carbon monoxide) can likewise mirror similar effects. Add organic and inorganic irritants, and the impact will likely have a synergistic, enhanced impact on building occupants. For instance, low levels of formaldehyde off-gassing from multiple building materials (e.g., kitchen cabinets and other engineered wood products) may be enhanced by low levels of hydrogen sulfide off-gassing from tainted gypsum board (e.g., Chinese drywall).

Environmental Movement Banishes Toxin Building Materials

In 2009, the International Living Building Materials Institute was created. It is a U.S.-based sustainable building, environmental impact certification organization with the mission of "transforming to a world that is socially just, culturally rich, and ecologically restorative" (International Living Future Institute 2015a,b). The components and requirements of the certification (see Table 3.10)

TABLE 3.10

International Living Building Certification Requirements

Petals	Imperative #	Imperatives Itemized
Place	01	Limits to Growth: *first half of Imperative 01 dealing with appropriate siting of buildings* (i.e., built on grayfields or brownfields, not on sensitive lands such as wetlands and farm land)
	02	Urban Agriculture (i.e., opportunities for agriculture appropriate to its scale and density using the Floor Area Ratio)
	03	Habitat Exchange (i.e., land away from the project site must be set aside in perpetuity through the Institute's Living Future Habitat Exchange Program or an approved Land Trust organization)
	04	Car-Free Living (e.g., bicycles, walking trails, and promotion of stairs instead of elevators)
Water	05	Net Positive Water (e.g., water retention tanks, recycled water, and wastewater treatment)
	06	Ecological Water Flow (i.e., all stormwater and water discharge, including gray and black water, must be treated on-site and managed either through reuse, a closed loop system, or infiltration)
Energy	06	*Net positive energy* (e.g., 105% renewable solar and wind with energy storage)
Health and Happiness	07	Civilized Environment (e.g., frequently occupied spaces with operable windows for fresh air and light)
	08	Healthy Interior Environment (e.g., smoking prohibited, fresh makeup air as per ASHRAE 62, and an indoor air quality test before and nine months after occupancy)
	09	Biophilia (i.e., designed to include elements that nurture the innate human/nature connection)
Materials	10	"Red List" (i.e., list includes 22 human, wildlife, and environmental "classes of chemicals," 777 actual chemicals)
	11	Embodied Carbon Footprint (i.e., carbon offset in the Institute's new "Living Future Carbon Exchange" or an approved carbon offset provider)
	12	Responsible Industry (i.e., advocate for the creation and adoption of third-party certified standards for sustainable resource extraction and fair labor practices)
	13	Living Economy Sourcing (e.g., local materials for reduced transportation costs)
	14	Net Positive Waste (i.e., strive to reduce or eliminate the production of waste during design, construction, operation, and end of life in order to conserve natural resources and to find ways to integrate waste back into either an industrial loop or natural nutrient loop)

(Continued)

TABLE 3.10 (*Continued*)

International Living Building Certification Requirements

Petals	Imperative #	Imperatives Itemized
Equity	15	Human Scale and Human Places (i.e., promote culture and interaction through such things as paved areas, street and block design, and building scale)
	16	Universal Access to Nature and Place (i.e., project may not block access to, nor diminish the quality of, fresh air, sunlight, and natural waterways for any member of society or adjacent developments)
	17	Equitable Investment (e.g., donation to charity or ILFI's Equitable Offset Program)
	18	Just Organizations (i.e., transparent disclosure of the business practices of major contributors such as architects, engineers, and contractors)
Beauty	19–20	*Beauty and spirit; inspiration and education*

Source: Adapted from International Living Future Institute. 2015a. A brief history of the living building challenge. *International Living Future Institute.* Accessed March 21, 2016. http://living-future.org/ilfi/about/history-0.; International Living Future Institute. 2015b. Two-part certification. *Living Future.* Accessed May 3, 2016. http://living-future.org/living-building-challenge/certification/certification-details/two-part-certification.

Note: *Net Zero Energy Building Certification Requirements*: Minimum of Imperative first half of 1; all of 6, 19, and 20 (https://living-future.org/net-zero/requirements).

Petal Certification: Minimum of three petals, one of which is water, energy, or materials.

Full Certification: All 20 imperatives with many preliminary audits and some 12-month audits after occupancy.

is a highly charged system for micromanaging choices such as a few limited, author-selected examples:

- Promotion of stairs as opposed to elevators
- On-site wastewater treatment facilities
- Dedicated renewable energy with energy storage
- Materials carbon offset in Institute's new "Living Future Carbon Exchange" or an approved offset provider
- Just and equitable pay practices

Although the concept is onerous and controversial—a highly charged, hot political topic—the group promotes green building materials and has developed a list of building material components the Institute seeks to banish ("not to be used in construction").

The "Red List" was excerpted from U.S. government agencies (e.g., EPA), the European Union Commission on Environment, and the State of California. The list is inclusive of material substances that are harmful not only to humans but also to other living creatures and/or the environment (e.g., climate change) as well. See Table 3.11 (Green Spec 2016).

TABLE 3.11

Red List of Environmental Toxic Substances

Environmental Toxic Substances	Building Materials That Contain Environmental Toxins
Alkylphenols	Plastics: Antioxidants, fire retardants, and UV stabilizers
Asbestos	Multiple building products prior to 1978
Bisphenol A	Plastics: Antioxidants
Cadmium	Plastic stabilizers and paint pigments
Chlorinated polyvinyl chloride (CPVC)	Plastic doors and windows
Chlorobenzene	High-boiling solvent in the manufacture of adhesives, paints and paint removers
Chlorofluorocarbons (CFCs) and Hydrochlorofluorocarbons (HCFCs)	Refrigerants for cooling
Chloroprene (Neoprene)	Rubber: Adhesives, sound insulation, gaskets, improvement of bitumen
Chlorosulfonated polyethylene	Roofing material, electric wire and cable sheathing, and paint
Chromium VI	Stainless steel and chrominated copper arsenate wood treatment
Formaldehyde	Multiple materials containing thermoset polymer (e.g., composite wood, insulation, spar varnish)
Halogenated flame retardants	Plastics (mainly XPS and EPS): Fire retardant
Hydrochlorofluorocarbons	Refrigerants for cooling
Lead	Roofing, flashing, paint, cable covering, and x-ray shielding in walls
Mercury	Florescent lights
Polychlorinated biphenyls (PCBs)	1929–1979 cable insulation, thermal insulation, floor finish, caulk, oil-based paint, plastics, and light ballasts[a]
Perfluorinated compounds (PFCs)	Fire retardant in paints
Phthalates	Plastics (flexible): Plasticizers
Polyvinyl chloride (PVC)	Multiple vinyl products
Polyvinylidene chloride (PVDC)	Plastic screens
Short chain chlorinated paraffins	Plastics: Plasticizer and flame retardants
Wood treatments containing creosote, arsenic, or pentachlorophenol	–
Volatile organic compounds (VOC) in wet-applied products[b]	–

Source: Adapted from International Living Building Institute. 2014. Living building challenge 3.0. *Living Future*, p. 44. Accessed May 3, 2016. https://living-future.org/sites/default/files/15-1215%20Living%20Building%20Challenge%203_0_forweb.pdf.

[a] EPA list of items containing PCBs.

[b] Wet-applied products (coatings, adhesives, and sealants) must have VOC levels below the South Coast Air Quality Management District (SCAQMD) Rule 1168 for Adhesives and Sealants or the CARB 2007 Suggested Control Measure (SCM) for Architectural Coatings as applicable. Containers of sealants and adhesives with capacity of 16 ounces or less must comply with applicable category limits in the California Air Resources Board (CARB) Regulation for Reducing Emissions from Consumer Products.

The list has been published under the heading of "Toxic Chemistry: Chemicals in Construction" by Green Spec, a UK builders' resource for green building information. Each toxic substance has been offered up a narrative. Although the primary objective is to target environmental concerns, product emissions and worker exposures are discussed as well.

The list includes plastics, plasticizers, and plastic additives—major components in today's construction world that have not typically been part of the discussion in regard to indoor air quality. It addresses asbestos, lead, PCBs, and mercury—toxic building materials components and their sources encountered in older buildings renovated and occasionally found in new construction materials.

As a final aside, a curious comment by the UK Lead Sheet Association regarding lead in building materials (Green Spec 2015):

> The lead industry in the UK promotes the "environmental benefits" of lead particularly in its BRE *Green Guide* ratings, its recyclability and its low carbon footprint.
>
> Lead is the "new" eco material that has been around for hundreds of years. It is used on a variety of buildings from flashings on domestic porches and cladding and roofing on modern commercial building, to the domes on heritage buildings and churches across the world.
>
> Now, in the BRE *Green Guide*, for the first time with a rating of A or A+, depending upon the use to which it is put; lead is a contemporary building material that has a key role to play in sustainability.

Apparently, it appears as though the BRE *Green Guide** regards "sustainability" of natural resources and the environment over lead toxicity. Very odd!

Summary

As the world of chemistry goes to warp speed, the information highway is woefully behind. SDSs address only "non-proprietary secret" regulated toxic chemicals as listed in the approximately 470 OSHA exposure limits and 600 ACGIH exposure limits. The California OSHA list of 750 toxic substances is not a required listing.

---

* The UK BRE *Green Guide* compares the environmental impact of different types of building specifications. It assesses building materials and components across their entire life cycle from cradle-to-grave, within comparable specifications, to give a rating from A+ to E, with A+ being least environmental impact.

California is the most restrictive of all U.S. standards in regard to indoor air pollution and green building emissions testing. In California, small chamber testing for VOCs is mandated for building materials in all state and school new construction. The California VOC testing criteria for new building materials has been adopted or adapted by other state, national, and international green movement programs.

Green product emissions testing, certification, and labeling are global. Although many countries participate, manufacturers seeking product certification are limited. In order to go-around the considerable expense of participating in these certification programs, manufacturers design their own labels claiming that their product is a "green product." In the North America, Green Seal and GREENGUARD certification have a clearly defined label. Anything otherwise is likely to be a ruse, not backed by globally recognized GEN third party testing.

Under the TSCA, EPA tracks toxic substances for the purpose of recording all chemicals that are manufactured or imported in the United States. Those found to be toxic are regulated. Yet, about 40 years after implementation of the act, the list of inventoried substance toxicological testing is severely lacking—with but nine toxic substances that are restricted or banned. In terms of building material emissions, the TSCA has published a standard that defines "no emissions" and "low emissions" of formaldehyde from composite wood products. Claims on SDSs should comply with the definition.

The information limitations are considerable. Chemicals are rapidly being introduced into the world market on a daily basis while toxicological studies, proceeding at a snail's pace, cannot keep up with even the chemicals used extensively in manufacturing. Full disclosure on MSDSs is frequently not forthcoming with reports of less than 6% of product MSDSs disclosing potentially hazardous ingredients (e.g., acetone in cleaning products). Furthermore, nontoxic chemicals can react with other substances in the air to form toxic substances such as nontoxic terpenes reacting with ozone to form toxic formaldehyde.

The interpretation of multiple chemical mixtures is unclear. Nontoxic chemicals can react with other substances in the air to form toxic substances such as terpenes reacting with ozone to form formaldehyde. Multiple VOCs can have a greater health effect on humans than single organic chemicals. Chemical irritants may be a combination of organic and inorganic substances, not just one such as formaldehyde emissions from composite wood products and hydrochloric acid emissions from heated vinyl building materials.

The worlds of chemistry and building materials are inseparable. As technology races forward at breakneck speeds—so does our need for vigilance!

Polymers are the part of the new world of chemical unknowns, contributors to most of today's product emissions. Understanding polymers is paramount to understanding building products and material emission sources. Background information on polymers and their unbounded additives is the topic of Section II.

References

American Chemical Society. 2016. "Home." *CAS*. October 3. Accessed October 3, 2016. https://www.cas.org/cas-home.

California DPH. 2010. "Standard Method for the Testing & Evaluation of VOC Emissions." *California Department of Public Health*. February. Accessed February 4, 2015. http://www.cdph.ca.gov/programs/IAQ/Documents/cdph-iaq_standardmethod_v1_1_2010%20new1110.pdf.

California OEHHA. 2014. "Acute, 8-hour and Chronic Reference Exposure Level (REL) Summary." *California Office of Environmental Health Hazard Assessment*. June 28. Accessed August 2016. http://oehha.ca.gov/air/general-info/oehha-acute-8-hour-and-chronic-reference-exposure-level-rel-summary.

CAS. 2016. "CAS REGISTRY—The Gold Standard for Chemical Substance Information." *CAS*. Accessed March 30, 2016. http://www.cas.org/content/chemical-substances.

Center for Effective Government. 2015. "Reducing Our Exposure to Toxic Substances." *Center for Effective Government*. Accessed February 11, 2016. http://www.foreffectivegov.org/reducing-chemical-exposure.

EDF. 1997. "Toxic Ignorance." *EDF*. Accessed February 15, 2016. http://www.edf.org/sites/default/files/243_toxicignorance_0.pdf.

Global Ecolabelling Network. 2016a. "List of Members and Genics Status." *Global Ecolabelling Network*. Accessed February 5, 2016. http://www.globalecolabelling.net/members_associates/map/index.htm.

———. 2016b. "What is Ecolabelling?" *Global Ecolabelling Network*. Accessed February 5, 2016. http://www.globalecolabelling.net/what_is_ecolabelling/index.htm.

Green Spec. 2015. "Lead." *Green Spec*. Accessed March 21, 2016. http://www.greenspec.co.uk/building-design/lead-health-envionrment.

———. 2016. "The 'Red list' of Building Materials." *Green Spec*. Accessed April 22, 2016. http://www.greenspec.co.uk/building-design/red-list-of-banned-toxic-construction-materials.

Hess-Kosa, Kathleen. 2011. *Indoor Air Quality: The Latest Sampling and Analytical Methods*. Boca Raton, FL: CRC Press.

International Living Building Institute. 2014. "Living Building Challenge 3.0." *Living Future*. Accessed May 3, 2016. https://living-future.org/sites/default/files/15-1215%20Living%20Building%20Challenge%203_0_forweb.pdf.

International Living Future Institute. 2015a. "A Brief History of the Living Building Challenge." *International Living Future Institute*. Accessed March 21, 2016. http://living-future.org/ilfi/about/history-0.

International Living Future Institute. 2015b. "Two-Part Certification." *Living Future*. Accessed May 3, 2016. http://living-future.org/living-building-challenge/certification/certification-details/two-part-certification.

Molkhave, Lars. 1991. "Volatile Organic Compounds, Indoor Air Quality and Health." *Indoor Air* 373.

OSHA. 2012. "Hazard Communication Standard: Safety Data Sheets." *OSHA Brief*. Accessed September 10, 2016. https://www.osha.gov/Publications/OSHA3514.html.

Steinemann, Anne. 2015. Volatile emissions from consumer products. *Air Quality, Atmosphere & Health* 8 (3): 273–281. Accessed February 9, 2016. http://www.drsteinemann.com/Articles/Steinemann%202015.pdf.

UL Environmental. 2016. "GREENGUARD Certification criteria for building products." *UL.* Accessed July 28, 2016. http://industries.ul.com/wp-content/uploads/sites/2/2014/09/GG_VOC_tables.pdf.

Section II

Polymers in Construction

4

Polymers in Our World

In 1941, Dr. Victor E. Yarsley, a British chemist stated:

> Let us imagine a dweller in the Plastic Age ... This "Plastic Man" will come into a world of color and bright shining surfaces ... a world free from mold and rust ... a world in which man, like a magician, makes what he wants for almost every need (Yarsley 1941, 57, 68).

In construction, plastics (i.e., polymers) have enabled man to reach for the stars—design a multitude of architectural shapes and forms, create inexpensive, structurally sound buildings, impart durable paints and coatings to surfaces, and provide exceptional weather protection to buildings. Light weight, strong, and inexpensive columns replace the more expensive stone columns. Styrofoam™ can be sculpted (e.g., EIFS). Polymer resin-impregnated wood products are used for structural support where heavier, more expensive steel beams would otherwise be required. Durable, weather resistant, inexpensive, and energy-efficient vinyl window frames would not be possible without polymers. Durable vinyl siding replaces less durable wood and asbestos cement siding. Inexpensive plastic plumbing and waste pipes replace steel and copper. Long lasting plastic roofing materials replace short-lived asphalt and stone. These examples are only the beginning—an insight into the world of polymers.

Plastics are affordable. As the price of copper pipe rises, PVC and PEX pipes rise to the occasion. PEX is not only durable. It is easier to install.

The life expectancy of plastics in construction is 35 years with wildly variable extremes. Plastic wallpaper has a life expectancy of 5 years, while PVC has a life expectancy of over 80 years. In the San Francisco earthquake of 1971, an estimated 50% of the city's water mains and service lines failed with one exception. The plastic PVC pipes survived while the steel, cement, and clay failed.

The significance of polymers in building materials is far reaching and growing by the day. As mankind seeks to minimize building costs and preserve nature's limited treasure-trove of resources, polymers have enabled humanity to reduce the utilization of our natural resources—lumber, steel, aluminum, copper, and natural rubber.

Plastics have enabled man to perform and produce products that would otherwise have limited mankind's horizons. Without polymers, high power electric transmission lines could not be protected and convey electricity to our homes. With polymers, we can mimic nature (e.g., beautiful, durable engineered wood flooring) at a reduced cost and enrich our lifestyle. Without polymers, modern architectural advances would not exist (e.g., sculpted building façades) and electronics (e.g., computers and cell phones) would be unattainable. Plastic buildings are easy to maintain and energy efficient. Yet, there is a downside to plastics in building materials.

Although polymeric building materials have become a coveted prize in our modern society, the miracle of plastics does not go unchallenged. The albatross of environmental doom-and-gloom foretells potential hazards associated with polymeric building materials. Beyond the ever looming public concerns regarding environmental plastic waste hazards harming wildlife and leaching into our streams, rivers, and underground water ways—polymers pose a potential threat to human health and wellbeing when used in building materials.

- Polymer building materials may emit, or off-gas, regulated toxins (e.g., formaldehyde) and unregulated irritates and/or toxic substances (e.g., plasticizers). The latter unregulated toxins get lost in the fog and require greater scrutiny. As the world of plastics is in a constant state of change, so are new polymer formulations.

- Polymer water containers (e.g., water retention tanks) and plumbing supply systems (e.g., PVC pipe) may leach plasticizers and additives into drinking water. There is considerable literature regarding leaching of plasticizers in plastic drink containers, baby bottles, and food containers. Alarm bells ring out, "Chemicals Are Found in Soft Plastic Products We Use Every Day." The media is rife with consumer plastic container concerns, but the building industry, as of the writing of this book, has yet to connect water contamination of plastic building water conveyances. Additives (e.g., lead and antimony) are also overlooked.

- Polymer building materials evolve potentially toxic gases during a fire. These gases may impact an occupant's ability to escape during a fire as well as a fire fighter's decision to don special PPE and environmental monitoring equipment.

As more and more plastics are used in building materials, the complex nature of polymer emissions requires attention. Environmental professional and building contractor vigilance is paramount.

The World of Polymers

Since the emergence of the industrial revolution around the 1800s, the world population has grown from 1 billion to over 7 billion in 2015 (Feldman and

Akovali 2005; Challis 2011). Our natural resources and agricultural production have been challenged. Mankind has scrambled to find solutions. World War I stretched our natural resources. The Great Depression was fraught with despair and diminished life styles. World War II placed a further burden on our way of living. Post World War II, however, witnessed a boom—in lifestyle and plastics. Without the discovery and advances in synthetic polymers, one must ask, "What would the world look like without plastic?"

In the nineteenth century, African elephants were being exploited for ivory that was used extensively in the making of billiard balls. With one million pounds of ivory consumed annually, the elephants approached extinction. Thus, there were fears of ivory shortages. In 1853, as rumors go, Phelal and Collander, a New York billiards ball manufacturer/supplier, offered a handsome fortune of $10,000 in gold to anyone who could come up with a suitable substitute. John Wesley Hyatt, an enterprising young journeyman printer, was inspired. He took the bait and began tinkering with chemistry in a shack behind his house. Finally, in 1870, Hyatt patented the first successful process for making plastic. The process involved cellulose nitrate (i.e., gun cotton) with camphor under heat and pressure, and he called the plastic "celluloid." Although celluloid billiard balls failed to perform, Hyatt and his brother set up a company to make celluloid denture plates that had previously been made from hard natural rubber (Nanns 2011). Celluloid was also widely acclaimed as an inexpensive replacement of ivory combs and introduced the first plastic film that revolutionized still photography and brought in the motion picture industry. However, its flammability made manufacturing dangerous and ultimately celluloid film was phased out in 1949 when safer polymers were discovered. Other products fell by the wayside as well with one exception. Today, one celluloid product remains unscathed by the advent of newer and better plastics. Celluloid table tennis balls are still in the game. Despite its limited success, celluloid was the spark that set off the avalanche of modern plastics.

By 1894, cellulose acetate rallied after celluloid at a snail's pace and was marketed as "safety film" in 1909 and was used extensively for waterproofing and stiffening fabric-covered airplane wings during World War I. Today, it is used in making tool handles and eye glass frames.

Casein plastics were introduced at the beginning of the twentieth century. Milk protein was precipitated by the enzyme rennin, the resin mix then molded under moderate heat and pressure. It was then hardened by soaking the natural polymer in formalin (5% formaldehyde). This is the first time formaldehyde was used in a polymerization process. Brightly colored casein buttons, belt buckles, and fountain pens are now collector's items.

Although man began experimenting with synthetic rubber in 1906, it was not until after World War II did he improve the quality to the point that it rivalled that of natural rubber. Wartime necessity became the impetus for the emergence of synthetic rubber on a large-scale basis when governments began building plants to offset natural rubber shortages. Synthetic rubber plants were built around the world post World War II, primarily in Europe, North America, and Japan. In 1960 the use of synthetic rubber surpassed that of natural rubber for the first time. Synthetic rubber has maintained the lead ever since. In construction, synthetic rubber is used in roofing, flooring, carpet backing, acoustic panels, and more.

In 1909, Leo H. Baekeland patented the first synthetic plastic and called it Bakelite™. Formulated by mixing phenol and formaldehyde, Bakelite was the first successful polymer to withstand the test of time. Although today it is referred to as phenol formaldehyde (PF), in the early 1900s, Bakelite was a household name—the only act in town. Initially, it was used for insulating electric wires and for automobile parts. By the Roaring Twenties, Bakelite jewelry became the rage, and the synthetic billiard balls replaced ivory. Households showcased Bakelite clocks, telephones, radios, bowls, and kitchenware. Post World War II, other plastics replaced many of the Bakelite products and have become high priced collector's items. Yet, PF retained its appeal in building materials. Due to its bonding strength, durability, and water resistance, PF composite wood products (e.g., exterior plywood and Oriented Strand Board [OSB]) have become integral with and coveted as important building materials.

In 1919, prompted by the success of PF, Hanns John of Czechoslovakia patented the process for formulating urea formaldehyde (UF). Although it was more costly than phenol formaldehyde, UF had unlimited color range. Its original uses mimicked that of Bakelite, and these too were short lived. In 1931, UF was used to make plastic laminate, and by the end of World War II, it became indispensable in the building industry. The price is less than that of phenolic resins, but it has poor water resistance and delaminates when wet. Today, UF resins are used in composite wood products that are not exposed to water damage. Some of the UF building materials developed in the twentieth century are plywood, medium density fiberboard (MDF), and foam insulation.

Urea formaldehyde foam insulation (UFFI) had been developed in Europe in the 1950s as an improved means of insulating difficult-to-reach cavities in house walls. It is typically made at a construction site from a mixture of UF resin, a foaming agent, and compressed air. When the mixture is injected into the wall, urea and formaldehyde unite and "cure" into an insulating foam plastic. During the curing process, formaldehyde was released, resulting in excessive exposures to formaldehyde. Occupants began complaining of eye, nose, and throat irritation, persistent coughing and respiratory distress, nausea, headache, and dizziness. During the 1970s, when concerns about energy efficiency led to efforts to improve home insulation, UFFI became an important insulation product for existing houses in North America. In Canada, most UFFI installations occurred between 1977 and the 1980s when it was banned in Canada. Canada also has a real estate disclosure requirement as to the presence of UFFI. Although research findings concluded that houses with UFFI showed no higher formaldehyde levels than those without, cured UFFI and other UF building products that are exposed to moisture can break down and, once again, release formaldehyde. In 1982, the U.S. Consumer Product Safety Commission banned UFFI use in residences and schools, and a year later the U.S. Court of Appeals struck down the law. However, because of the controversy, UFFI is not widely used in the United States. UFFI is still used in Europe where it was never banned and is considered one of the better types of "retrofit" insulations.

In 1933, melamine formaldehyde (MF) was developed as an intermediate between PF and PU. As it has outstanding clarity, heat/chemical/light stability, and fire/abrasion resistance, MF products surged in the post World War II years. Principal among these were decorative laminates (e.g., Formica®). There have been no reports of formaldehyde exposures associated with MF.

In 1835, the first plasticized polymer—polyvinyl chloride—was discovered quite by accident when H.V. Regnault observed that vinyl chloride changed form when exposed to sunlight. Yet, it remained a laboratory curiosity until 1932 when the Germans attempted to make PVC tubes with but marginal success. Then, there was a breakthrough. In 1935, one hundred years after the polymer was discovered, the first PVC pipes rolled off a German production line. Shortly thereafter PVC pipe was pressed into service for residential drinking water distribution and waste pipelines. In 1949, PVC pipe was introduced in North America, and the first PVC water distribution pipes were laid in the United States. Most of the PVC water lines, dating back to 1936, are still in service today. Still, dramatic advances in stabilizers and additives improved the PVC pipe that was already serviceable. By the 1960s, PVC pipe was also used for gas distribution, electrical conduit, chemical processing lines, and vent pipes. The stabilizers and additives may well alter the landscape of how we look at plasticized polymers. They are generally undefined by the manufacturers while posing a potential health problem as they leach out of the water pipes.

The 1930s saw a boom in polymer research and development not only in construction but throughout society. Plastic fabrics (e.g., nylon and polyester clothing) and faux leather (e.g., Naugahyde® boots and purses) were to become indispensable, low cost household items.

On March 27, 1933, two British research chemists, R.O. Gibson and E.W. Fawcett, discovered, quite by accident, the polyethylene polymer. Their story is a testament to "turning a lemon into lemonade." While autoclaving ethylene and benzaldehyde, the test container sprang a leak, and all of the pressure escaped. The result was a white, waxy substance that greatly resembled plastic. Upon carefully repeating and analyzing their research, the scientists discovered the polymerization process for polyethylene. Registered in 1936, polyethylene was to play a key supporting role for the British in World War II. Its first was used as an underwater cable coating and then as a critical insulating material for vital military applications in radar insulation. Polyethylene was a highly guarded secret until after the war. Subsequently, polyethylene has been used in the manufacture of electrical insulation, foam insulation, PEX water pipes, and moisture barriers.

Polycarbonate was discovered at Bayer by Dr. Hermann Schnell and his team in Bayer's R&D laboratories. The official patent was granted on October 16, 1953. Polycarbonate is clear as glass; it is unbreakable; and it is one-sixth the weight of glass. It is used for bulletproof glass and greenhouse covering.

According to Dorel Feldman and Guneri Akovali:

> After 1980 continuous growth was recorded with the development of a number of high performance polymers that could compete with traditional materials such as: polyetheretherketone, polyetherimide (1982), 4,6-polyamide (1987), syndiotactic PS (1989), metallocene polyolfins, polyphthalamide (1991), styrene-etheylene copolymer, syndiotactic PP in 1992 and nanocomposites (Feldman and Akovali 2005).

Research and development is on-going both in the polymer sciences as well as stabilizers and additives. The complex nature of polymer sciencecan be overwhelming, and the science is constantly evolving.

The Chemistry of Polymers

Plastic or polymer, "What difference does it make? Well, it makes a big difference—all plastics are polymers, and but not all polymers are plastics." Let's discuss the differences.

Polymers are high molecular weight molecules that consist of thousands, if not millions, of repeating smaller units. Most, not all, are organic, and some are inorganic. Polymers are either natural or synthetic.

Human DNA has over 20 billion constituent atoms, the most complex of all natural polymer. Shellac, wool, silk, and natural rubber are also natural polymers. Shellac is a high molecular weight resin that is secreted by the female lac beetle in the forests of India and Thailand, and resources are limited. Wool comes from sheep; silk comes from an insect larvae cocoon; and natural rubber is extracted from rubber tree. Most natural polymers are renewable yet limited, and they lack the resilience and durability of synthetic polymers.

Synthetic polymers are manmade polymers. Nylon, rayon, faux leather, latex, vinyl, Tyvek®, Styrofoam™, and latex are synthetic polymers that most people recognize in their everyday lives. Do you recognize them as synthetic polymers or as plastics? If the answer is "plastics," you are neither right nor wrong. The terms synthetic polymer and plastic are interchangeable.

The list of synthetic polymers is endless, and new polymer formulations are registered every day. Likewise, the list of synthetic building materials is ever growing, evolving daily. The science is complex and far reaching. Awareness of the basic science of synthetic polymers is necessary in order to understand the nature of the beast, to address the potential health impact of plastic building materials.

Polymerization of Synthetic Polymers

Polymerization is a means whereby monomers are processed. These monomers are bonded together by one, or a combination, of two main processes— condensation polymerization and addition polymerization.

In condensation polymerization, each step of the process is accompanied by giving up a monomer component, such as hydrogen gas or water, in order to link one monomer to another. The reaction usually requires heat and/ or pressure for polymerization to be completed. Polyester is a condensation polymer.

In addition polymerization, the monomers react to form a polymer without relinquishing monomer components. Addition polymerization is usually carried out in the presence of a catalyst. With an alkene monomer, polypropylene is an addition polymer.

Copolymerization is a series of polymerizations between two or more different monomers. Copolymerization renders larger, more complex polymers with higher melting points and more tensile strength than single monomers polymerization. For instance, ethylene is a single monomer, and polyethylene is used for fabrics; it has a low melting point and low tensile strength. Acrylonitrile, butadiene, and styrene are three different monomers that form acrylonitrile–butadiene–styrene (ABS) which is used in vinyl siding; it has a high melting point and high tensile strength.

The further the chemist delves into the latest and greatest processes, the more exotic the end product. The possibilities may lead us into unknown chemical components with unknown health effects.

The Polymeric Backbone

Most polymers are organic, possessing hydrogen and carbon, and most synthetic polymers have carbon in their backbone—the spine that glues them together. Many synthetic polymers have carbon backbones and other elements. For example, the nylon backbone is carbon and nitrogen. The backbone for polycarbonate bulletproof glass is carbon and oxygen (see Figure 4.1).

Some polymers, however, lack carbon in their backbone, and most of these have carbon in the branch units. They remain organic polymers with an inorganic backbone. For example, the backbone for polydimethylsiloxane (e.g., silicone caulk) is silicon and oxygen. The crosslink backbone for polysulfides (e.g., vulcanization of rubber) is sulfur, and the backbone for

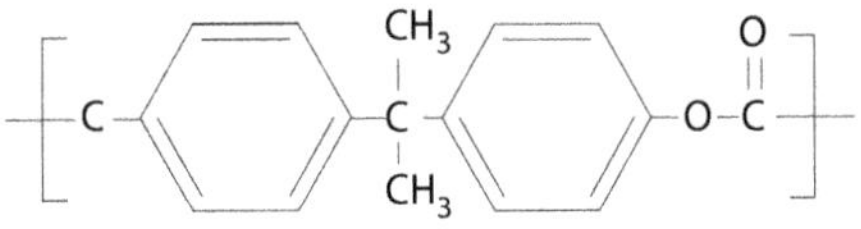

FIGURE 4.1
Polycarbonate polymer (carbon and oxygen backbone).

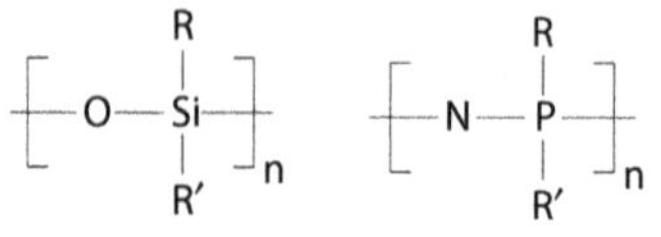

FIGURE 4.2
Silicone backbone (left); polyphosphazene backbone (right).

polyphosphazene (e.g., fire resistant expandable foam) is phosphorus and nitrogen. All contain carbon and hydrogen in the branches. See Figure 4.2.

The backbone confers properties that extend the possibilities. Research and development is expanding their sights into the exciting realm of backbone manipulation. The end polymers are exotic and poorly researched as to their effects on human health.

Thermoset Plastics versus Thermoplastic

Based on polymer properties, plastics are labeled as thermoset plastics and thermoplastic polymers. A vast amount of building materials contain thermoset plastics, while the thermoplastics outnumber the thermoset plastics in diversity.

Thermoset plastics are polymers that become permanently hardened when heated or cured. The curing process of thermosets causes a chemical reaction that creates permanent three-dimensional bonding. They have superior durability and will not change shape in extreme thermal and chemical conditions—the operative word being "change shape." In elevated temperatures and in the presence of water, thermoset plastics can and will degrade, reverting back to their monomers. Phenol–formaldehyde resins and polyisocyanurate foam are thermoset plastics.

Thermoplastic polymers become soft when heated and hard when cooled. Once polymerized, always polymerized. Thermoplastics do not degrade when heated. Thus, heating/reheating softens the polymer so it can be formed and/or reformed. Most thermoplastic polymers can be identified by their poly- suffix, such as polyester, polyethylene, and polyvinyl chloride. If there is a poly- suffix, it is likely a thermoplastic polymer, but not all thermoplastics have the poly- suffix. For example, ABS is a thermoplastic polymer.

Thermoset and thermoplastic polymers are somewhat predictable as to their emissions and thermal decomposition by-products. A vast majority, in terms of number, of polymers used in building materials are thermoplastic.

Molecular Arrangement and Properties of Thermoplastic Polymers

The molecular arrangement of thermoplastic polymers is linear and branched. Their physical properties are based on how they are arranged. The two basic molecular polymeric arrangements are amorphous or crystalline.

Amorphous polymers are randomly oriented and formless. They are generally transparent (e.g., glass-like). The chains are long and unorganized. When heated to above the glass transition temperature (T_g), amorphous polymers lose their rigid, glass-like properties becoming more viscous and rubber-like (e.g., able to stretch and deform without fracturing). Amorphous polymers with a T_g below ambient temperatures are referred to as elastomers. Plexiglass™ and bulletproof windows are amorphous thermoplastics.

Crystalline polymers are organized and arranged into distinct patterns. They are translucent or opaque. The chains are long and tightly organized. They are strong, stiff, chemically resistant, and stable. Their T_g is generally higher than that of amorphous polymers, and they tend to be more brittle. Sounds simple? There are no practical absolutes in polymer chemistry. Crystallinity makes polymers strong but reduces impact resistance. For example, polyethylene prepared under high pressure (5000 ATM) had high crystallinity (95%–99%) but were extremely brittle and easily shattered. Thus, most crystalline polymers are "semi-crystalline." To further confound the narrative, crystalline polymers, as they reach their melting point revert to amorphous polymers. Polypropylene and polyethylene moisture barriers are semi-crystalline thermoplastics.

The properties of thermoplastic polymers, outside molecular arrangement, are often manipulated by chemistry. Plasticizers and additives are key to manufacturers' production of quality plastics.

Plasticizers and Additives

Through the miracles of chemistry, plasticizers and additives can drastically change the landscape of a polymer. This is where the polymer chemist makes his mark, and companies covet their prize formulations. As plasticizers and additives are not bonded to the polymer, they are free molecules within the polymer matrix. They may leach out into liquids (e.g., lead additives in PVC plumbing) and/or be emitted into the environment (e.g., phthalate plasticizers in occupied buildings).

Due to their tremendous diversity and endless possibilities, many plastics and formulations have yet to be discovered—offering a fruitful bounty for research and development. The impact of these emerging plastics has yet to be unveiled!

Hazards of Polymers

The worldwide news media heralds the disasters of plastics, screams foul play, priming the pump of a misinformed public. Disinformation, if repeated frequently, becomes reality. Sometimes perceived reality obscures commonly accepted scientific findings. Our task is not to rule out but to delve in, seek the possibilities, and ferret out the probable.

An India blog, "Dangers of using plastics," assigns doom-and-gloom hazards to polymers for what are the health concerns of monomer(s). In just one of their bullet points, they claim, "polystyrene, the form of plastic used to make Styrofoam™ articles such as disposable cups and plates, it is believed, enters the body with food and accumulates in fat tissues. It can also cause irritation in the eyes, nose and throat." This is true of the monomer, not the polymer (Barwarchi 2013).

Some media publications are well researched but draw unsubstantiated inferences. In a Mother Earth News article, the author presents the generally accepted science while drawing speculative conclusions. "The latest scientific research has given us a lot of good reasons to think carefully about how we use plastics. The main concern with several types of plastic is that they contain endocrine disruptors—substances that, when taken into our bodies, alter normal hormonal function. Over the past several years, scientists and the media have struggled to find answers to mysteries such as precocious puberty, declining fertility rates in otherwise healthy adults, hyperactivity in kids, the fattening of America, and the persistent scourges of prostate cancer and breast cancer. Although multiple factors play a role in all of these conditions, one recurrent theme is the brew of endocrine disruptors infiltrating our lives" (White 2009).

On the other hand, some media publications are well researched and refer to unconfirmed information as "alleged." This leaves the door open on inferences. In a Seattle Times article, the writer states, "Polyvinyl chloride, also called PVC or vinyl, poses a particular hazard to health and the environment, according to these sources. Alleged problems with PVC include dioxin emissions when PVC plastics are manufactured or burned, and 'plasticizer' additives called phthalates … Studies have linked phthalates to problems including lowered sperm counts and smaller genitalia in males, says the ICEH. Items made from PVC include baby toys, bibs and teethers; cling wraps and food containers; and children's lunch boxes" (Watson 2007).

The science of polymers and associated hazards is ever evolving, ever expanding. Likewise, knowledge regarding the hazards of polymers is evolving. Although the jury has yet to render a verdict, some scientific studies are consensus conclusions. The primary areas of concern regarding polymer hazards are product emissions and combustion by-products.

Polymer emissions of "regulated" hazardous materials have received considerable attention, especially in regard to the more frequently encountered plastics in building materials. Yet, they are only the tip of the iceberg! Many known and unknown health hazards lurk in the shadows. From whence do they come?

- Unregulated toxic/carcinogenic monomers
- Unregulated unbound plasticizers

- Regulated polymer additives leaching from building materials
- Unregulated polymer additives leaching from building materials

Manufacturers are only required to list regulated "Hazardous Ingredients," not unregulated potentially hazardous ingredients. Evolving, exotic plastics may introduce toxic monomers that have neither been studied nor considered for listing as toxic substances. In accordance with the Globally Harmonized Safety Data Sheets (GH SDS), only regulated toxins/carcinogens must be identified and listed within the SDS.

Most, if not all, manufacturers also list plasticizers and additives, but disclosure goes no further. Clarity as to the type of plasticizer and additives is rarely disclosed. Most manufacturers covet their polymeric formulations and declare plasticizer and additive information to be "proprietary."

Although hazardous combustion by-products of polymer building materials are more widely studied, researched, and published than the emissions and leaching of polymers, the product manufacturers report "hazardous thermal decomposition products" dissimilar data from that which is reported by their competitors. For instance, one manufacturer may list carbon dioxide and carbon monoxide, while another will list carbon dioxide, carbon monoxide, hydrocarbons, and cyanide. There is no single source for information regarding combustion by-products. Furthermore, highly toxic dioxins (e.g., TCDD) are occasionally listed as hazardous thermal decomposition by-products, particularly as regards PVC products.

Information regarding polymer building material emissions and combustion by-products is constantly changing. Research is on-going. The list of toxic components is restricted "regulated" substances only, and some of the additives that may be regulated are being declared "proprietary." MSDSs are inconsistent from one manufacturer and another. As polymer research and development is on the rise, the information on health hazards of plastics in building materials must rise to the occasion. Beware of the possibilities!

Summary

Synthetic polymers had their start with a young journeyman printer who with a prayer, a shoe string, and a dream, developed and patented in 1870 the first successful process for making plastics. Others followed and some successes were accidental. The promise of replacing natural resources with chemically manipulated petrochemical by-products that were not only stronger and more resilient but more durable lead to the plastic revolution throughout the early 1900s. Only after World War II, however, did polymer building materials gain momentum. Today, everywhere you look, you see plastic consumer goods, clothing, and building materials; new formulations and polymers are developed with increased gusto as we catapult into the twenty-first century. Plastics are here to stay.

Whole books and research papers have been written regarding the chemistry of polymers—a chemist's carousel of delights. Thus, the purpose of the section on chemistry has been to introduce only the salient points of interest. Polymer chemistry influences emissions and thermal decomposition products.

- Most polymerization involves the condensation (i.e., subtraction) and/or addition of "monomers" to form high molecular weight plastics. Whereas their chemical make-up influences product emissions, there is very little known about the recently developed, more exotic chemicals, particularly about their health effects.

- The backbone of most polymers is carbon or carbon and another element (e.g., oxygen), and some are strictly inorganic (e.g., silicone; silicon, and oxygen). The backbone gets complicated, may potentially impact emission and thermal decomposition products.

- Thermoset plastics are permanently hardened when heated or cured (e.g., UF), and thermoplastic polymers become soft when heated and hard when cooled and can be recycled (e.g., polypropylene). Emissions and thermal decomposition chemistry are related to polymer type.

- Thermoplastic polymers are either formless amorphous plastics or organized semi-crystalline polymers, transparent and rigid or translucent/opaque and brittle. The properties of thermoplastics are altered by plasticizers and additives.

- Plasticizers and additives can drastically alter the properties of a plastic, and they are not bound within the polymer. They are free within the polymer matrix.

Beyond regulated components of plastics in building materials are a galaxy of unknowns, a fog of uncertainty. With the list numbering into the thousands, unregulated toxins fall under the radar. Unregulated, unbound plasticizers and additives—many of which are suspect health hazards—are potential emission health hazards. Some are regulated additives are a potential water health hazard. And polymer thermal decomposition and combustion by-product information is generally non-inclusive of plasticizers and additives. All is not well in "Plasticville"!

References

Barwarchi, Sify. 2013. *Dangers of Using Plastics*. Accessed February 13, 2015. http://food.sify.com/articles/Dangers_of_using_plastic-241445.

Challis, Dr. Tony. 2011. *People & Polymers*. Accessed February 3, 2015. http://www. plastiquarian.com/index.php?id=4&pcon=.

Feldman, Dorel and Guneri Akovali. 2005. "The Use of Polymers in Construction: Past and Future Trends." In *Polymers in Construction*, ed. Guneri Akovali, 13–16. Shawbury, Shrewsbury, Shropshire, UK: Rapra Technology Limited.

Nanns, Roy. 2011. John Wesley Hyatt (1837–1920). Accessed February 3, 2015. http:// www.plastiquarian.com/index.php?id=54.

Watson, Tom. 2007. "The Hazards in Our Plastics." *Home & Garden*. May 20. Accessed February 13, 2015. http://seattletimes.com/html/homegarden/2003703428_ ecoconsumer19.html.

White, PhD, Linda B. 2009. "Dangerous Plastics, Safe Plastics." *Mother Earth News*, August/September. Accessed February 13, 2015. http://www.motherearth-news.com/natural-health/dangerous-plastics-safe-plastics-zmaz09aszraw. aspx.

Yarsley, Victor E. and Edward Couzens. 1941. *Plastics*. Westminster, London: Penguin Books. http://www.epsomandewellhistoryexplorer.org.uk/Yarsley.html.

5

Plastics Commonly Found
in Building Materials

Many plastics are comprised of toxic, carcinogenic, and/or irritating monomers—no big surprise! The question is "When and under what conditions
are the polymer components most likely to impact human health?"

Polymer components may off-gas from building materials; they may
behave as surface contaminants; and they may leach into surrounding
media. Another overriding consideration in assessing the potential human
impact of plastics in buildings is combustion by-products.

Off-gassing, or emissions, is most likely to occur when the plastic or
plastic-containing building material is fresh off the production line. When
stacked, containerized, then stored and transported, the residual emissions
are likely to be trapped until they reach the end user. New product emissions
are generally limited in duration and dissipate faster in open unconfined
spaces. Within an enclosed building, the emissions are likely to be trapped
and recycled in the ventilation system. Multiple products emitting multiple
gases can add up. Whereas singular emission components may not impact
human health in an enclosed building, multiple gases may exasperate an
otherwise low level exposure to product emissions. Multiple emissions and
their additive or synergistic effect are oftentimes overlooked. Awareness of
the possibilities is paramount!

Once the new materials have off-gassed, their impact dissipates—until the
next wave. Conditions that can set off a second tsunami of emissions are
heat, high humidity, and water damage.

Heat is the greatest influencer of plastic emissions. The higher the temperature, the higher the probability of increased emissions. The main heat load
in a building is in the attic. This is certainly agreed to by most observers.
Well, how high can high get? Many agree that the temperatures in an attic
can reach as high as 160°F (71°C), particularly in poorly ventilated, poorly
insulated attics. Some places where the outside temperature reaches 120°F
(49°C) or 130°F (54°C), the attic temperatures in poorly ventilated, poorly
insulated attics can reach up to 190°F. Emissions from polymer building
materials in the attic will increase with temperature, and 190°F (97°C) is certainly the outer limit for most plastics. Elevated temperatures should also be
anticipated to building materials in direct sunlight, enclosed in uninsulated
enclosures, in direct contact with a heat source (e.g., fireplace), and in direct

contact with hot items (e.g., frying pans). Expect elevated levels of off-gassing with increased temperatures.

Water is most likely to impact plastics that have been processed by water condensation (e.g., extraction of water) in the polymerization process. This applies particularly to thermoset plastics such as urea formaldehyde. Moisture and water damage may result in emissions of plastic components, particularly those that have weaker monomer bonding. The likelihood for weaker bonded urea formaldehyde to off-gas formaldehyde is greater than that of PF. Sources of water damage extend beyond rain water. Indoor sources of water damage include (1) water heater leaks; (2) water faucet leaks in kitchens and bathrooms; (3) plumbing leaks; and (4) spills. Consider it all!

Beyond polymer emissions, plastic monomers may be released slowly from surfaces or the unbound semi-volatile plasticizers may leach out to the outer surface of the polymer building material and/or furnishings. Additives may also leach into plastic water containers and water supply systems. Plasticizers and additives are manufacturer formulated components of all thermoplastic polymers.

For clarity and ease of reference, the term plastics is used herein to differentiate synthetic resins (e.g., UF) and solid polymers (e.g., polycarbonate) from expanded foam (e.g., Styrofoam™) and elastomeric (e.g., Neoprene™) polymers. Emissions and combustion by-products from "plastics" used in building materials are discussed herein.

Thermoplastic Polymers

In terms of emissions and combustion by-products, thermoplastic polymers behave in a similar manner. Yet, they are different in the extent of research and information available, in the monomer components, in their properties and end use, and in the level of usage in building materials. Once formed, thermoplastic polymers can be melted down and reformed. The melting temperatures are generally higher than 212°F (100°C) and require artificial heating or burning to melt. Elevated temperatures within a building are not likely to be sufficiently high to melt thermoplastic polymers.

On the other hand, temperatures in a building may be high enough to cause a solid thermoplastic polymer to become rubbery, and in some cases the rubber-like state occurs below 73°F (23°C). This is referred to as the glass transition temperature (T_g). At the T_g, it becomes more likely for plasticizers and additives to be released from within the polymer matrix. For instance, the rubbery state for polyethylene and polypropylene are well below 32°F (0°C), and they are more likely to off-gas plasticizers and additives than many thermoplastic polymers that have a higher T_g value. The T_g for ABS is 176°F (80°C) to 257°F (125°C), and the T_g for PVC is about 186°F (85°C).

Each of the more prevalent thermoplastic polymers used in building materials is unique and worthy of further discussion. Below they are presented in alphabetical order.

Acrylonitrile–Butadiene–Styrene

Unregulated product emissions: See "Plasticizers" and "Plastic Additives."

Regulated product emissions: –

Combustion products: Carbon dioxide, carbon monoxide, and hydrogen cyanide

Widespread use in building materials: 1948 to present

ABS, an amorphous thermoplastic polymer, is produced by the polymerization of styrene and acrylonitrile in the presence of polybutadiene, a rubber polymer. The styrene-acrylonitrile copolymer gives the plastic rigidity, hardness, and heat resistance, and polybutadiene gives resilience and impact resistance to building materials. See Figure 5.1.

As it is similar to yet more resilient than PVC, ABS is used in many places where PVC is found. It is used in the manufacturing of drain-waste-vent pipe systems and window profiles. ABS building materials generally have additives (e.g., fire retardants, antioxidants, and UV protection), and they can also be blended with other polymers, such as PVC, to improve the resilience of the end product.

According to the Center for Fire Research, thermal decomposition of ABS results in the evolution of styrene within the early stages of a fire. Due to its flammability, styrene then provides "fuel to the fire." Combustion products are carbon dioxide, carbon monoxide, and hydrogen cyanide.

Styrene has a sharp, sweet odor that is detectable at 0.017 ppm, below the ACGIH TLV of 20 ppm. Thus, the presence of the styrene odor may serve as an early warning of exposures. Emissions from ABS building materials at temperatures below 158°F (70°C) have not been reported.

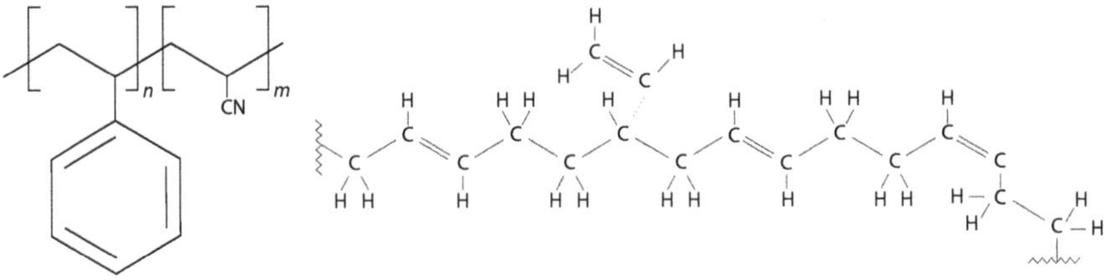

FIGURE 5.1
Polystyrene acrylonitrile (left); polybutadiene rubber (right).

Polycarbonate

Unregulated product emissions: See "Plasticizers" and "Plastic Additives."

Regulated product emissions: –

Combustion products: Carbon dioxide, carbon monoxide, and hydrocarbon
  fragments

Widespread use in building materials: Originated in 1958

Polycarbonates (PCs), an amorphous thermoplastic polymer, is most
commonly formed by the reaction of bis-phenol A with carbonyl chloride. Although it has exceptional strength (i.e., 250 times stronger than
glass), shatterproof (e.g., bulletproof), thermal stability, and good trans-
parency, PC building materials are limited to greenhouse glazing panels, dome lights, flat/curved window glazing, and bullet proof windows.
After polymerization, the end products are not reported to emit gases. See
Figure 5.2.

   According to most manufacturer MSDSs, the thermal decomposition products are carbon dioxide, carbon monoxide, and hydrocarbon fragments (e.g.,
charred particles with hydrocarbons adsorbed on the surface).

Polymethyl Methacrylate

Unregulated product emissions: See "Plasticizers" and "Plastic Additives."

Regulated product emissions: Methacrylate (trace amounts)

Combustion products: Carbon dioxide and carbon monoxide

Widespread use in building materials: Originated in 1936 (safety glass)

Polymethyl methacrylate (PMMA), a thermoplastic polymer, is an acrylate
polymer derived from the polymerization of acrylic acid esters and salts.
They are heat and impact resistant, have good clarity and UV resistance.
PMMA is used to make safety glass (e.g., Plexiglass™ and Lucite™), and it is
used in acrylic "latex" paints.

   PMMA is not only impact resistant, but it is also clearer than glass
and has been used in large aquariums such as California's Monterey Bay

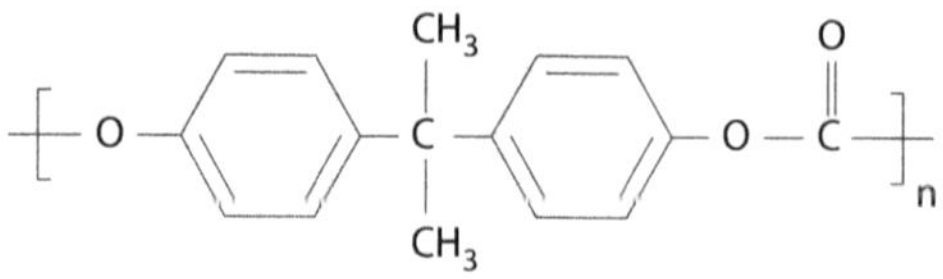

FIGURE 5.2
Polycarbonate.

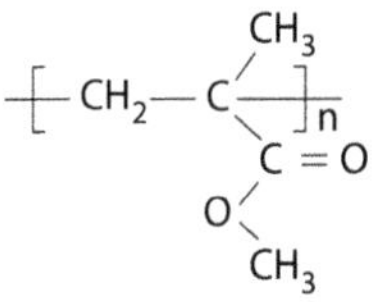

$$\left[CH_2 - C \underset{\underset{O \diagdown CH_3}{\overset{\displaystyle |}{\underset{\displaystyle C = O}{|}}}}{\overset{\displaystyle CH_3}{|}} \right]_n$$

FIGURE 5.3
Polymethyl methacrylate.

Aquarium—a large single sheet of clear plastic, 54-feet long, 18-feet high, and 13-inches thick. Some consumer uses, of interest, include contact lenses, dentures, and inert bone fillers.

In 1979, PMMA building materials held a 7% share of all plastics used in building materials. The PMMA monomer is methyl methacrylate. The monomer is highly irritating to the eyes, skin, and respiratory tract and has a sharp odor at levels (0.014 to 0.46 ppm) well below the ACGIH TLV (50 ppm). In building materials, PMMA is used in the production of break resistant glass (e.g., Plexiglass®) and acrylic paints. After polymerization, the end products are not reported to emit gases. See Figure 5.3.

The monomer, methyl methacrylate, is an irritant to the eyes, skin, and mucous membranes. It has a detectable sharp odor between 0.049 and 0.34 ppm, well within the ACGIH TLV of 50 ppm.

According to most manufacturer MSDSs, the thermal decomposition products are carbon dioxide and carbon monoxide. Some manufacturers include aldehydes and other hydrocarbons. Then, too, a Turkish chemistry professor claims that polymerized PMMA can contain trace amounts of its monomer methyl methacrylate in the final polymer, and it can also be a thermal decomposition by-product. Methyl methacrylate is an irritant to the eyes, mucous membranes, and skin.

Polyethylene

Unregulated product emissions: See "Plasticizers" and "Plastic Additives."

Regulated product emissions: –

Combustion products: Carbon dioxide and carbon monoxide

Widespread use: Late 1950s to present

Polyethylene (PE) is a thermoplastic polymer produced by the catalytic polymerization of ethylene gas at elevated temperatures and pressures. Although simple in structure, the properties are highly variable, based on the manufacturing process, plasticizers, and additives. The most frequently manufactured PEs are low-density polyethylene, high-density polypropylene (HDPE), and cross-linked polyethylene. See Figure 5.4.

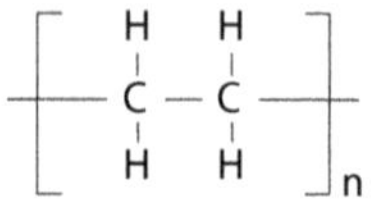

$$\left[\begin{array}{ccc} & \text{H} & \text{H} \\ \underset{}{\overset{}{-}} & \underset{|}{\overset{|}{\text{C}}} - \underset{|}{\overset{|}{\text{C}}} & \underset{}{\overset{}{-}} \\ & \text{H} & \text{H} \end{array}\right]_n$$

FIGURE 5.4
Polyethylene.

Low-density polyethylene (LDPE), a highly branched PE, has a low molecular weight, low density (0.910–0.940 g/cm³), and a low melting point (110°C). It is soft and flexible; it does not shatter; and it is used in wire and cable insulation.

HDPE has a relatively high molecular weight, high density (0.915 to 0.925 g/cm³), and high melting point (120°C). Without the benefit of plasticizers and additives, HDPE, a linear PE, is hard and brittle. HDPE is more thermostable than LDPE, requiring higher temperatures for thermal decomposition (Paabo and Levin 1987b; Akovali 2005). HDPE is used in chemical-resistant piping, natural gas distribution pipes, potable water pipes, storm drain pipes, and building vapor barriers.

Cross-linked PE (PEX) is a medium- to high-density polyethylene containing cross-linked bonds, creating stronger bond energy changing the PE from a thermoplastic to a thermoset polymer. It has a process dependent melting point between 110°C and 120°C. PEX has high impact resistance, high tensile strength, and is flexible. Europeans began using PEX around 1970; it was introduced in the United States in 1980. PEX use has been increasing ever since, replacing copper pipe in many applications, especially for hot water. Then, the rumor mill raised its angry head (WaWa 2005):

> A 2002 Norwegian Regional Food Control Authority report stated that "(v)olatile organic components were migrating from plastic pipes (HDPE, PEX and PVC) into drinking water," further stating that significant amounts of volatile organics were leaching into the test water from PEX pipes. The Norwegian study sparked a hail storm of inconclusive reports.
>
> Several California groups blocked adoption of PEX for a decade for concerns about toxins getting into the water, either from chemicals outside or inside the pipes such as methyl tertiary butyl ether (MTBE) and tertiary butyl alcohol. California eventually permitted PEX use in all occupancies. An environmental impact report and subsequent studies determined there were no causes for concerns about public health from PEX piping use.

Most MSDS regarding polyethylene limit the thermal decomposition products to carbon dioxide and carbon monoxide, and some include VOCs and/or aldehydes. Yet, the literature review by the National Bureau of Standards Center for Fire Safety found that thermal degradation products from polyethylene are largely dependent on temperature and oxygen availability.

> In the absence of oxygen, a mixture of alkanes and alkenes are produced.
> In the presence of oxygen, low molecular weight aldehydes (e.g., acrolein

and formaldehyde) and carboxylic acids are formed. Flaming combustion (e.g., destructive burning) resulted carbon dioxide and a variety of hydrocarbons. Yet, the results presented in the review show that the combustion products of polyethylenes are neither highly or unusually toxic (Paabo and Levin 1987a).

Polypropylene

Unregulated product emissions: See "Plasticizers" and "Plastic Additives."

Regulated product emissions: –

Combustion products: Carbon dioxide and carbon monoxide

Widespread use in building materials: Originated in 1957

Polypropylene (PP), a thermoplastic polymer, is produced by the polymerization of propylene (C_3H_6). In building materials, olefin fibers are used in indoor/outdoor carpeting, carpet backing, wallpaper, and house wrap (e.g., Tyvek®). In consumer products, some of the olefin fibers were used in cold weather gear (e.g., Thinsulate®), ropes, upholstery, disposable protective clothing, cigarette filters, and diapers. See Figure 5.5.

According to most manufacturer MSDSs, the thermal decomposition products are carbon dioxide and carbon monoxide. In one study, conducted by the U.S. Consumer Product Safety Commission, the pyrolysis products of PP between 200°C and 600°C produced oxygenated hydrocarbons, aromatic hydrocarbons, aliphatic hydrocarbons, CO, CO_2, and H_2O. As combustion temperature increased, oxygenated and aliphatic hydrocarbons decreased and aromatic hydrocarbons increased.

Polyvinyl Chloride

Unregulated product emissions: See "Plasticizers" and "Plastic Additives."

Regulated product emissions: Hydrogen chloride

Combustion products: Carbon dioxide, carbon monoxide, and hydrogen chloride

Widespread use in building materials: Middle 1950s to present; originated in 1940

PVC, also referred to as vinyl, is a thermoplastic polymer. It is comprised of long chains of polymerized vinyl chloride. It was first commercially

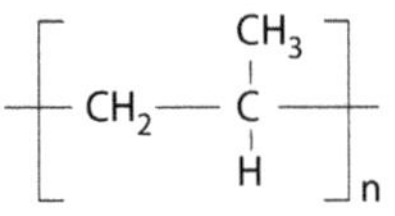

$$\left[CH_2 - \underset{\underset{H}{|}}{\overset{\overset{CH_3}{|}}{C}} \right]_n$$

FIGURE 5.5
Polypropylene.

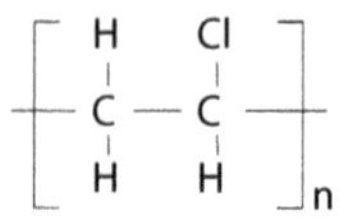

$$\left[\begin{array}{cc} \overset{\displaystyle H}{\underset{\displaystyle H}{\overset{|}{\underset{|}{C}}}} & \overset{\displaystyle Cl}{\underset{\displaystyle H}{\overset{|}{\underset{|}{C}}}} \end{array}\right]_n$$

FIGURE 5.6
Polyvinyl chloride.

introduced in the early 1930s in the United States and Germany, and by 1979, PVC building materials were contending worldwide with polyethylene. At that time, over 55% of all plastic building materials were comprised of PVC (Akovali 2005, p. 7). In a recent publication, worldwide PVC product usage is in excess of 34 million tons/year, 26% of which is pipe (PVC 2015). Uses of PVC in building materials include potable water pipes and fittings, waste drain pipes and fittings, roofing, window cladding, flooring (e.g., sheet vinyl), electrical wire and cables, trim, and molding. See Figure 5.6.

In its pure polymer form, PVC is rigid, strong, and fire/water resistant. Yet, due to its rigidity, PVC must be extruded or molded above 100°C, a temperature high enough to initiate chemical decomposition. For this reason, stabilizers are added (e.g., tin and lead) to minimize breakdown of the plastic. To further alter its properties, create a more flexible plastic (e.g., sheet vinyl tile), PVC is heated and mixed with plasticizers which in some cases are as high as 50% by weight of the end product polymer. The greater the amount of plasticizer, the greater the flexibility of the polymer. For example, vinyl rain coats have a high plasticizer content; they are strong, flexible, and water resistant. In some instances, however, plasticizer is not added such as it is in rigid PVC pipe. Rigid PVC is referred to as unplasticized PVC (UPVC).

Product emissions occur during degradation of PVC in the presence of heat and UV light. For example, vinyl window treatments are constantly exposed to heat and UV light. If there are insufficient stabilizers present, hydrogen chloride gas will be emitted; polymer discoloration (e.g., yellowing) will occur; and the plastic will deteriorate and lose its integrity. Stabilizers prevent this from happening. Yet, even stabilizers get bad press as stabilized PVC pipe has been implicated as a likely source of lead in potable water.

> Lead is used in 95 percent of PVC pipe in India, 86 percent in the Middle East and Africa, and 61 percent in South America. In Europe, by contrast, 29 percent of all PVC pipe systems use lead, while in North America, the figure is less than 1 percent, because nearly 100 percent of vinyl pipe systems use tin as a stabilizer (Jie 2013).

If this is not problematic enough, the more flexible materials (e.g., vinyl floor tiles) may degrade and release plasticizers that are bound within the polymer. The greater the degradation, the more plasticizer rises to the surface. If degradation becomes extensive, the semi-volatile plasticizer will collect on the surface of the polymer, leaving a sticky film, and the polymer will become hard and brittle and crack. See further discussion of plasticizers in Chapter 8, "Plasticizers."

The PVC monomer, vinyl chloride, has a detectable sweet odor of 10 ppm which is in excess of the ACGIH TLV of 1 ppm. According to a study, vinyl chloride levels are low (e.g., less than 0.2 ppm) when PVC is welded in a poor ventilation area (Williamson and Kavanagh 1987). Beyond this one study, vinyl chloride emissions have neither been reported nor found to be a problematic—once polymerization has been completed at the manufacturing facility. Vinyl chloride is strictly an occupational exposure hazard.

Due to its high chlorine content (57% chlorine by weight), the PVC thermal decomposition products—as reported in many SDSs—are hydrochloric acid and small amounts of aromatic (e.g., benzene) and aliphatic hydrocarbons (e.g., ethylene). Welding fume analyses have indicated levels of hydrogen chloride (1.0–3.5 ppm) in excess of the ACGIH standard—generated during normal welding (Williamson and Kavanagh 1987). Several sources claim that elevated temperatures due to amplified solar energy (e.g., increased by glass reflecting or magnifying the sun's rays) are likely to result in hydrogen chloride emissions which could cause eye and respiratory tract irritation. The combustion products are carbon dioxide, carbon monoxide, and detectable levels of hydrogen chloride.

Thermoset Plastics

The anticipation of the emissions and combustion products of thermoset plastics lives in a different realm from that of thermoplastic polymers. Thermoset plastics cannot be melted and reformed as thermoplastic polymers can. Furthermore, thermoset plastics do not melt nor do then they have a clear T_g. They do, however, break down in the presence of heat and/or water. The stronger the resin matrix, thermoset plastics are less apt to break down than the weaker matrices.

The list of thermoset plastics is short in terms of components, while usage in building materials is extensive. The more common thermoset plastics are discussed herein from the weakest to the strongest matrices.

Urea Formaldehyde

Unregulated product emissions: –

Regulated product emissions: Formaldehyde

Combustion products: Carbon dioxide, carbon monoxide, aldehydes, and organic acids

Widespread use in building materials: Middle 1950s to present; originated in 1931 (decorative laminates)

$$
\text{NH}_2 - \text{C} \overset{\displaystyle \text{O}}{\underset{\displaystyle \text{NH}_2}{\Big\langle}} \qquad \overset{\displaystyle \text{H}}{\underset{\displaystyle \text{H}}{\text{C}}} = \text{O}
$$

FIGURE 5.7
UF monomers—urea (left); formaldehyde (right).

UF, a thermoset resin, is formed through the reaction of urea with formaldehyde cross-linking to form a network of chains. They become permanently solid upon subjecting the polymerized PF to heat and/or pressure. See Figure 5.7.

Over 70% of manufactured UF is applied to wood products. The resin is used as the glue that bonds indoor, weather-protected particleboard, medium density fiberboard, hardwood plywood, and engineered lumber (e.g., laminated veneer lumber [LVL]). It is also used as an adhesive for veneer lamination. Due to its low cost, about 90% of all composite wood products are bonded with UF. Although its usage is extensive, UF products hold the dubious honor of contributing to formaldehyde exposures indoors due to building product emissions.

Awareness of UF building product's contribution to elevated levels of formaldehyde in manufactured homes became widespread in the 1980s. UF resin was used extensively in composite wood sub-flooring, cabinets, and wall panels. UF glue was used for carpet and veneer surface adhesion. Lawsuits ensued!

With public awareness on the rise, hysteria spread! Building materials in all forms of construction came into question. Indoor air quality professionals began focusing on formaldehyde "off-gassing" from building materials in residential and commercial buildings. Extensive investigations were launched.

Studies indicated that temperature, humidity, ventilation, and age of the building contributed to differences in formaldehyde levels. Emission rates were constant over the first 8 months after construction, then began to decline. In 1985, a U.S. study investigated formaldehyde levels in different types of housing.

> (The) study demonstrated that formaldehyde levels in 38 conventional U.S. homes averaged 40 parts per billion (ppb) with highs of 140 ppb. Nineteen apartments and 11 condominiums were also studied and had formaldehyde levels averaging 80 ppb and 90 ppb, respectively, with highs of 290 ppb. A more recent study of new homes found the geometric mean formaldehyde level was 34 ppb in manufactured homes and 36 ppb in site-built homes. This study suggested that formaldehyde concentrations in conventional homes have decreased greatly since the 1980s due to decreased use of plywood paneling and reduced emissions from the composite wood products used (CDC-FEMA 2009).

Subsequently, the manufacturers of UF composite wood products took action to reduce the levels of formaldehyde off-gassing. The effectiveness of the manufacturer's efforts to reduce formaldehyde off-gassing from composite wood products has been demonstrated by a 2005 U.S. EPA study. The study found a mean level of formaldehyde for mobile homes or trailers ranged from 15.5 to 24.7 ppb, a 50% reduction over the 1985 study. The strictest worldwide recommended exposure limit to formaldehyde in office buildings is the California Chronic Exposure Level. The limits are 23 ppb, based on the concept of "as low as reasonably achievable" (CDC-FEMA 2009). Based on the California recommended exposure limit, the most recent U.S. EPA findings of formaldehyde exposures in manufactured homes are acceptable. Yet, the studies did not account for the impact of temperature and humidity.

Elevated temperatures and humidity result in a significant increase in formaldehyde emissions from UF building materials. In a 1981 study, an increase in humidity from 30% RH to 75% RH results in about 60% increase in formaldehyde emissions. An increase in temperature from 77°F (25°C) to 104°F (40°C) at 75% RH results in a 420% increase in formaldehyde. A 2012 study extended the previous study. The more recent study assessed the increase in formaldehyde, over background levels, at 77°F (25°C) and 95°F (35°C) with increasing humidity, up to 100% RH. The increase in formaldehyde levels were three times and 10 times the background levels, respectively (Wescott et al. 2012).

The formaldehyde monomer is irritating to the eyes, skin, and respiratory tract. Its pungent odor is detectable by some at levels as low as 27 ppb (AIHA 1989). This level may not be detectable by the majority of people, and the consensus of the U.S. EPA and state agencies is that most people detect pungent formaldehyde gases at more elevated levels (0.8–1 ppm) than that of the WHO non-occupational exposure limit (100 ppb in 30 minutes) (WHO 2010, p. 142). The WHO limit alleges the guideline was developed to protect against sensory irritation in the general population, and prevent the effects on lung function as well as long-term health effects, including nasopharyngeal cancer and myeloid leukemia (WHO 2010, p. 142).

Most manufacturer of UF composite wood products list thermal decomposition products on their MSDSs as carbon monoxide, aldehydes, and organic acids. One manufacturer additionally lists hydrogen cyanide and polynuclear aromatic compounds.

Melamine Formaldehyde

Unregulated product emissions: –

Regulated product emissions: Formaldehyde

Combustion products: Carbon dioxide, carbon monoxide, aldehydes, and nitrogen oxides

Widespread use in building materials: Early 1970s; originated in 1933 (resin)

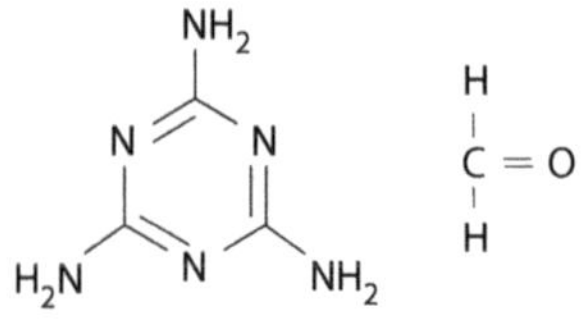

FIGURE 5.8
MF monomers—melamine (left); formaldehyde (right).

MF is a thermoset resin. It is formed through the reaction of phenol with formaldehyde cross-linking to form a network of chains. The MF resin becomes permanently solid upon subjecting the polymerized PF to heat and/or pressure. See Figure 5.8.

Due to its low formaldehyde emissions and resistance to water, MF is an intermediate between UF and PF. As it is accepted as a viable LEED replacement for UF, MF in building materials is expected to rise and may ultimately replace UF. Presently, MF resins are used in medium density fiberboard, particleboard, high pressure laminates (e.g., Formica), and engineered wood flooring.

Whereas UF products are reputed to off-gas formaldehyde, MF is not so reputed. This is not to say that MF will not release formaldehyde, but it has not been discovered to off-gas formaldehyde as yet. Formaldehyde has a detectable pungent odor of 27 ppb, below the ACGIH ceiling limit of 300 ppm (ACGIH 2013) and the WHO non-occupational settings limit (100 ppb in 30 minutes) (WHO 2010, p. 142).

The MSDSs indicated that the thermal decomposition products are carbon monoxide, aldehydes, or organic acids. One study presented in the *Journal of Polymer Science*, published in 1979, indicated that there is a considerable amount of formaldehyde given off at 200°C (Manley and Higgs 1973).

Phenol Formaldehyde

Unregulated product emissions: –

Regulated product emissions: Formaldehyde

Combustion products: Carbon dioxide and carbon monoxide

Widespread use in building materials: Early 1950s to present (residences); early 1970s to present (all building construction)

PF is a thermoset resin. It is formed through the reaction of phenol with formaldehyde cross-linking to form a network of chains. They become permanently solid upon subjecting the polymerized PF to heat and/or pressure. PF resins are used mostly in exterior plywood, OSB, laminates, and countertops. See Figure 5.9.

Cured PF is heat resistant (e.g., industrial heat shields), water/mold resistant (e.g., exterior sheathing in buildings), and chemically resistant (e.g., gasoline,

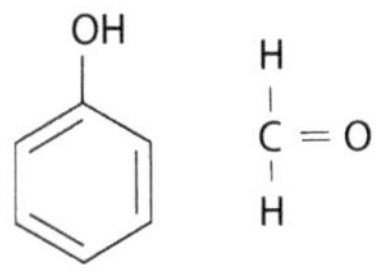

FIGURE 5.9
PF monomers—phenol (left); formaldehyde (right).

oil, and weak acids/bases). PF water/mold resistance makes PF wood products superior product to UF which deteriorates in high humidity and delaminates in water. However, as PF is two to three times the cost of UF, PF building materials are used as an exterior building material, and the less expensive UF building material is used indoors (e.g., kitchen cabinets).

Most of the phenolic building materials (e.g., exterior plywood, oriented strand board, and laminated composite lumber) are produced by the resole process. Resole products, used for gluing and bonding building materials, involves cross-linking phenol with an "excess of formaldehyde" in the presence of an alkaline catalyst (Kizilcan, 2013–2014). Thus, there is a process tendency for off-gassing of formaldehyde from resole processed PF. Although there are no reports of product emissions of formaldehyde from the finished products, the use of PF products within a building's interior spaces is rare. So, a "no report" of formaldehyde emissions may be due to the predominant use of PF building materials in exterior walls.

Formaldehyde has a detectable pungent odor of 27 ppb, below the ACGIH ceiling limit of 300 ppm (ACGIH 2013) and the WHO nonoccupational settings limit (100 ppb in 30 minutes) (WHO 2010, p. 142). A cautionary comment, "Although the consensus of polymer experts has been that, due to the superior bonding of PF resin, water resistance, and heat resistance, PF wood products do not off-gas formaldehyde, an excess of formaldehyde is used in resole formulation which may be residual after manufacturing." It is not likely to be problematic in exterior walls, but it may pose as a formaldehyde emission source indoors.

Most manufacturers of UF composite wood products list thermal decomposition products on their MSDSs as carbon dioxide and carbon monoxide only. One manufacturer, however, includes aldehydes, organic acids, hydrogen cyanide, and polynuclear aromatic compounds. In a research study conducted by a U.K. Fire Research Station study involving PF laminates, the findings were that phenolic components were identified at temperatures between 572°F (300°C) and 1022°F (550°C). The main phenolic components have been identified as phenol, o- and p-cresols, 2:4- and 2:6-xylenols and 2:4:6-trimethyl phenol. The maximum yields of the phenols are obtained at 460°C in nitrogen. In the presence of oxygen, however, phenol components are greatly reduced. What does this mean? In fires where the oxygen is greatly reduced, phenolic laminates could generate hazardous amounts of phenolic products (Woolley and Wadley 1970).

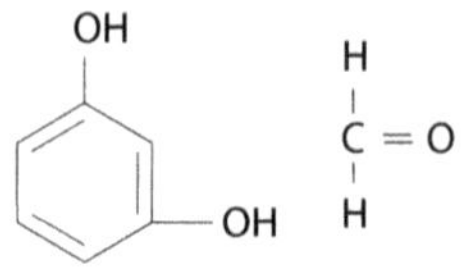

FIGURE 5.10
RF Monomers and polymer—resorcinol (left); formaldehyde (right).

Resorcinol Formaldehyde

Unregulated product emissions: –

Regulated Product emissions: Formaldehyde

Combustion products: Carbon dioxide and carbon monoxide

Widespread use in building materials: Patented in 1964

Resorcinol formaldehyde (RF) is a thermoset resin that is similar to PF, but has stronger adhesion and greater moisture resistance. It is, however, extremely expensive to process. The benefits of RF are not sufficient to warrant the added expense. Occasionally, RF is mixed with PF to form a phenol–resorcinol–formaldehyde (PRF) resin. See Figure 5.10.

Of the four thermoset resins discussed herein, RF is the least likely to off-gas formaldehyde. And it is the most expensive!

References

ACGIH. 2013. *Guide to Occupational Exposure Limits.* Cincinnati, Ohio: ACGIH.

AIHA. 1989. *Odor Thresholds for Established Occupational Health Standards.* Fairfax, Virginia: American Industrial Hygiene Association.

Akovali, Guneri. 2005. *Polymers in Construction.* Shawbury, Shewsbury, Shropshire, UK: Rapra Technology Limited.

CDC-FEMA. 2009. "Formaldehyde Exposure in Homes: A Reference for State Officials to Use in Decision-Making." Accessed January 28, 2015. http://www.cdc.gov/nceh/ehhe/trailerstudy/pdfs/08_118152_Compendium%20for%20States.pdf.

Jie, Steve Toloken Wang Zhan. 2013. "China's PVC Pipe Makers under Pressure to Give up Lead Stabilizers." *Plastics News.* Accessed January 19, 2015. http://www.plasticsnews.com/article/20130906/NEWS/130909958/chinas-pvc-pipe-makers-under-pressure-to-give-up-lead-stabilizers.

Kizilcan, Nigun. 2013–2014. "Experiment for Phenol–Formaldehyde Resin." In *Synthesis and Characterization of Macromolecules,* 91–93. Maslak, Istanbul: Polymer Science and Technology.

Manley, T. R. and Higgs, D. A. 1973. "Thermal Stability of Melamine Formaldehyde Resins." *Journal of Polymer Science: Polymer Symposia (Wiley),* 4 (3): 1377.

Paabo, Maya and Levin, Barbara. 1987a. "A Literature Review of the Chemical Nature and Toxicity of the Decomposition Products of Polyethylenes." *Fire and Materials*, 11: 68–69.

Paabo, Maya and Levin, Barbara. 1987b. "A Literature Review of the Chemical Nature and Toxicity of the Decomposition Products of Polyethylenes." *Fire and Materials*, 11: 55.

PVC. 2015. "How Is PVC Used?" *PVC*. Accessed February 1, 2015. http://www.pvc.org/en/p/how-is-pvc-used.

WaWa, Lady. 2005. "PEX Pipe and Water Quality." *Highwater Marks*. Accessed January 17, 2015. http://blog.highwaterfilters.com/pex-pipe-and-water-quality/.

Wescott, J., Spraul, B., Burn, M. 2012. "Formaldehyde Emissions from Composite Wood Products." Accessed January 28, 2015. http://www.whitehouse.gov/sites/default/files/omb/assets/oira_2070/2070_07092012-1.pdf.

WHO. 2010. "WHO Guidelines for Indoor Air Quality." *World Health Organization Europe*. Accessed May 18, 2015. http://www.euro.who.int/__data/assets/pdf_file/0009/128169/e94535.pdf.

Williamson, J. and Kavanagh, B. 1987. "Vinyl Chloride Monomer and Other Contaminants in PVC Welding Fumes." *AIHA Journal (American Industrial Hygiene Association)*, 48 (5): 532–436.

Woolley, W. D. and Wadley, A. I. 1970. "The Thermal Decomposition Products of Phenol-Formaldehyde Laminates Part 1. The Production of Phenol and Related Materials." *Fire Safety Science Digital Archive*. Fire Research Note #851. Accessed February 1, 2015. http://www.iafss.org/publications/frn/852/-1/view/frn_851.pdf.

6

Polymeric Foams

Throughout the ages, from cork and mud to rock wool and fiberglass, mankind has sought to insulate buildings against extremes in temperature, and the ultimate in insulation has arrived in the form of polymeric foams. The insulation value of mud brick (e.g., adobe) is R-0.03 per inch; cork is R-3 per inch; rock wool is R-2.5–3.7 per inch; and high density fiberglass batting is R-3.6–5.0 per inch. Without polymeric foams, we stop at the door to attainment of high R-value, quality insulation. Although used extensively in consumer products (e.g., upholstery, mattresses, and food containers), polymeric foams are primarily used as insulation material in buildings. An exception to the rule is carpet pads. Polymeric foam can have an insulation value as low as R-3.0 per inch and as high as R-10 per inch. For instance, polyethylene spray foam has an R-3.0. Polyurethane rigid panel insulation has an R-7.0–8.0 per inch, and silicone foam has an R-10.0 rating.

It is projected that by the end of 2015 the polymeric foams that will dominate the United States in terms of sales are polyurethanes and polystyrene. A much smaller portion (e.g., <25%) is anticipated for polyvinyl chloride, polyethylene/polypropylene, and others (BCC Inc. 2010a,b).

What differentiates plastics from polymeric foams? Simply stated, "A polymeric foam is a gas-filled plastic." In many cases the addition of a gas or gases can cause the plastic to expand as much as 100 times its original volume. The foaming process involves one, or a combination, of the following:

- Gaseous foaming—nitrogen, air, carbon dioxide, and/or an air/helium mixture
- Liquid foaming—hydrochlorofluorocarbons (HCFC), hydrofluorocarbons (HFC), and other chlorofluorocarbon substitutes
- Chemical foaming—mineral or chemical usually solids, that are able to decompose when heated to liberate large amounts of gas such as nitrogen, carbon dioxide, carbon monoxide, and hydrogen
  - The mineral foaming agents are generally salts (e.g., ammonium bicarbonate) and weak acids.
  - The chemical foaming agents are organic, consisting of one of several chemicals such as azo and diazo compounds, N-nitroso compounds, sulfonyl-hydrazides, azides, triazines, sulfonyl semicarbazides, urea derivatives, guanidine derivatives, and esters.

Foams are extruded, molded, or blown; they are open cell or closed cell. And in the end, the finished product is a solid–gas composite.

With all their positive contributions to energy efficiency in building, polymeric foams have received some bad press due to the toxic dalliances of a few. Each is up for review!

Foremost in Popularity

According to BCC Research, polyurethane foam excels in sales in the United States by a huge margin—reflecting its popularity in terms of cost, insulation R-value, confidence, and reliability. Polyurethane/polyisocyanurate foam sales are double that of polystyrene sales. Polystyrene foam sales are three times that of polyvinyl chloride foam (BCC Research 2010a,b). These numbers refer to overall sales volume—consumer products and building materials. In consumer products, many of the polyurethane (PU) foams are used in bedding, upholstery, and packaging, and the polystyrene (PS) foams are found in floatation devices and Styrofoam® food storage containers. Most of the PU and PS foams building materials are used in rigid insulation panels and spray foam sealants.

Polyurethane Foam

Unregulated product emissions: "Plastic Additives"

Regulated product emissions: TDI, MDI, and HDI

Combustion products: Carbon dioxide, carbon monoxide, oxides of nitrogen, and hydrogen cyanide

Widespread use in building materials: 1970s to present

PU foams, thermoset plastic polymers, are formed by reacting an isocyanate (e.g., MDI) with a polyol (e.g., glycerol). See Figure 6.1. They are polymerized in the presence of foaming agents and additives (e.g., catalysts, stabilizers, UV protection, fungicides, colorants, and flame retardants), and the end product is about 97% gas, an excellent thermal insulation material (Feldman 2005, pp. 246–247).

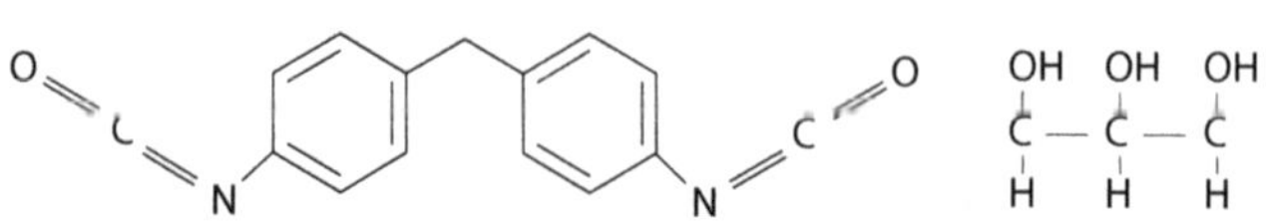

FIGURE 6.1
Polyurethane monomers: methylene diisocyanate (left); glycerol (right).

PU foam can be flexible or rigid, open or closed cell which depends on: (1) the ratio and chemical makeup of the starting components; (2) foaming agent(s); and (3) type of technology (Feldman 2005, p. 246). Most PU foam has an isocyanate/polyol ratio of around 100:1; the higher the ratio, the more brittle the product. For rigid, closed cell foam, the most common foaming agents are HCFCs. Whereas flexible, open cell foams are used mostly in consumer products (e.g., cushions, furniture foam padding, and mattresses), rigid, closed cell PU insulation is extensively used in construction (e.g., faux stone panels). The PU rigid panels (CFC/HCFC expanded) insulation has a very high insulation value—R-7.0 to 8.0 per inch. In construction, PU foam is used less extensively in flexible, open cell noise insulation and in expandable rigid sealants.

Icynene MD-C-200™ is a spray-on, pour fill semi-rigid "open cell PU foam" that seems to be gaining popularity despite its expense and controversial concerns (e.g., soaks up water and may be prone to infestation by wood-boring insects) (Blog Spot 2008). The manufacturer claims "no detectable emissions" and an environmentally safe water-based blowing agent—the likely reasons for its popularity.

As for probable emissions from other PU foam, free isocyanates are likely to be residual in the end products. Isocyanates include toluene diisocyanate, (TDI), methylene diphenyl diisocyanate (MDI), and hexamethylene diisocyanate (HDI). As they have no odor, isocyanate exposures offer no warning—other than symptoms. The symptoms of exposure to MDI are eye, nose, and upper respiratory irritation, and MDI can cause dermal sensitization. Although less frequently used in the manufacture of PU foams, TDI symptoms are more serious than that of MDI. TDI is a strong irritant of the eyes, mucous membranes, and skin, and it is a potent respiratory tract sensitizer. HDI symptoms are similar but to a lesser degree to that of TDI.

According to most manufacture MSDSs for PU foam, the hazardous thermal decomposition products are carbon monoxide, carbon dioxide, oxides of nitrogen, and hydrogen cyanide. Other manufacturers also list free isocyanates, acetaldehyde, and acrylonitrile.

Polyisocyanurate Foam

Unregulated product emissions: "Plastic Additives"

Regulated product emissions: MDI

Combustion products: Carbon dioxide, carbon monoxide, oxides of nitrogen, and hydrogen cyanide

Widespread use in building materials: 1970s to present

Polyisocyanurate (PIR) foam, a thermoset plastic, is similar to polyurethane in chemical composition. See Figure 6.2. Yet, due to processing differences, PIR foam has greater bond strength which makes it more brittle, and it is chemically and thermally more stable than PU foam. The foil faced PIR rigid

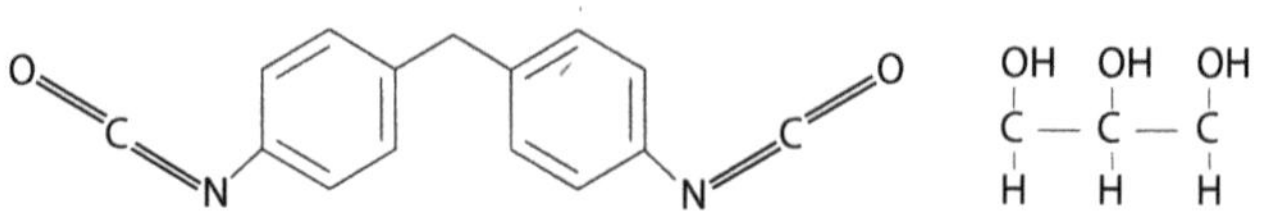

FIGURE 6.2
Polyisocyanurate monomers (similar to polyurethane): methylene diisocyanate (left); glycerol (right).

panel (pentene expanded) insulation has an insulation value approaching that of PU—R-6.8 per inch. The PIR isocyanate/polyol ratio ranges between 200:1 and 500:1. The higher the ratio, the stronger the bonds, the more brittle the foam. The brittleness of PIR foam is key to the rationale for using the less brittle PU foams in construction.

Refer to preceeding "Polyurethane Foam" for emissions and combustion products.

Polystyrene Foam

Unregulated product emissions: "Plastic Additives"

Regulated product emissions: –

Combustion products: Carbon dioxide and carbon monoxide

Widespread use in building materials: Originated in 1959

Polystyrene foam, also referred to as expanded polystyrene (EPS) foam, is a rigid, closed cell thermoplastic polymer which is expanded up to 40 times the original volume of polystyrene beads (Figure 6.3). In construction, EPS foam is molded (e.g., pipe insulation) or extruded (e.g., Styrofoam® panels). The foaming agents may be one, or a combination, of hydrocarbons and/or halocarbons. Halocarbons impart flame retardant properties to the EPS—a desirable characteristic in extruded foam panel insulation. The foaming agent for molded pipe insulation is a hydrocarbon. The end product is about 98% gas (e.g., air), and the PE insulation values range from R-3.85 per inch for low density molded EPS to R-3.6–5.4 per inch for EPS.

Thermal insulation is the sole function of EPS in building materials. It's all in the air, all in the R-values. EPS is used in interior insulation, exterior

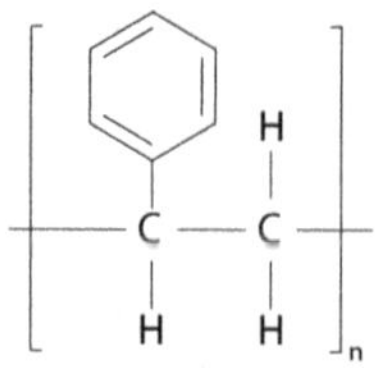

FIGURE 6.3
Polystyrene.

wall insulation (e.g., EIFS), roof deck insulation, masonry wall insulation, and pipe insulation. Recycled EPS is further used as the support in insulated concrete forms (ICF). Also high density PU foam is used in simulated wood.

Now, a paradox deserves a microcosm of attention. After having reviewed "multiple" Safety Data Sheets, I must ask, "Why does a product that is referred to as polystyrene foam insulation adhesive have neither styrene nor polystyrene in its ingredients?" The answer is yours to ponder!

Although most manufacturers claim that there is no off-gassing from EPS after production, the U.S. EPA claims, "Indoor air is the principal route of styrene exposure for the general population. Average indoor air levels of styrene are in the range of 1–9 $\mu g/m^3$ (i.e., 2 ppb or 0.002 ppm), attributable to emissions from building materials, consumer products, and tobacco smoke" (U.S. EPA 2013). There are other claims of "possible" residual styrene monomer after production (Akovali 2005b, p. 436). The levels of anticipated styrene emission from EPS are anticipated to be very low, well within the strictest of exposure limits.

Styrene has a sharp, sweet odor detectable at 0.34 ppm (U.S. EPA 2013), below the ACGIH exposure limit of 20 ppm. Emissions from PS building materials at temperatures below 158°F (70°C) have not been reported.

During a fire, most PS foam burns with a luminous yellow, sooty flame and sweet odor, slowly melting and bubbling until the ignition source has been removed—at which time it ceases to burn. Although to every rule there is an exception. Wherein a flammable hydrocarbon (e.g., pentane) has been used as a blowing agent, the flammable gases trapped in the air pockets will continue to burn. A fire retardant (i.e., brominated hydrocarbons) added during the foaming process will retard the spread of fire. The combustion products of polystyrene not foamed with a halocarbon are carbon dioxide and water, whereas those that have been foamed with halocarbons produce halogen and acid halogen gases in fires exceeding 430°C (Owens Corning 2007). As lumber burns between 282°C and 480°C, the toxic combustion gases will not be generated until a fire is well out of control and temperatures not survivable.

Minor Contributors

Polyvinyl chloride foam is one-third the United States volume in sales of polystyrene, one sixth that of polyurethane foams. Polyethylene is only slightly less in sales volume than polyvinyl chloride (BCC Research 2010a,b). In consumer products, PVC foam is used in road signs and billboards, while PE foam is used in floatation devices, sheet packaging, and foam padding for delicate instruments (e.g., camera case liners). In construction, PVC is used in mostly in thermal insulation, and PE foam is used in door and window backer rods and wrapped pipe insulation. Due to the lack of information and limited use in construction, PE foam is only mentioned.

Polyvinyl Chloride Foam

Unregulated product emissions: (See "Plasticizers" and "Plastic Additives.")

Regulated product emissions: Hydrogen chloride (detection)

Combustion products: Carbon dioxide, carbon monoxide, and hydrogen chloride

Widespread use in building materials: Originated in 1970s

Polyvinyl chloride (PVC) foam is a unique rigid, closed cell thermoset polymer that has thermoplastic properties. What does this mean? Well, PVC foam is actually a mixture of PVC and polyurea—a three dimensional copolymer matrix (Diab 2015). See Figure 6.4. The end product has good mechanical characteristics (e.g., tensile, compressive, and shear strength), resistance to termites and bacterial growth, good fire stability, and resistance to crumbling under impact or vibration.

During manufacturing, PVC and polyurea are mixed with additives and a chemical foaming agent (e.g., azobisformamide) under controlled conditions and expressed into a mold. Once the mold is filled, the mix is placed into a large press and heated. Upon completion, the solid material emerges from the mold. The solid material is then subjected to a hot water bath at which time it expands. As the hot expanded foam cools, it cures to a solid form. Azobisformamide foaming agent is an unregulated toxin (see "Plastic Additives"). For more flexible products (e.g., foam flooring), plasticizer is added.

In construction, PVC foam is used in nonstructural thermal insulation, rigid water/frost insulation, exterior/interior rigid wall panels, and flexible coated fabric flooring (Akovali 2005a, p. 5, 245). As a thermal insulator the insulation value is highly variable ranging from R-3.6–6.3 per inch (MatWeb 2015).

The PVC monomer, vinyl chloride, has a detectable sweet odor (10 ppm) in excess of the ACGIH standard (1 ppm). Vinyl chloride is not reported to be a problematic emission after polymerization.

Due to its high chlorine content (57% chlorine by weight), PVC decomposes at relatively low temperatures to give a high yield of hydrochloric acid into the air along with a wide variety of minor organic compounds. At higher temperatures (e.g., combustion), CO_2 and CO are generated. Most reports also claim detectable levels of hydrogen chloride as a by-product of PVC combustion.

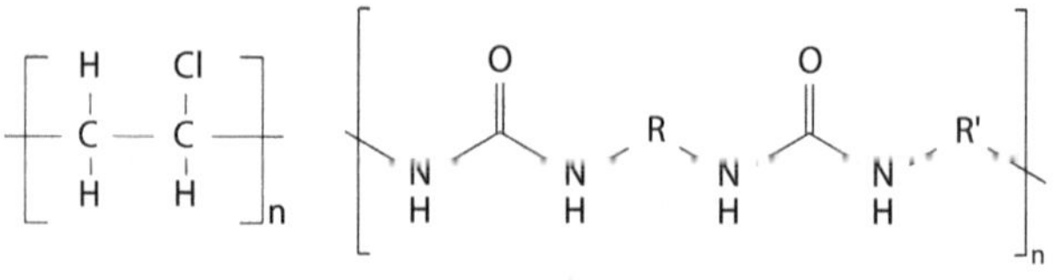

FIGURE 6.4
PVC foam components: polyvinyl chloride (left); polyurea (right).

Polyethylene Foam

Unregulated product emissions: "Plastic Additives"

Regulated product emissions: –

Thermal decomposition products: Carbon dioxide and carbon monoxide

Widespread use in building materials: Unknown

Polyethylene (PE) foam is processed by melting low-density polyethylene (LDPE) with additives and blowing a gas foaming agent (e.g., carbon dioxide) into the molten polymer. The mix is extruded, and it expands about 30 times the unprocessed LDPE volume. The end product is a closed cell foam that has excellent buoyancy, superb strength, and tear resistance, is shock resistant, flexible, impervious to mold, and chemically/grease resistant. These properties are good for consumer products but of limited use in construction—backer rod for sealing around exterior doors and windows and pipe insulation.

In a 2010 Fin Pan Pro Panel Foam Backer Board MSDS, the Chinese manufacturer claims flame retardant additives as "proprietary," while there is a slight hydrocarbon odor. It is not clear as to the component source of the odor. "Hazardous decomposition products" are more extensive than that of unfoamed PE plastic.

> Does not normally decompose. Evolution of small amounts of hydrogen halides occurs when heated above 250°C. Under high heat, non-flaming conditions, small amounts of aromatic decomposition products depend upon temperature, air supply and the presence of other materials. Hazardous decomposition products may include and are not limited to ethyl benzene, aromatic compounds, aldehydes, hydrogen bromide, hydrogen chloride, hydrogen fluoride, polymer fragments, and styrene.

The combustion by-products are likely due to the fire retardant, and thermoplastic polymers are likely to snuff themselves out when a propagation fire source is removed. Emissions are likely due to additives, and combustion products are likely due to additives as well.

Infamy Worthy of Mention

At this moment and space in time, there is one thermoset polymeric foam that has entered the halls of infamy. Yet, it still remains on the United States and European markets. Although attempts to ban UFFI from these countries were unsuccessful, the material is presently banned in Canada.

Urea Formaldehyde Foam

Unregulated product emissions: "Plastic Additives"

Regulated product emissions: Formaldehyde

Thermal decomposition products: Carbon dioxide, carbon monoxide, aldehydes, and organic acids

Widespread use in building materials: 1970s–1980s

UF foam is a mixture of urea formaldehyde resin (Figure 6.5), an acidic foaming agent, and a propellant (e.g., air). UF foam, predominantly used as a "retrofit" insulation, is open cell, friable, and easily damaged. For this reason, UF foams are produced and applied on site. The UF resin pre-polymer stock, water solution, and a hardener-surfactant solution are sprayed from a portable foaming unit. The application mix also includes a catalyst and foaming agents. The foam is sprayed, or pumped, into wall cavities, attics, and other structural voids within a building. Curing usually takes about 24 hours—if formulated properly.

If not formulated properly, UF foam insulation deteriorates and loses its insulating capacity. Its expected insulation value is around R-4.0–4.6 per inch. Well, UF formulation is only the tip of the iceberg!

During the application and curing process, UF foam off-gases formaldehyde indoors. In the 1970s, there were efforts to provide tighter, more energy efficient building, and formaldehyde gases had a greater opportunity to get trapped and retained in the occupied spaces. Building occupants began to complain about irritating odors and adverse health effects. Then, a demon came forth in the name of "formaldehyde."

> In the early 1980s, the media went on a feeding frenzy, and lawyers found another target for litigation. In 1982, the U.S. Consumer Product Safety Commission banned the sale of UF foam insulation. Shortly thereafter a law prohibiting its sale was enacted, later to be struck down in 1983 because there was "no substantial evidence clearly linking UF foam insulation to health complaints." The seed of discontent had been planted, and UF foam insulation is not widely used in the USA today. It was banned in Canada in 1980—after some 10,000 homes had been insulated with UF foam insulation. Yet, it not only remains in use throughout Europe today, but UF foam insulation is still considered an excellent retrofit insulation

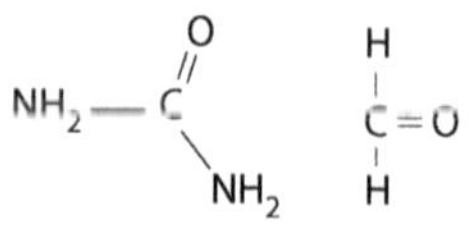

FIGURE 6.5
Urea formaldehyde monomers—urea (left); formaldehyde (right).

by Europeans. Clearly, differences in opinion prevail to this day! So, let's take a closer look at formaldehyde off-gassing from UF foam insulation.

There is no question that during application and curing, formaldehyde is generated. Yet, all are not created equal. The methods in which UF foam was mixed and sprayed on site were highly variable, and the amount of post application off-gassing was variable.

Over time, the off-gassing dissipates, and formaldehyde off-gassing lessens with time. There are claims alleging that months or, at the most, a few years after application, formaldehyde is no longer detected emanating from the foam insulation. Yet, this claim has not been substantiated.

The rate at which formaldehyde off-gasses is also dependent upon temperature and humidity. The higher the humidity and higher the temperature, the greater the off-gassing. Water damage and high moisture content in a UF foam insulated wall cavity may result in formaldehyde emissions—years after application.

Formaldehyde is irritating to the eyes, skin, and respiratory tract. Its pungent odor is detectable by some at levels as low as 27 ppb. This level may not be detectable by the majority of people, and the consensus of the U.S. EPA and state agencies is that most people detect formaldehyde are more elevated levels (0.8–1 ppm) that exceed the WHO nonoccupational settings limit (100 ppb in 30 minutes) (WHO 2010, p. 142). The WHO limit alleges the guideline was developed to protect against sensory irritation in the general population, and prevent the effects on lung function as well as long-term health effects, including nasopharyngeal cancer and myeloid leukemia (WHO 2010, p. 142).

Most manufacturers of UF composite wood products list thermal decomposition products on their MSDSs as carbon monoxide, aldehydes, and organic acids. One manufacturer additionally lists hydrogen cyanide and polynuclear aromatic compounds.

Summary

Polymeric foams have allowed modern structures to reach new heights in energy efficiency. Whereas fiberglass batts have an insulation value of R-3.1–4.3 per inch, polyurethane rigid panels can attain a value of R-8 per inch.

In the 1970s, UF spray foams were used to insulate older homes which had poor or no insulation. Formaldehyde emissions became a political football, and UF retrofits became a pariah. Open cell PU (e.g., Icynene®) has since provided a smidgen of hope to replace UF spray foam. Yet, it is not without the dark cloud of controversy. So, in the meantime, building material polymeric foams are predominantly closed cell "insulation panels."

Product emissions from polymeric foams are fairly consistent in that slight residual monomers "may" off-gas detectable monomers. Monomer off-gassing has not been confirmed! However, plastic additive emissions are likely wherein the additives are chemicals—particularly chemical foaming agents. Also, plasticizers can off-gas from limited usage polymeric foams such as PVC foam flooring. Most, but not all, of the plasticizers and plastic additives are unregulated, and many are potentially irritants/toxins. Product manufacturer identification of the unregulated irritant/toxins is not required.

Combustion by-products of all the polymeric foams are carbon dioxide and carbon monoxide. Additionally, the PU/PIR foam thermal decomposition products include oxides of nitrogen and hydrogen cyanide, and PVC foam includes hydrogen chloride—chlorine in its PVC monomer.

References

Akovali, Guneri. 2005a. *Polymers in Construction*. Shawbury, Shewsbury, Shropshire, UK: Rapra Technology Limited.

Akovali, Guneri. 2005b. Possible health issues related to plastics construction materials and indoor atmosphere. In *Polymers in Construction*, ed. Guneri Akovali, Shawbury, Shewsbury, Shropshire, UK: Rapra Technology Limited.

BCC Research. 2010a. "Polymeric Foams." *BCC Research*. August. Accessed February 2015. http://www.bccresearch.com/market-research/plastics/polymeric-foams-pls008g.html.

BCC Research. 2010b. "Polymeric Foams." *BCC Research*. August. Accessed August 2015. http://www.bccresearch.com/market-research/plastics/PLS008F.html.

Blog Spot. 2008. *Icynene Dangers*. Accessed March 4, 2015. http://icynenedangers.blogspot.com/.

Diab. 2015. "PVC Foam." *Net Composites*. Accessed February 26, 2015. http://www.netcomposites.com/guide/pvc-foam/91.

Feldman, Dorel. 2005. "Plastics and Polymer Composites: A Perspective on Properties Related to Their Use in Construction." In *Polymers in Construction*, ed. Guneri Akovali. Shawbury, Shewsbury, Shropshire, UK: Rapra Technology Limited.

MatWeb. 2015. "DIAB Klegecell® R 45 Rigid, Closed Cell PVC Foam Core Material." *Mat Web*. Accessed February 26, 2015. http://www.matweb.com/search/datasheet.aspx?matguid=9bc6c71838cd4c4e93a4e502647a7db5.

Owens Corning. 2007. "Celfort(R) Extruded Polystyrene Insulation." *Material Safety Data Sheet*. Toledo, OH, May 21.

U.S. EPA. 2013. "Styrene." *Technology Transfer Network—Air Toxics Web Site*. October 18. Accessed February 25, 2015. http://www.epa.gov/ttn/atw/hlthef/styrene.html.

WHO. 2010. "WHO Guidelines for Indoor Air Quality." *World Health Organization Europe*. Accessed May 18, 2015. http://www.euro.who.int/__data/assets/pdf_file/0009/128169/e94535.pdf.

7

Elastomeric Polymers

An elastomeric polymer is a synthetic rubber whose evolution is a story to inspire the imagination and to herald appreciation of the complex evolution of rubber.

> When Christopher Columbus revisited Haiti on his second voyage, he observed some natives playing ball. Columbus' own men had brought their Castilian wind-balls to play with in idle hours. However, they found that the balls of Haiti were incomparably superior toys; they bounced better. These high bouncing balls were made from a milky fluid, the consistency of honey, which the natives harvested by tapping certain trees and then cured over the smoke of palm nuts.
>
> In 1736, a French astronomer was sent by his government to Peru to measure an arc of the meridian. He brought home samples of the milky fluid and reported that the Indians used it for lighting. He wrote that it burned without a wick very brightly and that the Indians made shoes from it which were waterproof. The Indians collected the gummy fluid from trees in pear-shaped bottles on the necks of which they fasten wooden tubes. Pressure on the bottle sends the liquid squirting out of the tube, so they resemble syringes. Their name for the fluid, he added, was cachuchu or caoutchouc. Thirty-four years later, an English writer wrote about a different use for the tree gum and a new name. A stationer accidentally discovered that it would erase pencil marks ... by rubbing. Thus, it came to be called that which we refer to today as "rubber" (Bellis 2015).

About the year 1820, American merchants sailing between Brazil and New England, used rubber as ballast on their voyage home, "dumping the rubber" on the wharves as they returned to Boston. One enterprising merchant brought home for sale five hundred Brazilian indigenous, native rubber shoes. Although they were thick, clumsy, and heavy, they were an overnight success. A few years later, half a million pairs were imported and sold—annually. As the Brazilian shoe success was on the rise, New England entrepreneurs glommed onto the gravy train. They bid against one another along the wharves for used rubber ballast, and manufactures produced "North American-made rubber shoes."

Europeans developed waterproof rubber fabrics which became an overnight success. But the thrill wore off in short order as the fabrics began to melt in the summer heat, and they became rigid/brittle in the winter freeze. Then, in 1842, Goodyear developed a process called vulcanization. Problem solved! The vulcanized rubber retained its elasticity in extreme temperatures. Subsequently, rubber found another niche that ultimately lead to production of the universally coveted rubber tires.

The first synthetic rubber, styrene–butadiene copolymers, was launched in 1875. At the time, it marginally duplicated natural rubber. In 1882, the isoprene polymer, a natural rubber equivalent, was developed. This time it did replicate natural rubber, but the production process was expensive. Then, out of necessity, the Germans industrialized the process during World War I. But in the U.S. synthetic rubbers were put on the back burner until World War II. Without synthetic rubber, the U.S. war efforts would have been severely compromised. Thus, the driving force for synthetic rubber was borne in the bowls of war. Post World War II, the world market exploded, and elastomer research and development have since been on the rise.

An elastomer is a polymer that can be stretched, twisted, and deformed and when release will reconfigure back into its original form. Synthetic polymers withstand temperature extremes. They can withstand high pressures and bounce back. They can be twisted into a pretzel and bounce back. They can be abused and bounce back. They are contortionists, and they don't readily burn. Synthetic elastomers are truly a polymer like none other!

Currently, however, elastomeric polymers in construction represent a very small piece of the pie. Synthetic rubber is used in low pitch rubber roof coverings as well as in construction adhesives, sealants, and roofing asphalt. Beyond the polymerization process, plasticizers (e.g., diisooctyl phthalate) and additives (e.g., fillers) are a component of the formulation. Rubber adhesives and sealants are suspended in organic solvents. Plasticizers, additives, and organic solvents remain unbound, available for release into the environment!

Butyl Rubber

Unregulated product emissions: Eye, skin, and respiratory tract irritation

See also "Plasticizers" and "Plastic Additives"

Regulated product emissions: –
Combustion products: Carbon dioxide and carbon monoxide
Widespread use in building materials: 1940s to present

Butyl rubber, a thermoset elastomer, comprises of 98% polyisobutylene and 2% isoprene. See Figure 7.1. Although the terms butyl rubber and polyisobutylene are often used interchangeably, polyisobutylene is not a

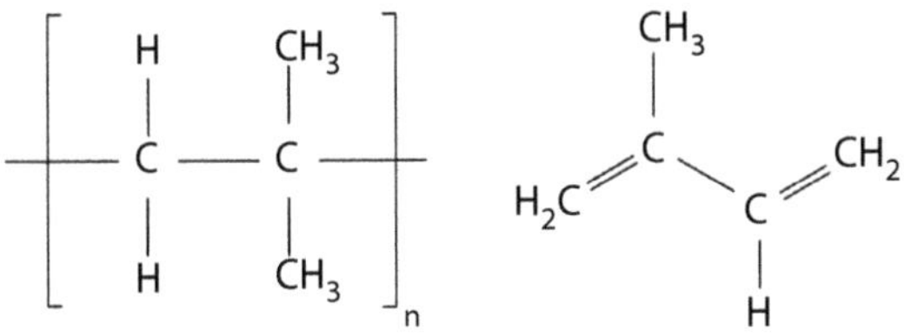

FIGURE 7.1
Butyl rubber components: polyisobutylene (left); isoprene (right).

rubber—without isoprene (which provides for crosslinking during vulcanization). The polyisobutylene monomer is isobutylene, and the polymerization process is a high-energy exothermic process. Thus, polymerization requires extreme cold (e.g., −100°C) to control the reaction rate. In construction, butyl rubber is used as rubber sealing tape, caulk, and as a recycled rubber "additive" in roofing asphalt.

Although the monomers, isobutylene and isoprene, are nontoxic, they are flammable. The butyl rubber sealants and adhesives are, however, suspended in VOCs (e.g., up to 35% by weight). The VOC content varies by manufacturer; and the toxic potential of the organic components should be assessed on the basis of published information (e.g., MSDS). Nontoxic fillers (e.g., kaolin and limestone) may be as high as 70% by weight in butyl sealants. Plasticizers and other additives, however, cannot be ruled out. Flammable organic vapors emissions are likely.

Chlorosulfonated Polyethylene

Unregulated product emissions: Eye, skin, and respiratory tract irritation

See also "Plasticizers" and "Plastic Additives"

Regulated product emissions: Carbon tetrachloride

Combustion products: Carbon dioxide, carbon monoxide, nitrogen oxides, and hydrogen cyanide

Use in building materials: 1965 to present

Chlorosulfonated polyethylene (CSPE) rubber, also referred to as Hypalon®, is the odd one! See Figure 7.2. It is comprised of a thermoplastic polyethylene backbone that is converted to an elastomer by chlorosulfonation and chlorination—the addition of chlorine and sulfur dioxide. Once this

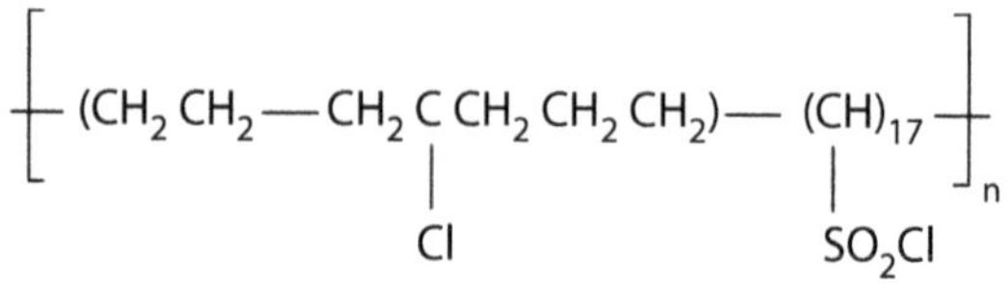

$$\left[(CH_2\,CH_2 - CH_2\,\underset{\displaystyle Cl}{C}\,CH_2\,CH_2\,CH_2) - \underset{\displaystyle SO_2Cl}{(CH)}_{17} \right]_n$$

FIGURE 7.2
Chlorosulfonated polyethylene.

process has been completed, the polymer can no longer be recycled—a characteristic attributable only to thermoplastic polymers. The end product has a chlorine content of 27%–45% and a sulfur content of 0.8%–2.2%. Owing to the presence of chlorine, CSPE is fire, oil, and microbe resistant. It also resists the effects of ozone, UV light, and inorganic acids (e.g., acid rain), and it retains its elasticity at a temperature of –31 to 284°F (–35 to 140°C). In building materials, CSPE is used predominantly in low pitch roof membranes.

Reports and MSDSs are limited regarding CSPE. In a China originated MSDS, chlorosulfonated polyethylene contains 96% CSPE, about 2% talc (no asbestos), 0.2% carbon tetrachloride, 0.02% chloroform. They claim all components (including the carbon tetrachloride and chloroform) are "not hazardous" (WestMoonint Corp. LTD 2010). Yet, carbon tetrachloride and chloroform are indeed regulated toxins and listed by the IARC as possible human carcinogens. WestMoont further disclosed the following:

> Some individuals with specific sensitivities may injure their eyes, nose, throat slightly … (when they are exposed to) … fumes for a long time …. Decomposition can be as low as 150 degrees C (302°F) … HAZARDOUS DECOMPOSITION PRODUCTS: carbon tetrachloride is liberated on standing; evolution is accelerated with heat. Decomposition products (at elevated temperatures in excess of 150°C) include carbon monoxide, hydrogen chloride, sulfur dioxide, and hydrocarbon oxidation products including organic acids, aldehydes, and alcohols … In combustion case(s), (CSPE) emits toxic fumes of carbon oxides … nitrogen (oxides) and HCN (hydrogen cyanide) (WestMoonint Corp. LTD 2010).

Relying on the limited information available, and reading between the lines, one might readily conclude irritant/toxic carbon tetrachloride emissions are likely, especially in the newer CSPE products.

At temperatures in excess of 150°C (302°F), thermal decomposition products of CSPE include carbon monoxide, hydrogen chloride, sulfur dioxide, and hydrocarbons (e.g., organic acids, aldehydes, and alcohols). Combustion products are carbon dioxide, carbon monoxide, nitrogen oxides, and hydrogen cyanide.

Ethylene Polypropylene Diene Monomer Rubber

Unregulated Product Emissions: Eye, skin, and respiratory tract irritation

See also "Polymer Additives"

Regulated product emissions: –

Combustion products: Carbon dioxide, carbon monoxide, and hydrocarbons

Used in building aterials: 1965 to present

Ethylene polypropylene diene monomer (EPDM) rubber, a thermoset elastomeric terpolymer, is produced by the polymerizing ethylene, propylene, and a small percentage of diene. In commercial use rubbers, the diene represents 4%–5% by weight and the ethylene represents 30%–70% by weight of the polymer. The higher the ethylene content, the more the hardness and tensile strength. Solid, as opposed to liquid, EPDM is vulcanized with sulfur and/or other equivalent accelerators.

EPDM rubber can be stretched up to 400%; it does not tear, crack, or split; it retains its flexibility at temperatures between –50°C and 150°C; and it has a service life up to 30 years (Akovali 2005, pp. 78–85). Its primary use is in the consumer market (e.g., synthetic tires). In the building industry, however, EPDM rubber is used extensively in the production of low cost, high-quality roof membranes—more than 55% of all flat/low slope roofs installed today. It is also used to make safety mats, to waterproof retaining walls, foundations, and structures (e.g., basements).

According to manufacturers, EPDM roof membrane components are up to 50% fillers (e.g., carbon black or kaolin clay) and up to 35% additives which are often proprietary (Carlisle Syn Tec 2013).

The ethylene monomer is flammable but not toxic, and the polypropylene is inert. The most commonly used diene components in the manufacture of EPDM are 1,4-hexadiene, dicyclopentadiene, and ethylidene norbornene. Although not regulated by OSHA, the dienes may cause eye, skin, and respiratory tract irritation. According to NIOSH, dicyclopentadiene may cause irritation to eyes, skin, nose, throat; incoordination, headache; sneezing, cough; and skin blisters. Dienes are crystalline solids and have a "disagreeable camphor-like odor." As a low level component in rubber, diene emissions are unlikely from EPDM rubber.

According to many manufacturers, combustion products include carbon dioxide, carbon monoxide, and hydrocarbons. Some include alcohol. Some include sulfur dioxide. Some say "corrosive and toxic fumes"—not itemizing them. Some fail to report, commenting there is no information available. Others claim chlorine gas with no identified source of chlorinated compounds—likely originating from an unnamed additive. These itemized combustion products, as reported by different manufacturers, demonstrate inconsistencies in data disclosures.

Neoprene Rubber

Unregulated product emissions: Skin and respiratory tract irritant

See also "Polymer Additives"

Regulated product emissions: Chloroprene (possible but unlikely)

Combustion products: Carbon dioxide, carbon monoxide, small amounts of sulfur dioxide, and trace amounts of hydrogen cyanide

Widespread use in building materials: 1931 (invented); 1960s to present

Neoprene, also referred to as polychloroprene, is a synthetic rubber produced by the polymerization of chloroprene. See Figure 7.3. Its primary use in building materials has been roof membranes. Neoprene exhibits good chemical stability, and maintains flexibility over a wide range of temperatures (i.e., −58°F to 248°F; −50°C to 120°C). According to the MSDSs, stabilizers and other additives are common, and combustion by-products may include chlorinate and aliphatic hydrocarbons.

The monomer chloroprene is a regulated chemical and listed by the IARC as a possible human carcinogen. Chloroprene's sweet ether-like odor threshold (15 ppm) is 15 times greater than the NIOSH recommended limit (1 ppm), one and a half times the ACGIH recommended limit (10 ppm), and less than the OSHA limit (25 ppm).

Once polymerized, chloroprene emissions are possible, but not likely. As additives are not part of the polymer matrix and not bound within the polymer complex, additives may pose an emissions hazard. The MSDSs for neoprene indicate skin and respiratory tract irritation associated with product handling in confined spaces, and the odor is described as rubber-like which is the odor description for chloroprene in the ACGIH Odor Thresholds Manual (AIHA 1997).

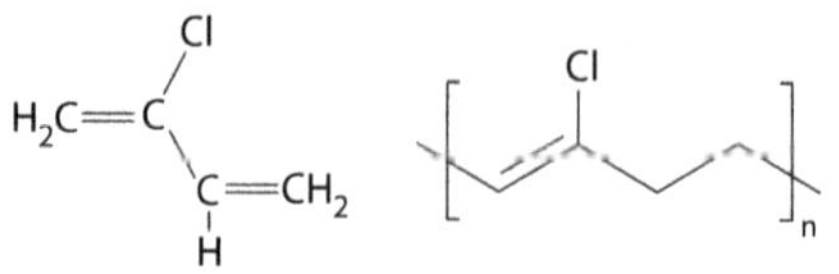

FIGURE 7.3
Neoprene rubber: chloroprene monomer (left); polychloroprene (right).

Polyurea/Polyaspartic Elastomer

Unregulated product emissions: Eye, skin, and respiratory tract irritant.

See also "Polymer Additives"

Regulated product emissions: –

Combustion products: Carbon dioxide, carbon monoxide, formamide, and isocyanic acid

Widespread use in building materials: 1990s to present

Polyurea, a thermoset elastomeric copolymer, is the product of a two part reaction between isocyanates and amines—as monomers and/or partial/fully polymerized monomer units. The extent of monomer polymerization in Part A and Part B is highly variable, based on different manufacturer formulations. Some use polymerized isocyanates in Part A and unpolymerized amines in Part B. Some use polyisocyanates (with less than 1% monomer) in Part A and polyether polyol in Part B. There is inconsistency amongst manufacturers as well as incomplete disclosure—trade secrets are *jealously guarded.*

Polyaspartic is an "aliphatic" polyurea (see Figure 7.4), comprised of aliphatic isocyanates (e.g., HDI) and amines (e.g., ethylenediamine). On the other hand, polyurea can be comprised of aliphatic or aromatic isocyanates (e.g., MDI).

All forms of polyurea resist water, and many polyureas have a tensile strength of up to 6000 psi with a flexible elongation rate up to 500%. They are strong and stretchy—ideal for coatings and self-healing paints.

Spray-applied polyurea coatings started slowly in the 1990s, but as its superior qualities were recognized, polyurea in waterproof coatings catapulted into the twenty-first century with a big bang. The coatings are puncture proof, impact resistant, chemical resistant, do not emit VOCs, and tolerate extreme temperatures (i.e., –20°C to >50°C). In addition to these qualities, polyaspartic resins are UV resistant and do not yellow with time as do the other polyurea resins.

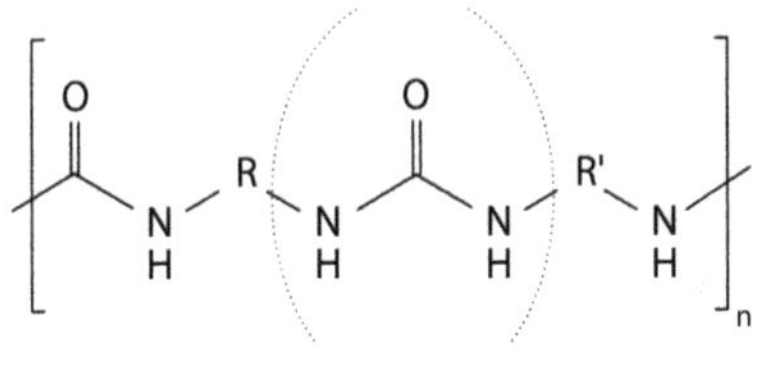

FIGURE 7.4
Polyurea.

Polyurea coatings have a long service life (e.g., 25–30 years) and a rapid curing rate which can be anywhere from a few seconds up to 120 minutes—depending on the mix and catalysts. With such spectacular properties, how could it fail? Polyurea waterproof coatings are on the rise! Expect to see polyurea products used with greater frequency in construction.

In buildings, polyurea coatings/paints are used to waterproof and protect concrete (e.g., floors), steel (e.g., beams), wood (e.g., decks), and other polymers (e.g., foam mats). It can be used as an encapsulant (e.g., radon barrier), water barrier (e.g., roof protection), joint fill and caulk (e.g., expansion joints), crack-resistant caulk and sealants.

MDI, HDI, and ethylene diamine are regulated monomers that are allegedly tied up in the polymerization process at the factory. Yet, due to the variability in manufacture reporting and formulations, the regulated monomers are due a modicum of attention. The odor threshold for methylene biphenyl isocyanate (MDI), a polyurea monomer, is 400 ppb which exceeds the ACGIH TLV of 5 ppb and the OSHA Ceiling Limits of 200 ppb. The odor description is vague—no odor, musty odor, and solvent like odor. Ethylene diamine, a polyamine monomer, has an ammonia-like odor at 10 ppm which is the same as the ACGIH TLV of 10 ppm. Shortly after application, emissions are possible, but highly unlikely—due to the pre-polymerized components and to rapid curing times.

According to a Marquette University research publication, the thermal decomposition products of polyurea in the presence of flame retardants (i.e., ammonium polyphosphate and expandable graphite) were carbon dioxide, carbon monoxide, formamide, isocyanic acid, simple hydrocarbons, and complex char (Awad and Wilkie 2010, p. 11).

Silicone

Unregulated product emissions: See "Plasticizers" and "Polymer Additives"

Regulated product emissions:–

Combustion products: Silicon dioxide (silicone); carbon dioxide, carbon monoxide, and silicon dioxide (organosilicon)

Widespread use in building materials: 1970s to present

Silicone, an inorganic elastomer, is a polymer composed of silicon, oxygen, and hydrogen. Silicon rubber is generally nonreactive, stable, and retains its flexibility in temperatures ranging from −55°C to 300°C. In buildings, silicone is used as a waterproof sealant. Sometimes organosilicon polymers are also referred to as silicone and are often used in conjunction with inorganic

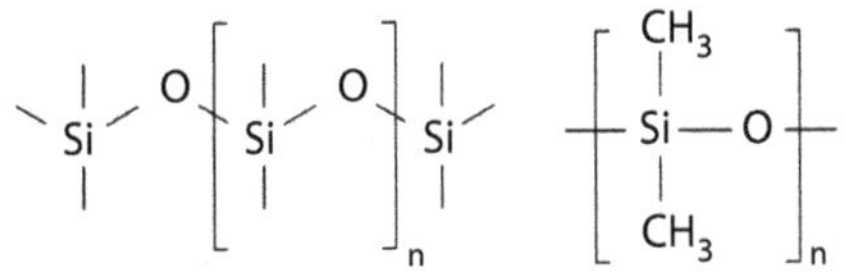

FIGURE 7.5
Silicone elastomer (left); organosilicon elastomer (right).

silicone (e.g., 2%–4% organosilicon). See Figure 7.5. All silicones are inert, nontoxic, and nonflammable. The combustion products of organosilicon include amongst the usual organic combustion products—formaldehyde at 300°C.

Summary

Elastomers were borne of synthetic rubber and morphed into other forms particularly in construction. The most frequently encountered rubber building material is EPDM waterproof membranes used on flat/low slope pitch roofs, foundations, and retaining walls. Other elastomers are used as high quality, durable sealants/caulk and as components of high quality, self-healing protective coatings (e.g., paints).

Although there are in some cases of regulated monomers, emissions of known toxins possible, but unlikely after product curing has been completed. Curing may have been completed at the factory. It may be instantaneous upon application, or the monomers are nontoxic.

With the exception of silicone, there is a consensus of the manufacturers claiming "eye, skin, and respiratory tract irritation" associated with each of the elastomers—despite any identified hazardous/toxic components. Unregulated components do not require identification on the MSDSs. Plasticizers and additives are rarely, if ever, identified. Components are ear marked "trade secret" or "proprietary."

References

AIHA. 1997. *Odor Thresholds for Chemicals with Established Occupational Health Standards*. Fairfax, VA: AIHA.

Akovali, Guneri. 2005. *Polymers in Construction*. Shawbury, Shewsbury, Shropshire, UK: Rapra Technology Limited.

Awad, Walid and C. Wilkie. 2010. "Investigation of the Thermal Degradation of Polyurea: The Effect of Ammonium Polyphosphate and Expandable Graphite." *e-Publications@Marquette*. March 14. Accessed April 10, 2015. http://epublications. marquette.edu/cgi/viewcontent.cgi?article=1040&context=chem_fac&sei-redir=1&referer=http%3A%2F%2Fwww.bing.com%2Fsearch%3F.

Bellis, M. 2015. "Early History of Rubber and Rubber Products." *About Money*. Accessed April 15, 2015. http://inventors.about.com/cs/inventorsalphabet/a/ rubber_2.htm.

Carlisle Syn Tec. 2013. *MSDS: Mule-Hide EPDM Sheeting*. Mule-Hide Products, Carlisle, PA.

WestMoonint Corp. LTD. 2010. *MSDS: Chlorosulfonated-Polyethylene*. WestMoonint (HK), Chengdu, Sichuan, China.

8

Plasticizers

Oh, that new car smell—pleasurable or malodorous? Off-gassing from automotive plastics is described by many as "leathery" or "plasticky." The odor becomes especially pungent after a car has been sitting in the hot sun for a few hours. This plastic-like pungent odor is attributable by many to plasticizers volatizing from a hot plastic dashboard, vinyl seat covers, door panels, and of the numerous plastic components in a car. In the evening, the plasticizers condense out of the inside air of the car and form an oily coating on the windshield. Can this not occur in buildings wherein the use of plastics in building materials is on the rise?

All plasticizers are liquid, semi-volatile organic compounds which "may volatize" (i.e., vaporize) at normal temperatures. Increased heat increases the amount of plasticizer evaporation. Thus, the new car effect in buildings is certainly possible.

As they migrate to the surface of plastic component(s), plasticizers evaporate, and the void allows migration of more plasticizer to the surface and more evaporation. The rate of loss is dependent upon plasticizer type, temperature, material thickness, humidity, and exposure time. Higher molecular weight plasticizers have a slower loss rate than lower molecular weight plasticizers. Loss of plasticizer in plastic building materials can be observed by one or a combination of the following:

- Reduction in weight
- Slight reduction in thickness
- Increase in strength
- Decrease in flexibility
- Increased brittleness

For example, a loss of plasticizer in automotive dashboards and vinyl seat covers results in brittleness and cracking. In building materials, a loss of plasticizer may be more subtle such as hardening of spray-on foam sealant or panel insulation both of which are hidden in the wall cavity or above the ceiling.

Plasticizers are added to synthetic polymers to alter the physical properties of an otherwise rigid, brittle plastic, rendering the plastic softer with increased flexibility and impact resistance. They are used predominantly in the manufacture of PVC plastic wherein the plasticizer is captured in the polymer matrix. However, plasticizers may also be used in other thermo-plastic polymers (e.g., ABS, polycarbonate, polystyrene, and acrylics).

About 80% of all plasticizers used today in plastic manufacturing are phthalates. Most, if not all, are organic esters. Some are benzoates (plasticizers used in rubbers, caulks, inks and the treatment of lice), biodegradable epoxidized ethers, sulfonamides, organophosphates, and glycols/polyether. With the exception of benzoates, most of the non-phthalate plasticizers are used predominantly in consumer products—not in building materials.

Most Prominent Plasticizers

There is no consensus on the percent phthalates used in building materials, and there are certainly no published projections on phthalate plasticizers used in building materials. There is, however, a range of 80%–90% of all plasticizers used in building materials.

Phthalates

Unregulated toxic/irritant product emissions: Eye, skin, and URL irritation

Regulated product emissions: Dibutyl phthalate (OSHA PEL: 5 mg/m^3); Di(2-ethylhexyl) phthalate (OSHA PEL: 5 mg/m^3)

Combustion products: Carbon dioxide and carbon monoxide

Widespread use: 1960s to present

Phthalates, also referred to as phthalate esters, comprise approximately 80% of the plasticizer additives in the United States. Phthalates are semi-volatile organic chemicals that comprise up to 50% by weight of a plastic wherein the phthalate is an unreacted component of the plastic. The phthalate component is mixed within the polymerized plastic and readily subject to release, or off-gassing, in enclosed high temperature environments.

According to several manufacturers, phthalates may cause irritation of the eyes, skin, and upper respiratory tract (i.e., nose and throat). A few also claim other possible health effects such as labored breathing, allergic contact dermatitis (e.g., cosmetics or plastic watch bands) with the sensation of burning, and central nervous system depression. One manufacturer claimed chronic exposures may cause headache, nausea, dizziness, and vomiting. Another claimed that elevated exposure levels to plasticizers in general, in the workplace have led to pain, numbness, weakness, and spasms in the extremities. While there are no published odor thresholds, phthalate odor is described by manufacturers in obtuse terms such as "mild characteristic odor" or "mild ester odor." The NIOSH Pocket Guide states that diethyl phthalate has a slight aromatic (e.g., sweat) odor.

Dibutyl phthalate and di(2-ethylhexyl) phthalate are both OSHA regulated at a PEL of 5 mg/m³. In an article entitled "Health Hazards in Nail Salons," OSHA identifies DBP as a health hazard that can cause nausea and irritated eyes, skin, nose, mouth, and throat (OSHA 2016). NIOSH also recommends a limit of 5 mg/m³ for the phthalates, and their limits are based on eye, upper respiratory tract, and stomach irritation. The remaining commonly used phthalates in building materials are not regulated. The regulated phthalates are often declared "proprietary" ingredients by the manufacturers, and the listing of nonregulated phthalates is not required on Safety Data Sheets. Knowledge of the plasticizers can aide one in assessing potential emissions and health impact.

Phthalate plasticizers impart a softness and flexibility to plastic which would otherwise be hard and brittle. Most phthalate plasticizers are added to PVC plastics. Other plastics, as opposed to polyvinyl chloride, to which phthalates are added include polyvinyl acetate, rubbers, cellulose plastics, and polyurethane.

Phthalate plasticizers are produced by reacting phthalic anhydride with alcohol(s). See Figure 8.1. The alcohols can be one or a combination of alcohols ranging from methanol to tridecyl alcohol. Plasticizers made of low molecular weight alcohols such as ethanol (e.g., diethyl phthalate) tend to be more toxic than plasticizers made up of the higher molecular weight alcohols, and the lower molecular weight phthalates are more volatile. Based on reduced volatility, the higher molecular weight alcohol components are less likely to off-gas, and they retain their flexibility.

As plasticizers are added to plastics, they are not chemically bound to the polymer. They are contained within. Thus, plasticizers can readily be expected to off-gas into the environment, and the phthalate content in the plastic, as high as 50% by weight, will readily off-gas more. And it goes without saying that indoor air concentrations are higher than outdoor air concentrations.

The more volatile, low molecular weight semi-volatile phthalates (e.g., dimethyl phthalate and diethyl phthalate) are likely to off-gas higher concentrations in indoor air as compared to the less volatile, higher molecular weight phthalates (e.g., di(2-ethylhexyl) phthalate. Elevated room and

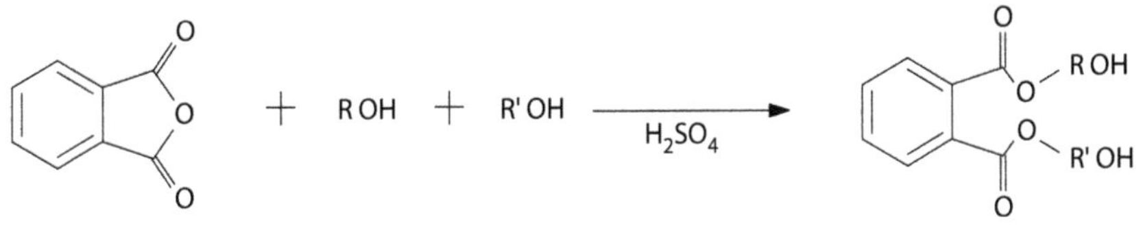

FIGURE 8.1
Formulation of phthalate esters: phthalic anhydride + alcohol(s) = phthalic ester.

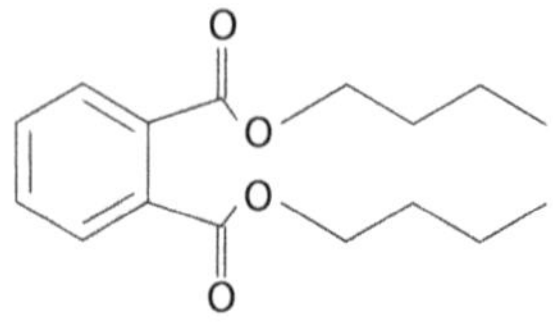

FIGURE 8.2
Dibutyl phthalate (MW: 278 g/mol; BP:240°C).

surface temperatures will also result in greater off-gassing. In other words, a low molecular weight phthalate-enhanced polymer that is exposed to solar heat will off-gas more plasticizer than the same plastic in the interior of an air conditioned space. Yet, let's not get fixated on off-gassing as a source of phthalate exposures. Dibutyl phthalate (Figure 8.2) is a low molecular weight phthalate. It is the most likely of the plasticizers to vaporize from plastic building materials.

Phthalates may be encountered in dust from surface abrasion, scraping, and rubbing. Airborne phthalate-containing dust can thus contribute to the off-gassed phthalates. Phthalates are also used in glues, caulk, and paint which may be overlooked as a contributing source of phthalate off-gassing. Sources of building material exposure to phthalates may be multiple. Yet, it doesn't end with building materials.

Phthalates are used as solvents in cosmetics such as perfume, eye shadow, moisturizers, nail polish, hair spray, and liquid soap. Phthalate plasticizers are used in children's toys, shower curtains, vinyl rain coats, automobile interiors, and medical devices. Plastic food containers have high phthalate levels that can leach into the food—especially when the container is put in a microwave to cook food.

Some noteworthy studies, research, and legal restrictions regarding phthalates follow:

> A mystery for some and a thorn in the side of plastic recycling facilities is the illusive fact that polyethylene terephthalate plastic, used to make drink containers, does not have phthalate plasticizers. Terephthalate polymer is a polymerized phthalate and phthalate plasticizers are chemically free. Yet, a number of studies have encountered free phthalate chemicals in bottled water and sodas. One suggested hypothesis is that plastics with phthalate plasticizers are a component of the some of the mix of recycled plastics that are ultimately used in plastic drink containers.
>
> A 2012 Swedish study of children found that phthalates in polyvinyl chloride flooring was taken up into their bodies, showing that children can ingest phthalates not only from food but also by breathing and through the skin.
>
> The phthalate di(2-ethylhexyl) phthalate has been detected on firefighter personal protective clothing. Yet, animal studies conclude cardiac effects and testicular cancer (Alexander and Stuart Baxter 2014, p. D47)–adverse health outcomes observed in firefighter populations.

The study suggests firefighters follow hazmat decontamination principles at the scene of structural fires (Alexander and Stuart Baxter 2014, pp. D43–D48).

The United States, European Union, and Canada have banned and/or restricted the manufacture and sale of polyvinyl chloride plastics which contain specified low molecular weight phthalates. Many of the Japanese car manufacturers have voluntarily eliminated polyvinyl chloride plastics in their car interiors.

In building materials, phthalate plasticizers are found in vinyl flooring (e.g., tiles and sheet vinyl), plumbing and sewage pipes, carpeting, adhesives, enamels, various foams, electrical wiring/cable insulation, caulk, and synthetic leather wall coverings. They are used as an adhesive for porous building stones (e.g., sandstone) and as a primer for paper coated drywall. See Table 8.1.

Di(2-ethylhexyl) phthalate (DEHP) is the most widely used phthalate plasticizer, worldwide. It is inexpensive and used predominantly as a softening agent in PVC plastics. The content is 1%–40% of unreacted chemicals, the higher the DEHP content, the softer the plastic (EcoUS 1989). Thus, you may reasonably anticipate more off-gassing from the softer plastics. Building materials where DEHP are utilized include floor tiles, wall coverings, electrical wire, and cables. DEHP is also used as a plasticizer in rubber, cellulose, and styrene (U.S. EPA 2012).

TABLE 8.1

Common Uses of Specified Phthalates

Chemical Name[a]	Abbreviation	Common Uses
Di-n-butyl phthalate	DBP	Adhesives, carpet backings, and paints (CDC 2014)
n-Butyl benzyl phthalate	BBP	Vinyl foam which provides a subsurface padding for floor tiles and sheet vinyl, and it is used in artificial leather which may be applied as a wall covering (Wikipedia 2014)
Diisoheptyl phthalate	DIHpP	Vinyl floor tiles and sheet vinyl flooring
Di(2-ethylhexyl) phthalate	DEHP	Floor tiles, wall coverings, and electrical wire and cables. DEHP is also used as a plasticizer in rubber, cellulose and styrene (EcoUS 1989; U.S. EPA 2012)
Diisononyl phthalate	DINP	Floor tiles, sheet vinyl, electrical wire and cables, roofing materials, rubbers, paints, lacquers, sealants, and sheet waterproofing
Butyl benzyl phthalate	BBzP	Vinyl tiles, traffic cones, food conveyor belts, artificial leather, and plastic foams
Diisodecyl phthalate	DIDP	Insulation of wires and cables
Dioctyl phthalate	DOP	Flooring materials and carpets

[a] Listing is lowest to highest molecular weight phthalates that are used in building materials.

Di(2-ethylhexyl) phthalate and diisononyl phthalate (DINP) content in PVC plastics is 0.005%–50%. Once again, the softer the plastic, the greater the anticipated off-gassing of plasticizer. DINP is used in floor tiles, sheet vinyl, electrical wire and cables, roofing materials, rubbers, paints, lacquers, sealants, and sheet waterproofing.

Presently, there are no legal restrictions on the use of phthalates in building materials. Environmental professionals may reasonably anticipate product emissions and airborne/surface dust exposures with contributing exposures from other sources.

Lesser Plasticizers

After the phthalates come all others. Most are ester-based plasticizers, as are phthalates, and all building material plasticizers are hydrocarbons.

Trimellitates

Unregulated toxic/irritant product emissions: Slight eye irritation

Regulated product emissions: –

Combustion products: Carbon dioxide and carbon monoxide

Widespread use: Infrequently used in building materials

Trimellitates are 1,2,4-benzenetricarboxylic acid with side chain esters ranging from C8 to C10. Although thought to be similar to phthalates in toxicity, they have a higher molecular weight and lower volatility. For example, tris-2(ethylhexyl) trimellitate (Figure 8.3) has a molecular weight of 547 and boiling point of 414°C, and dioctyl phthalate has a molecular weight of 390 and a boiling point of 340°C. Thus, it is highly unlikely that trimellitates will volatize in elevated temperatures, short of a raging fire. Trimellites are used predominantly and almost exclusively in high temperature PVC cables (ExxonMobil 2001). Due to low volatility and limited use, emissions are highly unlikely in indoor environments.

Trimellitates will cause slight eye irritation only, and building material emissions are extremely unlikely. The combustion products are carbon dioxide and carbon monoxide.

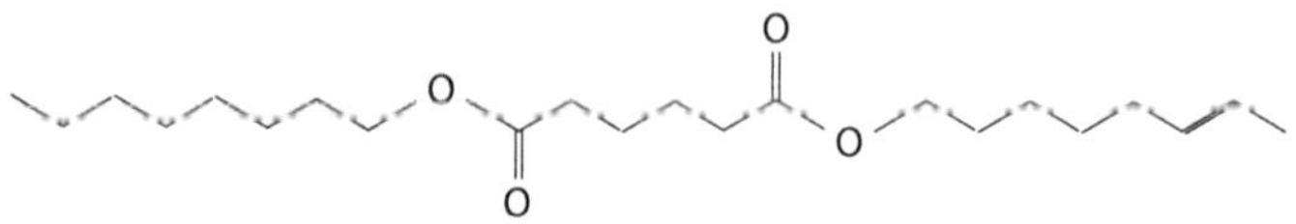

FIGURE 8.3
Tris(2-ethylhexyl) trimellitate (546 g/mol; BP: 414°C).

Adipates

Unregulated toxic/irritant product emissions: Eye, skin, and lung irritation
Regulated product emissions: –
Combustion products: Carbon dioxide and carbon monoxide
Widespread use: Infrequently used in building materials

Adipates are esters of adipic acid ($C_6H_{10}O_4$). They are slightly volatile—with a high molecular weight and a boiling point of less than 250°C. As a plasticizer, adipates impart flexibility and low temperature and are UV resistant. The most commonly used adipate plasticizer is dioctyl adipate (Figure 8.4). Adipates are generally used along with phthalate plasticizers to improve cold resistance to plastics—cold resistant plastic sheeting, cable, electric wires, paints, varnishes, faux leather, and outdoor, exposed water pipes. Due to their slight volatility, products containing adipates used in indoor environments may potentially emit adipates, especially where there are slightly elevated temperatures.

Adipates are eye, skin, and respiratory irritants, and they have a slight "aromatic" odor. The combustion products are carbon dioxide and carbon monoxide.

Sebacates

Unregulated toxic/irritant product emissions: Severe eye, skin, gastrointestinal, and lung irritation
Regulated product emissions: –
Combustion products: Carbon dioxide and carbon monoxide
Widespread use: Infrequently used in building materials

Sebacates are esters of sebacic acid ($C_{10}H_{18}O_4$). They are semivolatiles, requiring elevated temperatures to vaporize. The mostly commonly used sebacate plasticizer is dibutyl sebacate. Dibutyl sebacate (Figure 8.5) is used as a cold and UV resistant plasticizer used in PVC, polyethylene resins, rubber, and adhesives. Due to their semivolatility, sebacates emissions are unlikely in indoor environments unless the product is exposed to high temperatures.

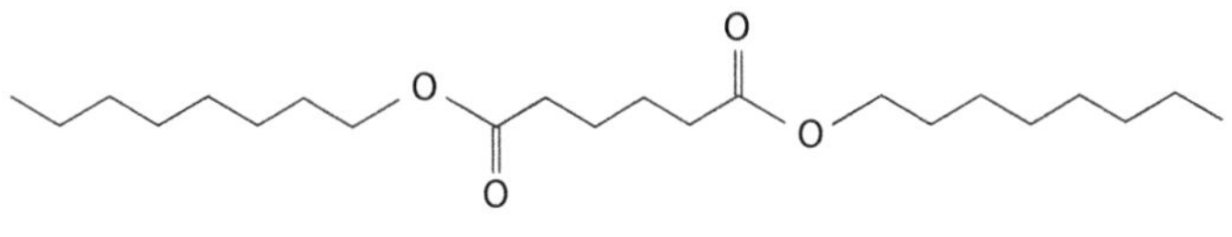

FIGURE 8.4
Dioctyl adipate (370 g/mol; 214°C).

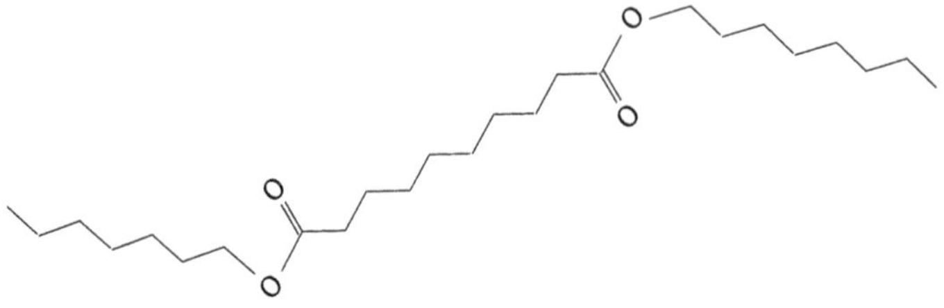

FIGURE 8.5
Dibutyl sebacate (314 g/mol; BP: 344°C).

Sebacates are severe eye irritants and they can cause skin, gastrointestinal, and respiratory irritation. Dibutyl sebacate may cause drowsiness, diarrhea, and possible allergy hypersensitization. Sebacates have a "mild" or no reported odor. The combustion products are carbon dioxide and carbon monoxide.

Maleates

Unregulated toxic/irritant product emissions: Eye, skin, and lung irritation
Regulated product emissions: –
Combustion products: Carbon dioxide and carbon monoxide
Widespread use: Infrequently used in building materials

Maleates are esters of maleic acid ($C_4H_4O_4$). They are slightly volatile and require elevated temperatures to vaporize. The most commonly used maleate is dioctyl maleate (Figure 8.6). Due to their slight volatility, maleate emissions are potential contributors to poor indoor air quality, especially where there are elevated temperatures. Dioctyl maleate is used in paints, varnishes, and adhesives.

Maleates are moderate eye and skin irritants, and potential respiratory tract irritants, and they have a pleasant ester-like odor. The combustion products are carbon dioxide and carbon monoxide.

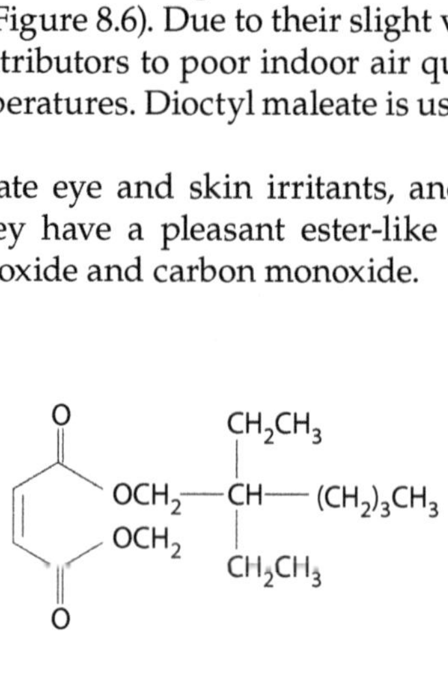

FIGURE 8.6
Dioctyl maleate (340 g/mol; BP: 229–239°C).

Benzoates

Unregulated toxic/irritant product emissions: Eye, skin, and lung irritation

Regulated product emissions: –

Combustion products: Carbon dioxide and carbon monoxide

Widespread use: –

Benzoates are esters of benzoic acid. They are nonvolatile and require elevated temperatures to vaporize. The most commonly used benzoate plasticizer is oxydipropyl dibenzoate (Figure 8.7). It is used in adhesives, caulks, and sealants.

In pure form, benzoates are moderate eye, skin, and respiratory tract irritants, and they have a pleasant ester-like odor. The combustion products are carbon dioxide and carbon monoxide.

Chlorinated Paraffins

Unregulated toxic/irritant product emissions: Eye, skin, and lung irritation

Regulated product emissions: –

Combustion products: Carbon dioxide, carbon monoxide, and hydrogen chloride

Widespread use: 1930s

Chlorinated paraffins (CP) are chlorinated straight chained hydrocarbons. They are highly complex mixtures, comprised mainly of n-paraffins of varying chain lengths, ranging from C_{10} to C_{30+}, and a large chlorine component, ranging from 35 to 70% by weight. In building materials, CPs are used as secondary plasticizers in PVC polymers as well as in paints, adhesives, sealants, caulks, and rubber.

The medium chain chlorinated paraffins (MCCP; C_{14-17}) represent the largest production and use of CPs in North America. An example of MCCP is 2,5,6,7,8,11,15-heptachloroheptadecane (Figure 8.8). The long-chain chlorinated paraffins (LCCP; $C_{>17}$) are secondary in the hierarchy of chlorinated paraffin production in the United States. And last, but not least, are short-chain chlorinated paraffins (SCCP; C_{10-13}), less commonly

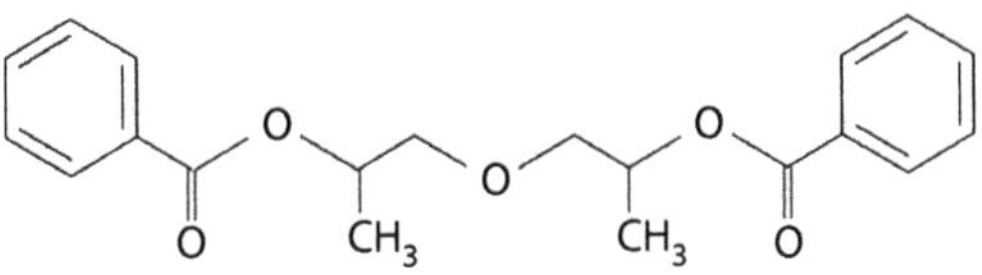

FIGURE 8.7
Oxydipropyl dibenzoate (342 g/mol; BP: 232°C).

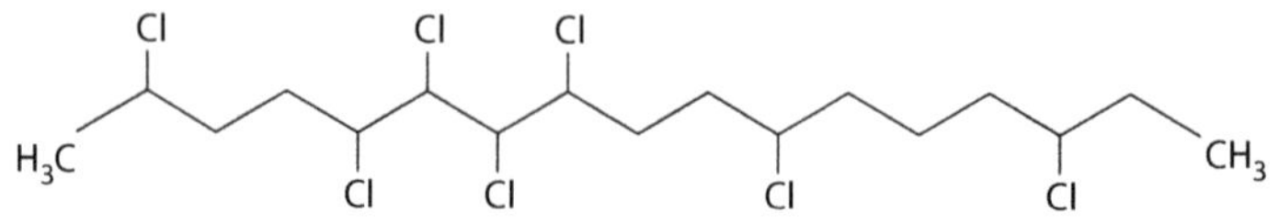

FIGURE 8.8
2,5,6,7,8,11,15-Heptachloroheptadecane (52% Cl by weight; 471 g/mol; BP: >500°C).

used and most environmentally controversial. An example of SCCP is 2,3,4,6,7,8-hexachlorodecane (Figure 8.9). CPs range from viscous liquids to waxy solids!

Although not regulated, MCCP and LCCP are eye, skin, and lung irritants. Yet, due to their high molecular weight and high boiling points, they are not likely to contribute to poor indoor air quality.

On the other hand, SCCP emissions from building materials are possible at elevated temperatures. SCCP are an irritant; their emissions are not regulated; and the U.S. EPA has placed SCCP on a "short list of worrisome chemicals."

> The chemicals, called short-chain chlorinated paraffins, persist in the environment, accumulate in human breast milk, can kill small aquatic creatures and travel to remote regions of the globe … (They are) primarily used as coolants and lubricants in metal forming and cutting … Although Europe has restricted use of SCCPs, their manufacture is growing in China and possibly in India, raising concerns that worldwide exposure levels for people and wildlife might be increasing. China's production of the chemicals has increased 30-fold in fewer than 20 years … In an unprecedented use of the 1976 Toxic Control Substances Act, the EPA in December placed short-chain chlorinated paraffins on a list of four chemicals that may pose unreasonable risks to health and the environment. In its action plan, the EPA announced its intentions to investigate and manage those risks, possibly restricting or banning future use of SCCPs in the United States (Mergel 2011).

Although eye, skin, and lung irritants, chlorinated paraffin emissions from building materials are unlikely. The combustion products are carbon dioxide, carbon monoxide, and hydrogen chloride.

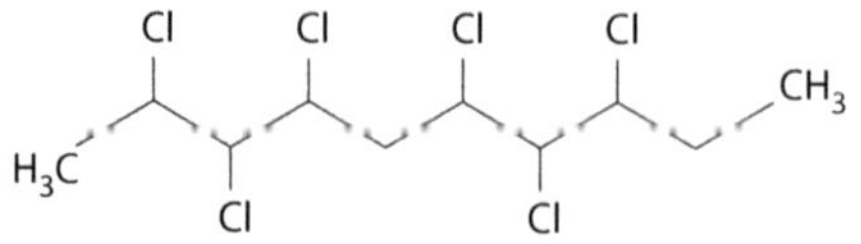

FIGURE 8.9
2,3,4,6,7,8-Hexachlorodecane (61% Cl by weight; 345 g/mol; BP: 412.0 ± 13.0°C).

Summary

A vast majority of plasticizers used in building materials are ester-based and slightly volatile to semi-volatile. They are readily released from within the polymer matrices, and vaporize more readily when the material is exposed to elevated temperatures. The higher the temperature, the more the plasticizer vaporizes, and the more the plasticizer vaporizes, the greater the emissions.

Although most plasticizer emissions are not regulated, all building material plasticizers can cause eye, skin, and respiratory irritation. Elevated temperatures can cause plasticizers to off-gas from plastic building materials and impact the health of building occupants. These same plasticizers are rarely, if ever, listed on the manufacturers MSDS "Hazardous Components List." Disclosure of components is mandated only when the substances are regulated. To date, however, there have been no published reports of toxic/irritant "encounters with plasticizers"—either for lack of awareness or failure to connect the dots. Plasticizer contributions to building-related health problems remain an unknown!

References

Alexander, Barbara M. and C. Stuart Baxter. 2014. "Plasticizer Contamination of Firefighter Personal Protective Clothing—A Potential Factor in Increased Health Risks in Firefighters." *Journal of Occupational and Environmental Health*, D43–D48.

CDC. 2014. "ToxFAQs™ for Di-n-butyl phthalate." *Agency for Toxic Substances & Disease Registry*. August 12. Accessed August 10, 2014. http://www.atsdr.cdc.gov/toxfaqs/TF.asp?id=858&tid=167.

EcoUS. 1989. "Toxicological Profile for Di(2-ethylhexyl)phthalate." *EcoUS*. Accessed March 9, 2015. http://www.eco-usa.net/toxics/chemicals/di_2-ethylhexyl_phthalate.shtml.

ExxonMobil. 2001. "High Production Volume Chemical Challenge Program." *EPA Publications*. December 13. Accessed March 11, 2015. http://www.epa.gov/hpv/pubs/summaries/trime/c13468tp.pdf.

Mergel, Maria. 2011. "Chlorinated Paraffins." *Toxipedia*. March 23. Accessed March 18, 2015. http://www.toxipedia.org/display/toxipedia/Chlorinated+Paraffins.

OSHA. 2016. "Health Hazards in Nail Salons." *OSHA Safety and Health Topics*. Accessed March 9, 2015. https://www.osha.gov/SLTC/nailsalons/chemical-hazards.html.

U.S. EPA. 2012. "Basic Information About Di(ethyhexy)phthalate in Drinking Water." *EPA*. December 13. Accessed January 24, 2015. http://water.epa.gov/drink/contaminants/basicinformation/di_2-ethylhexyl_phthalate.cfm.

Wikipedia. 2014. "Benzyl Butyl Phthalate." *Wikipedia*. August 12. Accessed March 9, 2015. http://en.wikipedia.org/wiki/Benzyl_butyl_phthalate.

9

Plastic Additives

The crowning glory that gives polymers their coveted properties is the additives. Plastics and rubber can be massaged and cajoled to perform in ways that nature might readily envy.

With stabilizers, plastics can resist the test of time, environmental degradation, and the ravages of mold—if so desired. On the other hand, plastics can be made to biodegrade, return to nature—if so desired.

Fire retardants enhance our chances of escaping a fire. A plastic that has been treated with a fire retardant does not support combustion like wood. Wherein wood is a tinder box, fire retardant treated plastic will melt.

Fillers extend a plastic and lower costs without compromising the integrity of a polymer. Then come the amazing reinforcing fillers! They are used to strengthen building structures without adding weight and enhance the opportunity for creative designs.

Foaming agents convert solid plastics into air-filled foams for use in energy saving thermal foams that excel over all natural products such as mineral wool and the ever infamous asbestos. But the benefits are not restricted only to thermal insulation. Polymer foams provide good impact protection for flooring and structural support. Mostly as a consumer product but sometimes as a building material, the most commonly encountered use of polymer foams is that of cushioning—mattresses and pillows.

Polymer additives are not part of the plastic matrix yet may comprise as much as 50% by weight of the plastic. They are not contained. The additives are available and are potential building material emitters—irritants, regulated toxins, and inert, unregulated nontoxins. They are not to be overlooked!

Stabilizers

Stabilizers are polymer additives that enhance and extend a polymer's ability to resist environmental stresses—transcend the foibles that impact nature. They include, but are not limited to, antioxidants, ultraviolet light stabilizers, and heat stabilizers.

Antioxidants

Unregulated toxic/irritant emissions: –
Regulated product emissions: –
Combustion products: Carbon dioxide and carbon monoxide
Widespread use in building materials: Unknown

Some polymers (e.g., polyethylene) and elastomers oxidize, degrade, and age due to photo-oxidation when exposed to the outdoors. To prevent these unwanted effects, small amounts of antioxidant(s) are added during the polymerization process. At the top of the polymer antioxidant list are hindered phenols followed by secondary-aromatic amines, alkyl phosphites, and thioesters. See Figure 9.1. All have low toxicity, are nonirritating, and unregulated.

Ultraviolet Light Stabilizers

Unregulated toxic/irritant emissions: Severe eye, skin, and lung irritation
Regulated product emissions: –
Combustion products: Carbon dioxide and carbon monoxide
Widespread use: Unknown

High-energy ultraviolet (UV) light can degrade and sever chemical bonds in plastics and polymeric adhesives, paints, and sealants. The process is referred to as photodegradation. This results in cracking, chalking, color change (e.g., yellowing), and loss of desirable physical properties. Polymer damage is like "sunburn" only, unlike human skin, polymers cannot regenerate—polymer damage is irreversible.

In the manufacturing process, the addition of "UV absorbers" minimizes the predictable damage to polymers anticipated to be exposed to solar and/or artificial UV light. The most commonly encountered UV absorbers are 2-(2-hydroxyphenyl)-benzotriazoles (Figure 9.2) and 2-hydroxy-benzophones. Concentrations normally range from 0.05%–2% by weight, with some applications up to 5%, and most UV absorbers are solids. Although they are eye and skin irritants, the more common UV absorbers are unlikely to off-gas from building materials. See Table 9.1.

FIGURE 9.1
Sterically hindered phenol and phosphite-based compound.

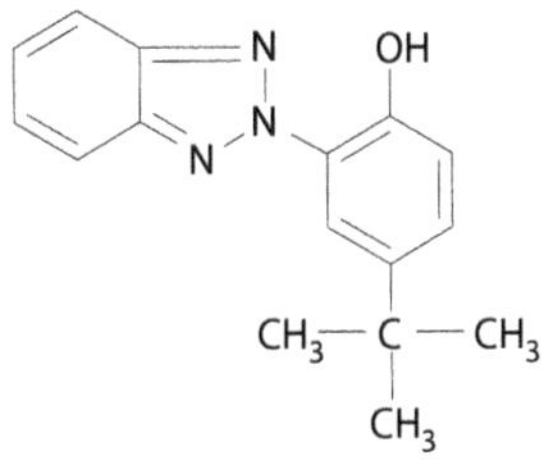

FIGURE 9.2
2-(5-Tert-butyl-2-hydroxyphenyl) benzotriazole (MW: 267 mol/g).

TABLE 9.1

Most Common UV Stabilizers

UV Stabilizer	Form	Excessive Exposure Symptoms	Typical Plastic
2-(2-Hydroxyphenyl)-benzotriazoles	Solid	Extreme eye and skin irritation	Polycarbonate
2-Hydroxy-benzophones	Solid	Serious eye and skin irritation	Polyvinyl chloride

Note: HALS volatile liquid serious eye irritation, skin, and respiratory tract irritation polypropylene and polyethylene.

Although there is little to no toxicity information available on other UV absorbers (e.g., hydroxyphenyl-s-triazines and oxalanilides), they are not likely to be components of plastic building materials.

One of the oldest ways to protect polymers from UV light is the use of pigments (e.g., carbon black and iron oxides). They are applied to the surface of plastics as a UV blocking agent. Carbon black and iron oxides are low in toxicity. They are nonirritating and unregulated.

Free radical "scavengers," as opposed to UV absorbers, stabilize polymers. These scavengers, referred to as hindered-amine light stabilizers (HALS), are almost exclusive derivatives of 2,2,6,6-tetramethylpiperidine (Figure 9.3). Due to their volatility, HALS may contribute to polymer emissions and poor indoor air quality. Although unregulated, HALS may cause serious eye irritation, skin, and respiratory tract irritation.

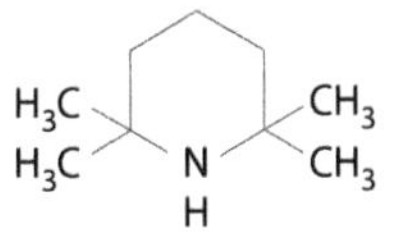

FIGURE 9.3
2,2,6,6-Tetramethylpiperidine (MW: 141 mol/g; BP: 152°C).

Heat Stabilizers

Unregulated toxic/irritant emissions: –

Regulated product emissions: "Potential leaching of lead and/or cadmium into drinking water."

Combustion products: Carbon dioxide, carbon monoxide, and hydrogen chloride

Widespread use in building materials: 1930s to present

Heat stabilizers are used to circumvent degradation of chlorine-containing polymers—mainly polyvinyl chlorides. Untreated PVCs tend to degrade, lose chlorine atoms—in the form of hydrogen chloride—from the polymer matrix when heated. Degradation accelerates in the presence of hydrogen chloride. So, once degradation has started it becomes self-propagating. The end is total deterioration of the plastic with considerable loss of mechanical strength with signs of discoloration, deformation, cracking, electrical insulation performance, and brittleness. Building materials typically treated are water/sewage pipes, vinyl window profiles, and electrical conduits.

To avoid PVC degradation during high temperature production as well as solar and artificial heat stress, metal salts are added as heat stabilizers during the formulation process—at a rate of 1%–1.5% by weight (Plastemart 2003). Since the 1930s, lead salts have held the dubious distinction of extending the service life of PVC products, and many of the lead treated PVC pipes remain in service to this day—having endured the test of time and earthquakes.

With advent of the new millennium, PVC manufacturers worldwide expressed concern for possible lead leaching from potable water PVC pipes. Whether real or imagined, their concerns led to a "voluntary" phase out of lead stabilizers in PVC pipes in Europe and North America. Other global manufacturers have also committed to a phase out, but there have been no overall country commitments. See Table 9.2.

TABLE 9.2

Production of Lead in PVC Pipe by Country

Country	PVC Potable Water Pipe with Lead Component (%)
North America	1
Europe	29
South America	61
Middle East and Africa	86
China	91
India	95

Source: Adapted from Zhan Jie, W. 2013. *Plastics News.* Accessed March 23, 2015. http://www.plasticsnews.com/article/20130906/NEWS/130909958/ chinas-pvc-pipe-makers-under-pressure-to-give-up-lead-stabilizers.

U.S. EPA Maximum Limits in Drinking Water
Lead (Pb): 15 μg/L
Cadmium (Cd): 5 μg/L
Barium (Ba): 2000 μg/L

Most manufacturers do not list additives in the "Hazardous Ingredients Section" of their MSDS. So, if lead in PVC pipes should—in the future—be found to leach lead into the drinking water, buyer beware! And if you live in a country in which manufacturers do not use lead stabilizers, most builders purchase the least expensive product which generally is "Made in China"— 91% of which contain lead. Once again, the amount of lead that may leach out into potable water from 1% to 1.5% lead-containing PVC water pipe has yet to be determined.

Alternative metal salt stabilizers include cadmium, barium, calcium, zinc, and organic tin. See Table 9.3. Cadmium is more toxic than lead, and barium has a very low toxicity. Although, lead, cadmium, and barium are all regulated by the U.S. EPA under the Safe Drinking Water Act, the reference is mostly to groundwater contamination, not leaching from PVC pipes, and manufacturers claim that heat stabilizers are "safe since there is no leaching of stabilizer to the water once the initial surface layer of stabilizer has been washed off by flushing" (Sevenster 2015). Calcium, zinc, and organic tin are relatively nontoxic. In an effort to stave off litigation, many manufacturers choose calcium, zinc, and organotin heat stabilizers over the less expensive lead.

Other than leaching, PVC building product emissions are not likely unless lead- or cadmium-containing PVC dust is generated during installation. This then becomes a worker exposure consideration, and if not properly cleaned up, the dust may become airborne and ultimately entrained in the HVAC system only to be distributed throughout the building. Although airborne

TABLE 9.3

Heat Stabilizers by PVC Building Material

PVC Product	Pb	PB/Ba/Cd	Ca/Zn	Ba/Zn	Organic Sn	K/Zn
Pipes	++	−	++[a]	−	++[b]	−
Fittings	++	−	−	−	+[a]	−
Profiles (e.g., windows)	++	++	+	−	−	−
Cable coverings	+	−	++	−	−	−
Flooring	−	−	−	++	+	++
Wall coverings	−	−	(+)	++	+	++

Source: Adapted from Sevenster, A. 2015. *PVC*. Accessed March 23, 2915. http://www.
         pvc.org/en/p/organotin-stabilisers.
Note: ++ Major use; + Minor Use.
[a] PVC potable water pipe.
[b] USA, not Europe, PVC potable water pipe.

PVC dust that has less than 2% lead or cadmium may not pose a serious health hazard, awareness is prudent.

Flame Retardants

Most untreated plastics are highly flammable—self propagating fuel sources. For instance, combustion of polystyrene building products breaks down the polymer into its monomer, flammable styrene. Once ignited, polystyrene will add fuel to the fire.

Although not able to stop a self-perpetuating fire, flame retardant additives in plastics will reduce flammability, impede burning, and/or self-extinguish. Then, too, the halogenated and polycarbonate plastics do not support combustion at ambient oxygen levels—lack of oxygen is the limiting factor. See Table 9.4. Yet, if a halogenated or polycarbonate polymer has been plasticized, the free, unbound plasticizer is flammable at ambient oxygen levels. For this reason, all plasticized plastics are treated with flame retardants.

Polymers that pose potential fire hazards must be modified through the use of one or a combination of flame retardant additives. The means by which fire propagation is managed are fairly basic concepts: (1) vapor phase inhibition; (2) solid phase char-formation; and (3) quench and cool.

In vapor phase inhibition, additives disrupt the formation of monomer gases that fuel the fire. See monomers and fire ratings in Table 9.4. Vapor

TABLE 9.4

Polymer Flammability Information

Polymer	Approximate LOI[a] Percent (Reilly and Beard 2009)	Combustion By-Product Monomer(s)	NFPA Fire Rating
PE	17.3	Ethylene	4
PMMA	17.3	Methyl methacrylate	4
PP	17.6	Propylene	4
PS	18	Styrene	3
ABS	18.5	Acrylonitrile–butadiene–styrene	3;3;3
PC	24	Phenol and phosgene	2;0
F-PVC	24.5	Vinyl chloride	4
R-PVC	42.5	Vinyl chloride	4

Source: Adapted from Reilly, T. and A. Beard. 2009. *Fire Retardants and Their Potential Impact on Fire Fighter Health*. Gaithersburg, MD: Clariant.

Note: F-PVC: Flexible PVC (plasticizer); R-PVC: Rigid PVC (no plasticizer).

[a] The limiting oxygen index (LOI) is the minimum concentration of oxygen that will support combustion of a polymer. Any polymer that has a LOI above 21% does not have sufficient oxygen in normal air to support combustion.

phase fire retardant additives are typically brominated/brominated organic compounds (e.g., polybrominated diphenyl ethers). In solid phase char-formation, a carbon "charred" barrier is formed, insulating the polymer, hindering the release of additional monomer gases—fuel. The more commonly used char-formation additives are phosphorus and nitrogen-based compounds (e.g., phosphate esters; melamine). Quench and cool involves the use of an endothermic (i.e., heat absorbing) reaction with the subsequent release of water molecules which, in turn, cool the polymer—slow combustion. This is usually accomplished by inorganic compounds (e.g., aluminum oxide hydrate). Emissions from polymerized brominated organic compounds, nitrogen bases esters, and inorganic compound fire retardants are unlikely.

Halogenated Flame Retardants

Unregulated toxic/irritant emissions: Possible, but unlikely

Regulated product emissions: –

Combustion products: Carbon dioxide, carbon monoxide, and hydrogen chloride/bromide

Widespread use in building materials: 1970s

Halogenated flame retardants act as "vapor phase additives." They inhibit, or slow down, combustion of treated materials by disrupting the formation of flammable components (e.g., monomers). Most halogenated flame retardants are brominated organics; very few are chlorinated organics; and some are both brominated and chlorinated organics. Many are additive, as opposed to reactive, and some have a phosphate component. See Table 9.5. They are most commonly used in plastics, textiles, and electronics.

TABLE 9.5

Examples of Halogenated Flame Retardants

Polymer	Flame Retardant Content (%)	Flame Retardant	Physical State	Molecular Weight
Polyamides (e.g., Kevlar and Nomex)	13–16	Decabromodiphenyl ether; brominated polystyrene	Solid	958 g/mol
Polypropylene (e.g., carpeting)	5–8	Decabromodiphenyl ether; propylene dibromo styrene	Solid	958 g/mol
Polyethylene (e.g., vapor barrier)	5–8	Decabromodiphenyl ether; propylene dibromo styrene	Solid	958 g/mol
Polycarbonates (e.g., bullet proof windows)	4–6	Brominated polystyrene	Solid	>1000 g/mol
Polystyrene Foam (e.g., insulation panels)	0.8–4	Hexabromocyclododecane	Solid	642 g/mol

In recent years, halogenate flame retardants have received worldwide scrutiny in terms of environmental impact—groundwater, surface water, and soil contamination. By extension, some would also suggest air exposures. Although toxicological data is limited, many of the halogenated fire retardants have been found guilty by association with structurally similar compounds. Many are linked to cancer and reproductive/developmental health effects. Disposal of consumer products containing halogenated flame retardants is implicated as the source of environmental contamination.

> In the European Union . . . the use of certain BFRs [brominated flame retardants] is banned or restricted; however, due to their persistence in the environment there are still concerns about the risks these chemicals pose to public health. BFR-treated products, whether in use or waste, "leach" BFRs into the environment and contaminate the air, soil and water. These contaminants may then enter the food chain where they mainly occur in food of animal origin, such as fish, meat, milk and derived products (EFSA 2014).

Building materials containing halogenated flame retardants have also come under attack—particularly polystyrene foam insulation. The flame retardant HBCD may no longer be produced or used. In May 2013, 160 countries at the UN Chemical Conference in Geneva decided that HBCD should no longer to be produced.

> Ultimately, the findings were unequivocal. (Thirty) years after they were first produced industrially and used around the world, the expert committee of the Stockholm Convention has classified HBCDs as POPs [persistent organic pollutants] and thus laid the foundations for a global ban. The resolution was formally passed on 9 May 2013 and comes into effect after a transition period of approximately one year (Empra 2013).

In 2008, the California Environmental Contaminant Biomonitoring Program Scientific Guidance Panel reported that halogenated flame retardants have "been found in house dust, indicating that they are being released from products." The alarm bells go off again!

The looming question is, "Can halogenated flame retardant emissions pose a health hazard to building occupants?" According to prominent manufacturer MSDSs, the more prominent halogenated flame retardants (see Table 9.6) can cause eye, skin, and lung irritation, and they are not regulated. As most, if not all, halogenated flame retardants are solids and possess high molecular weights, they are not likely to off-gas from building materials. Yet, as they have been reported in residential dust, the potential impact of halogenated flame retardants on indoor air quality remains unclear.

TABLE 9.6

Percent Content of Brominated Flame Retardants in Plastics

Polymer	Flame Retardant Content (%)	Flame Retardant
Polyamides (e.g., Kevlar and Nomex)	13–16	Decabromodiphenyl ether; brominated polystyrene
Polypropylene (e.g., carpeting)	5–8	Decabromodiphenyl ether; propylene dibromo styrene
Polyethylene (e.g., vapor barrier)	5–8	Decabromodiphenyl ether; propylene dibromo styrene
Polycarbonates (e.g., bullet proof windows)	4–6	Brominated polystyrene
Polystyrene foam (e.g., insulation panels)	0.8–4	Hexabromocyclododecane

Inorganic Flame Retardants

Unregulated toxic/irritant emissions: Possible, not likely

Regulated product emissions: Antimony and molybdenum dust (possible, not likely)

Combustion products: –

Widespread use in building materials: Unknown

Most inorganic flame retardants are "quench and cool" retardants that release water molecules when heated—quenching the fire. The most widely used inorganic flame retardant is hydrated aluminum oxide (e.g., aluminum trihydroxide; Al $(OH)_3$). Although inexpensive, greater than 60% of hydrated aluminum oxide is added to a polymer in order to generate sufficient water to quench a fire. Magnesium hydroxide is used in polymers which have higher processing temperatures.

Many inorganic flame retardants are used as synergists to enhance the performance of other flame retardants or to suppress smoke formation. Antimony oxides (e.g., antimony trioxide), sometimes in conjunction with zinc borate, are used with halogenated fire retardants to slow the release of gases and the impact of a fire. Less frequently used as a halogenated fire retardant are molybdenum compounds (e.g., molybdenum trioxide).

Borates are used as flame retardants in cellulose (e.g., boric acid treated paper insulation) and plastics (e.g., zinc borate). Zinc borates are used in conjunction with other inorganic flame retardants (e.g., aluminum trihydroxide and magnesium hydroxide). Boron compounds serve a dual function by not only releasing water in a fire, but they form a glassy coating which protects the surface in a similar fashion to charring. Beyond smoke suppression, zinc compounds also promote char formation.

Of all the inorganic flame retardants, only the antimony oxides (as Sb) and molybdenum compounds (as soluble Mo) are regulated. Antimony trioxide is an irritant to the eyes, skin, and mucous membranes. Although occupational exposures exposure levels to Sb and Mo are highly unlikely in indoor air quality, antimony trioxide is a suspect carcinogen. Subsequently, according to the MSDS from Fisher Scientific, some U.S. state governments require warning labels on all products containing antimony trioxide.

> California prop. 65: This product contains the following ingredients for which the State of California has found to cause cancer, birth defects or other reproductive harm, which would require a warning under the statute: Antimony trioxide California prop. 65: This product contains the following ingredients for which the State of California has found to cause cancer which would require a warning under the statute: Antimony trioxide Pennsylvania RTK: Antimony trioxide Massachusetts RTK: Antimony trioxide TSCA 8(b) inventory: Antimony trioxide SARA 313 toxic chemical notification and release reporting: Antimony trioxide CERCLA: Hazardous substances: antimony trioxide

Molybdenum trioxide may cause eye and skin irritation and is designated low in toxicity. As they are solids with a boiling point in excess of 1000°C, antimony trioxide and molybdenum trioxide are highly unlikely to off-gas from building material. Yet, based on residential air testing and due to the large amount of these inorganic flame retardants added to plastics, antimony and molybdenum dust cannot be ruled out—surface and/or air.

All other inorganic flame retardants are eye, skin, and upper respiratory tract irritants—at various levels of severity. They are solid and not likely to pose an emissions problem unless the polymer is pulverized or abraded and not cleaned up after construction.

All inorganic flame retardants are solids and are nonflammable. And many have a boiling point in excess of 1000°C. There are no reported combustion products.

Phosphorus-Based Flame Retardants

Unregulated toxic/irritant emissions: Possible, not likely

Regulated product emissions: Triphenylphosphate dust (possible, not likely)

Combustion products: Organics—carbon dioxide and carbon monoxide
   Ammonium polyphosphate—ammonia and inorganic acids

Widespread use in building materials: Unknown

Phosphorus-containing flame retardants cover a wide range of inorganic and organic compounds, both reactive (e.g., chemically bound within the plastic matrix) and additive (e.g., physically integrated during the mixing process). Chemically bonded retardants are not likely to result in product emissions whereas those that are physically mixed will pose a greater possibility of off-gassing or dust deposits. The most important phosphorus-containing flame retardants are phosphate esters, phosphonates, and phosphinates.

Additive phosphate esters are used primarily in polyphenylene oxide/high impact PS and PC/ABS blends as well as PVC and PC (e.g., triphenylphosphate). Other applications are for phenolic resins (e.g., phenol formaldehyde) and coatings.

Reactive phosphates (i.e., phosphate salts), phosphonates (i.e., phosphonic acid esters), and phosphinates (i.e., salts of phosphoric acid) flame retardants are used predominantly in flexible polyurethane foams (e.g., mattress foam). These flame retardants are an environmentally friendly alternative to organic halogens.

Expandable, intumescent flame retardants are used as coatings to protect combustible materials (e.g., plastics, wood, and steel). This is achieved by combining inorganic acids (e.g., ammonium polyphosphate) with binders and other noncombustible components. They are found in rigid and flexible polyurethane foams, PP, PE, phenolics, epoxies, and coatings for textiles. Ammonium polyphosphate combustion products are ammonia and inorganic acids.

Less frequently encountered, red phosphorus is used in glass fiber reinforced polymers (i.e., polyamide, polyethylene, polyurethane foam, and thermosetting resins). Red phosphorus, used in safety matches, is an adjunct flame retardant. It can cause eye and skin irritation; it is not regulated; and it is environmentally friendly.

Phosphate esters can cause severe eye and skin irritation and respiratory irritation, and triphenylphosphate is regulated—albeit at low toxicity, a low level of concern. They are solids and unlikely to pose an emission hazard. However, additive phosphate ester dust accumulation, although highly unlikely, cannot be ruled out.

There is little or no toxicity data on the other phosphorus-based flame retardants. As for the limited data available, health hazards to humans are minimal to nonexistent. The other phosphorus-based flame retardants are also environmentally friendly.

Fillers

The term plastic filler refers to solid inert extenders and reinforcing fibers that are incorporated into a plastic matrix. Inert fillers increase the volume and decrease the cost. Other properties affected include an increase in density, reduced shrinkage, increased hardness, and increased heat deflection temperature. Some fillers include sand, silica, glass spheres, calcium carbonate, carbon black, graphite, mica, talc, and clay. The amount of filler required is based on the polymer and process.

A polypropylene filled with 40% calcium carbonate would have a diffusivity approximately 3.5 times greater than that of neat PP. This means it will take much less energy from the extruder drive and will transfer energy much faster than the neat PP in relation to the mass of the output. This means lower motor load and quicker, more uniform melting than neat PP at the same output (lb/hr) (Frankland 2011).

Although regulated and a suspect human carcinogen, respirable crystalline silica in sand and silica extenders are unlikely to present an emission health hazard, because the fillers are bound within a plastic matrices. However, cutting, grinding, and abrading during construction may result in airborne crystalline silica dust exposure to workers.

Reinforcing fiber fillers are added to strengthen a polymer—increasing the tensile, compressive, and shear integrity of a plastic. Other properties imparted to a polymer with RF fillers include an increase in heat deflection, reduction in shrinkage, increase in modulus, and improve the creep behavior. In buildings, the most commonly used RF fillers are glass fibers and carbon fibers. As a plastic filler, glass fibers are by far preferred over the more expensive carbon fibers. Building materials in which fiber-reinforced plastics (FRP) are used include, but are not limited to, non-rust rebar (e.g., foundations), large water/wastewater main pipes, and building materials requiring light weight, unique designs (e.g., domes, columns, window/door frames, and moldings).

If "not" larger than 5 microns with a width to length ratio of 3 or more and a diameter of less than 3 microns, glass and carbon fibers are considered by OSHA to be a nuisance dust and may pose a physical irritation hazard only—if not bound within a plastic matrix. Especially after having been bonded to a matrix, glass and carbon fibers are generally quite large and not likely to pose an airborne asbestos-like health hazard.

Bound within a polymer matrix, fillers are highly unlikely to pose an emission or dust hazard to building occupants. Once again, however, construction dust is another story. It may pose a worker exposure hazard to respirable crystalline silica.

Foaming Agents

Foaming agents, also referred to as blowing agents, are additives which create gaseous microcellular pockets within a plastic matrix. The plastic may be thermoplastic or thermoset plastic matrices, and the end product may be rigid or flexible. The more flexible the polymeric foam, the greater the plasticizer content.

In building materials, rigid foams are used in thermal insulation (e.g., walls and pipes) and load-bearing isolation (e.g., foundations, roof equipment support, tilt-up/precast panel, and tank/column isolation). Although used predominantly in consumer products (e.g., mattresses and cushions), flexible foam is used in building materials in shock absorber materials (e.g., carpet pads, wood/carpet backing, and floor mats) and acoustical materials (e.g., ceiling and wall noise absorbers).

There are two general classes of foaming agents—physical and chemical. The physical foaming agents are gaseous or liquid, and the chemical are usually solids.

Physical Foaming Agents

Physical foaming agents are either gaseous or liquid. Generally accepted gaseous forms include nitrogen, air, carbon dioxide, and an air–helium mixture. Nitrogen and air are inert, nontoxic, and nonflammable, but they have a low diffusivity and poor mixing ability in most polymers. Carbon dioxide is nontoxic and nonflammable, but it is a greenhouse gas. The air–helium mixture is nontoxic and easily controlled, but it is flammable. Clearly, the gaseous forms of foaming agents are less than ideal.

Halogenated aliphatic hydrocarbons (e.g., trichlorofluoromethane) possess all the desirable characteristics—nontoxic, nonflammable, and good mixing ability. So, this would make them ideal as physical foaming agents with but one exception. They are greenhouses gases!

In 1987, the United Nations Montreal Protocol on Substances that Deplete the Ozone Layer was an agreement between the United States, Canada, and other countries to ban and phase out select HCFCs.

> In developed countries the production and consumption of halons formally ended by 1994, several other chemicals (such as CFCs, HBFCs, carbon tetrachloride, and methyl chloroform) were phased out by 1996, methyl bromide was eliminated in 2005, and HCFCs are scheduled to be completely phased out by 2030. In contrast, developing countries phased out CFCs, carbon tetrachloride, methyl chloroform, and halons by 2010; they are scheduled to phase out methyl bromide by 2015 and eliminate HCFCs by 2040 (Encyclopaedia Britannica 2015).

All CFC-11s (i.e., trichlorofluoromethane) were banned in 1995, and they were replaced with other HCFCs and HFCs. In the same year that CFC-11s were banned, the U.S. EPA published a phase out schedule for all HCFCs with all HCFCs banned and phased out by 2030. Manufacturers and researchers launched into a search for a replacement with zero ozone depletion potential. In 2014, Dow Chemical announced a "zero ozone-depleting foaming agent technology" for extruded high-density polystyrene foam. This was a start—change is on the way. The ideal physical foaming agent is knocking at the door to progress.

Chemical Foaming Agents

Chemical foaming agents are predominantly solids that are able to decompose at polymer melt temperatures, liberate large amounts of gas—carbon dioxide, carbon monoxide, and hydrogen. After foaming agent decomposition and gas formation, the resultant residue becomes part of the polymer matrix.

Mineral foaming agents (i.e., salts and weak acids) release gas by thermal dissociation or chemical decomposition. The most important of these is ammonium bicarbonate which has a very low decomposition temperature (i.e., 60°C) and does not leave a residue. See Figure 9.4. Then come sodium bicarbonate and sodium borohydride with an activator; they leave a residue. See Figure 9.5. Water has also been used as a foaming agent—in the form of steam.

Organic foaming agent, on the other hand, liberate nitrogen gas. Many have a maximum gas liberation temperature close to the flow temperature of their polymer matrix, and they can be mixed uniformly with other additives. The down side of organic foaming agents is not only their high cost but the toxicity of some of the foaming agents and the potential liberation of ammonia gases. See Table 9.7 for some of the plastics and foaming agents and organic foaming agents (Akovali 2005, p. 32; Chemcalland21 2015; Luebke and Weisneer 2002).

The organic foaming agents include azo and diazo compounds (Figure 9.6), *N*-nitroso compounds (Figure 9.7), sulfonyl-hydrazides (Figure 9.8), azides (Figure 9.9), and triazoles/tetrazoles (Figure 9.10) (Akovali 2005, p. 242). Those not listed herein are either research or of limited use components (not widely accepted and/or used).

Most PVC foam is made with chemical foaming agents such as azobis-formamide (also azodicarbonamide). The amount of the solid added to the polymer mix is about 1%–2% by weight (Akovali 2005, p. 245).

Although they may be unregulated toxins and flammable, the organic chemical agents only pose a health and flammable hazard to the manufacturer's workers during processing. The liberated gases and residue that are retained in the foam—after processing—are not reported to pose an emission health hazard. On the occasion where ammonia is generated during processing, this will degrade the end product which will likely be rejected as defective by manufacturers. Otherwise, ammonia would not only degrade the polymer but pose a regulated toxic off-gassing concern. However, ammonia off-gassing is not likely from the final foam product.

$$NH_4CO_3 + Heat \longrightarrow NH_3\uparrow + H_2O\uparrow + CO_2\uparrow$$

FIGURE 9.4
Decomposition of ammonium bicarbonate.

$$2NaHCO_3 + Heat \longrightarrow Na_2CO_3 + H_2O\uparrow + CO_2\uparrow$$

FIGURE 9.5
Decomposition of sodium bicarbonate.

TABLE 9.7

Organic Foaming Agents and Liberated Gases

Class of Organic Chemicals	Foaming Agent	Building Material Polymers	Liberated Gases in Order
Azo and diazo compounds	Azodicarbonamide	PVC and rubber	Nitrogen, carbon dioxide, and carbon monoxide or (ammonia)[a,b]
Nitroso compounds	N,N'-dinitroso-pentamethylenetetramine	PVC and rubber	Nitrogen, carbon dioxide, and carbon monoxide or (ammonia)[b]
Sulfonyl-hydrazides	p-Toluenesulfonyl hydrazide	PVC, PP, LDPE, and PS	Nitrogen, carbon dioxide, and carbon monoxide or (ammonia)[b]
Azides	4'-Oxydibenzenesulfonyl hydrazide	PVC, PP, and LDPE	Nitrogen, water[a]
Triazols and tetrazoles	5-Phenyltetrazole	PC and rubber	Nitrogen, carbon dioxide, carbon monoxide, (ammonia)[b]

Note: (ammonia): Ammonia will be liberated if insufficient heat.

[a] Luebke, G. and M. Weisneer. 2002. *Blowing Agents and Foaming Processes 2002 Conference Proceedings*, 31. Shawbury, Shrewsbury, Shropshire: Rapra Technology Limited.

[b] p-Toluenesulfonyl hydrazide. http://chemicalland21.com/specialtychem/perchem/p-TOLUENESULFONYL%20HYDRAZIDE%20%28BLOWING%20AGENT%29.htm.

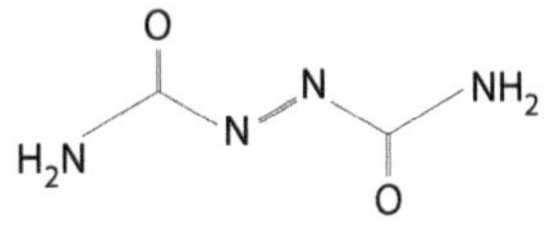

FIGURE 9.6
Azodicarbonamide (azo foaming agent).

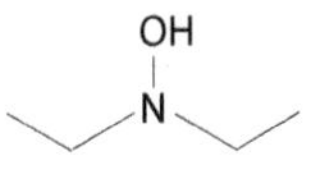

FIGURE 9.7
N,N-diethylhydroxylamine (nitroso foaming agent).

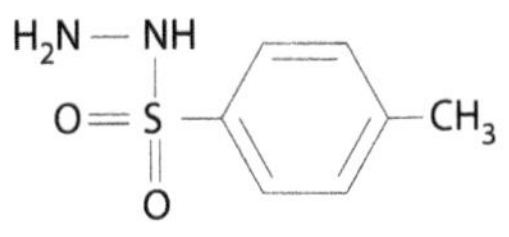

FIGURE 9.8
p-Toluenesulfonyl hydrazide (sulfonyl-hydrazide foaming agent).

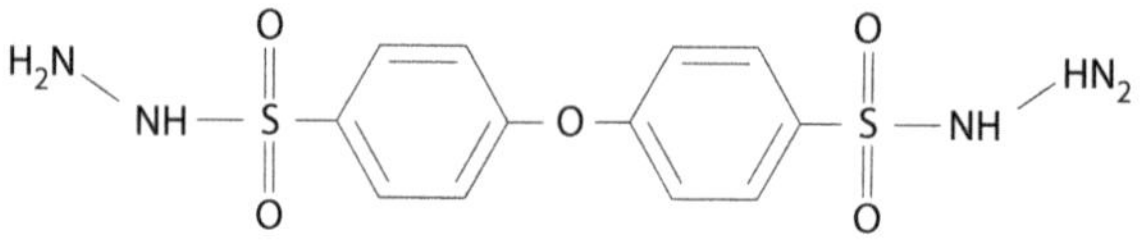

FIGURE 9.9
4'-Oxydibenzenesulfonyl hydrazide (azide foaming agent).

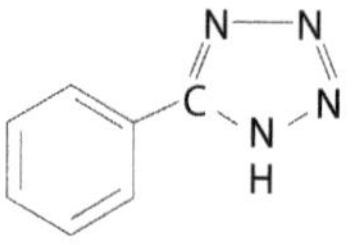

FIGURE 9.10
5-Phenyltetrazole (tetrazole foaming agent).

Summary

The contribution of unbound additives to a polymer matrix is significant when assessing potential emissions and off-gassing from building materials. Polymeric additives are rarely, if ever, addressed in indoor air quality evaluations. Yet, the volume of these additives and their potential health effects can truly be significant.

Stabilizers may contribute as much as 5% by weight to a polymer. Many of the UV light solid stabilizers cause eye and skin irritation, yet they are unlikely to off-gas. Whereas hindered amine light stabilizer is a volatile liquid that can cause serious eye irritation as well as skin and respiratory tract irritation. All heat stabilizers are solid and predominantly used in PVC pipes. That said, lead and cadmium are not only toxic but also regulated in the air and in drinking water. The concern here is not for toxic emissions but toxic leaching from PVC water pipes into potable water. At this time, the impact is speculated, not substantiated with supporting data.

Fire retardants may contribute as much as 16% by weight to a polymer (Akovali 2005, p. 245). Although they are solid and unlikely to pose an emission hazard, most halogenated flame retardants are being phased out worldwide. The inorganic flame retardants are solid as well, but toxic, and regulated antimony and molybdenum may pose a dust hazard. Phosphorus-based stabilizers are solid, and building material emissions are unlikely.

Fillers may contribute as much as 40% by weight to a polymer (Akovali 2005, p. 245). With one exception, extension fillers are nontoxic. The exception is respirable crystalline silica, an airborne, long-term regulated toxin. Although not likely to be airborne once bound within a resin, silica components may, but are not likely to, pose a dust hazard.

Foaming agents may expand a polymer greater than 40 times the polymer's original volume with a 97% gas component, and wherein a solid foaming agent is introduced into a mix, it may contribute 1%–2% by weight (Akovali 2005, p. 245). Foaming agents are nontoxic after processing but halogenated hydrocarbon foaming agents are not only the most ideal but they have been designated contributors to "greenhouse gases." They are being phased out worldwide.

The more complex the addition, the more complex the assessment. There truly is a convoluted theater of if-and-maybe. If cut or abraded during construction, additives may be released and become airborne. If physically damaged after construction, additives may be released and become airborne. If subjected to physical abuse and wear-and-tear, additives may be released and become airborne. If irritant volatile additives are attached to airborne dust, sampling for vapors may result in incomplete sample information. In conclusion, don't be blind-sided by a singular focus. The devil may be in the dust!

Plasticizers and other additives are encountered not only in those which are typically identified as plastics but in an ever increasing array of building materials which have polymer components. Plastics are the new norm, the staple of modern construction.

In the next section, typically, most commonly encountered, building materials are identified by function with an explanation of their components and associated health hazards. Improper installation of these building materials may also contribute to indoor air pollution—even having a devastating impact on the health of the occupants. Some of the more prominent construction oversights are mentioned as well.

References

Akovali, Guneri. 2005. *Polymers in Construction*. Shawbury, Shrewsbury, Shropshire, UK: Rapra Technology Limited.
Chemcalland21. 2015. "p-Toluenesulfonyl Hydrazide." *Chemicalland 21*. Accessed April 2, 2015. http://chemicalland21.com/specialtychem/perchem/p-TOLU-ENESULFONYL%20HYDRAZIDE%20%28BLOWING%20AGENT%29.htm.
EFSA. 2014. "Brominated Flame Retardants." *European Food Safety Authority*. March 11. Accessed March 26, 2015. http://www.efsa.europa.eu/en/topics/topic/bfr.htm.
Empra. 2013. "Worldwide Ban on Flame Retardant." *Science Daily*. August 26. Accessed March 26, 2015. http://www.sciencedaily.com/releases/2013/08/130826105752.htm.
Encyclopaedia Britannica. 2015. "Montreal Protocol International Treaty." *Encyclopaedia Britannica*. Accessed March 31, 2015. http://www.britannica.com/EBchecked/topic/391101/Montreal-Protocol.
Frankland, Jim. 2011. "How Fillers Impact Extrusion Processing." *Plastics Technology*. November. Accessed April 2, 2015. http://www.ptonline.com/columns/how-fillers-impact-extrusion-processing.
Luebke, Gunther and Marcel Weisneer. 2002. "Development of New High Temperature Chemical Foaming Agents." In *Blowing Agents and Foaming Processes 2002 Conference Proceedings*, 31. Shawbury, Shrewsbury, Shropshire: Rapra Technology Limited.

Plastemart. 2003. "PVC Additives Grow Steadily." *Technical Articles & Reports on Plastic Industry*. Accessed March 23, 2015. http://www.plastemart.com/upload/Literature/PVC_additives_grow_steadily.asp.

Reilly, T. and A. Beard. 2009. "Additives used in Flame Retardant Polymer Formulations: Current Practice and Trends." *Fire Retardants and Their Potential Impact on Fire Fighter Health*. Gaithersburg, MD: Clariant. https://www.nist.gov/sites/default/files/documents/el/fire_research/2-Reilly.pdf

Sevenster, Arjen. 2015. "Organotin Stabilizers." *PVC*. Accessed March 23, 2015. http://www.pvc.org/en/p/organotin-stabilisers.

Zhan Jie, Wang. 2013. "China's PVC Pipe Makers Under Pressure to Give Up Lead Stabilizers." *Plastics News*. September 6. Accessed March 23, 2015. http://www.plasticsnews.com/article/20130906/NEWS/130909958/chinas-pvc-pipe-makers-under-pressure-to-give-up-lead-stabilizers.

Section III

Building Materials by Function

10

Overview of Building Material Components

In today's world of plastics and advanced technology, assessing building material emissions and combustion products is like attempting to control a menagerie of monkeys. Once you think you have a handle on it, you encounter the unexpected. The task is difficult but not impossible. So, let's get started.

Function versus Type

Building materials are typically identified by type, not by function. Type is a general description of products that serve multiple functions. It does not address trade differences in terms of product composition (e.g., mineral wool insulation versus mineral wool ceiling tiles). It does not encompass degradation challenges (e.g., insulation in wall cavities versus insulation in air conditioning units). It does not discern the differences between interior and exterior products (e.g., exterior wood siding versus interior bead board). So, let's look at a few of these differences.

- The framing and carpenter trades use lumber and composite wood products. Sounds simple enough! The type is lumber and composite wood. But not so fast! Framing is accomplished with treated/untreated lumber and composite "structural" wood and sheathing. Wood cabinets are comprised of "fine lumber" and veneered composite wood products. Framing materials are enclosed, and cabinets are interior. The type of building material is the same. Yet, the composition and accessibility is different.

- The plumbing and window installation trades use vinyl building materials. All vinyl additives are not the same; exposure environments are not the same; and degradation is not the same. The most common material used in plumbing is polyvinyl chloride pipe which is generally enclosed, not exposed to solar heat and UV light. On the other hand, window frames comprised of vinyl are often interior/exterior and exposed to solar heat and UV light. Additives used in PVC pipe and vinyl windows are also different. While the type of building material is the same, the composition is different. Environmental conditions are different, and solar conditions differ.

- The insulation and HVAC trades use insulation. Insulation contractors install fiberglass insulation, and HVAC contractors install fiberglass duct board. Loose-fill fiberglass insulation is installed in wall cavities/attics. It is generally enclosed, not subject to air disturbances. Rigid fiberglass duct board is, however, subjected to considerable air movement—air that is delivered to the occupied spaces on the interior of a building. Once again, materials are different, and conditions are different.

Thus, a discussion of "building materials by function" allows for extended conversations regarding composition, location, environmental influences, and special conditions.

Composition Gone Wild

The "toxic stew of chemicals" in building materials is on the rise. As of January 2016, more than 106 million unique organic and inorganic chemical substances and more than 66 million sequences were CAS-registered worldwide. About 15,000 chemicals are added on a daily basis (CAS 2016). Toxicological information is severely lacking!

As the world of chemistry races out of control and technology reaches for the stars, health hazards associated with building materials are becoming more and more elusive. Products are comprised of the reputed, long standing toxins of the past, poorly understood polymers of today, and those in between.

Toxins of Old

Asbestos, lead, and sand/quartz (e.g., crystalline silica) are long standing toxic building materials that, in some respects, allowed advances in construction and in civilizations—not unlike that of today's polymers. Some of the advances were sand and limestone bricks dating back to ancient times, the Roman lead potable water supply systems, asbestos fireproof roofing. Yet, today, they are the scourge of construction. Builders must manage building materials with caution, comply with federal, state, and local regulations, and remediate asbestos-containing and lead-based paint containing building materials. Airborne dust and water exposure health hazards, asbestos and lead are environmental and occupational health hazards whereas crystalline silica is an occupational hazard. Not one emits gases. So, you query, Why even mention these health hazards?

Unbeknownst to most builders, asbestos and lead is still manufactured and sold for use in building materials to this day. Such asbestos and/or lead-containing building materials may or may not be legal. They may or may not be reported. As for sand/quartz, OSHA regulated crystalline silica is in more building materials than builders are generally aware. For this reason, building materials containing asbestos, lead, and/or sand/quartz are identified and briefly discussed herein.

Polymers Run Amuck

As compared to the toxins of old, today's synthetic polymers are plastics run amuck. Used extensively in the later part of the twentieth century, there have been limited toxicological studies performed. Thus, polymer off-gassing and associated health hazards are not easily predictable. New polymers, plasticizers, and additives are being introduced daily. It is difficult, if not impossible, to track new polymer formulations and new polymer-containing building products.

There is a modicum of understanding regarding thermoset polymers, most of which are formaldehyde-based resins such as urea formaldehyde and phenol formaldehyde. With their increased use, formaldehyde emissions have a low odor threshold and may cause eye, skin, and respiratory tract irritation at low levels. Formaldehyde is regulated. Many of the formaldehyde emission sources are generally composite wood products such as plywood, pressboard, and MDF. Composite wood use in the manufacture of wood cabinets, laminated flooring, OSB flooring is but the beginning. Spray-applied urea formaldehyde foam insulation (UFFI) has also been implicated along with other building materials such as HVAC air duct board. Yet, these are only the tip of the iceberg. Look around your home, your business. Try and identify all building materials both visible and hidden that may emit formaldehyde gases! Then, test your knowledge. You may be surprised at how many products in your indoor environment emit formaldehyde!

Thermoplastic recyclable polymers (e.g., PVC) are more difficult to assess, not because of the polymer, but because of the plasticizers and plastic additives that are not bound within the polymer. Many of the plasticizers are irritants of the eyes, skin, and respiratory system, similar health effects to those of formaldehyde. Yet, they are not regulated. Listing of plasticizers is not required on an SDS. Some of the additives are regulated non-emitters (e.g., lead) that may, however, pose an environmental problem or ingestion health hazard such as when the polymer is used for the conveyance of potable water. On the other hand, UV light stabilizers are highly irritating to the eyes, skin, and respiratory tract. UV stabilizers are unregulated and found in most thermoplastic polymers such as vinyl windows. Some fire retardants are regulated toxins (e.g., antimony trioxide), and some are unregulated irritants of the eyes, skin, and upper respiratory tract (e.g., zinc borate). Formulations are protected information and constantly changing. Awareness of the possibilities and the possible health effects of plasticizers and polymer additives in plastic building materials is a good start.

The Brew of Organic Toxins

Consumers boast, "We eat only organically grown food!" Builders proclaim, "No volatile organic chemicals in my building materials!" There is an obvious discrepancy in perceptions and in the definition of organic.

The more popular definitions of organic follow:

- A compound that has carbon atoms—with exclusions
  - Carbon-containing alloys (carbon steel)
  - Metal carbonates (e.g., calcium carbonate; limestone)
  - Simple oxides of carbon (e.g., carbon monoxide)
  - Cyanides (hydrogen cyanide)
  - Allotropes of carbon (e.g., graphite)
- Any compound that has carbon and hydrogen atoms
  - This definition excludes carbon tetrachloride and many other carbon only compounds that are considered organic compounds by toxicologists and industrial hygienists
- All compounds containing carbon typically found in living systems
  - Generally, anything made from living systems, such as cloth, fuels, or wood
  - Organic foods are grown with no fertilizer except the organic compounds found in plants and animals

For purposes of this book, the first definition shall apply when referencing organic compounds.

The terms "volatile organic compounds" is another subject to ponder. All chemicals are gas, liquid, or solid. Methane is a gas, and formaldehyde is a gas that dissolves in water. Low molecular weight organics are typically gases. Gaseous organic compounds are by some considered volatile organic compounds, but technically, they are not. Typically, volatile organic compounds are liquids or solids that change directly from solid form to a vapor without becoming a liquid such as dry ice which changes from solid carbon dioxide directly to a vapor.

Volatile organic compounds are organic chemicals that readily evaporate at room temperature. They are released into indoor air in homes from a variety of products. Analytical laboratory and/or government agencies have different ways to define volatile organic compounds and differentiate them from semi-volatile organic compounds—generally based upon analytical capabilities which vary from one laboratory to the next.

- According to a prominent U.S. laboratory, key characteristics for defining VOCs are (Test America 2011)

- Vapor pressure or boiling point
- Aqueous solubility
- As the vapor pressure increases, a compound's volatility increases. As its aqueous solubility decreases, it also becomes more volatile than an aqueous solution. A brief example follows:

  1,2,4-Trichlorobenzene is about the highest boiling compound routinely included in a volatiles analysis. Its boiling point is listed as 214.4° Celsius, and it is water insoluble. Nitrobenzene's boiling point is 210.9° Celsius, but it is soluble in water. It does not purge (or process) with sufficient efficiency. So, it is not considered a volatile.

- According to Health Canada, VOCs have boiling points roughly in the range of 50–250°C (122–482°F). By this definition, organic compounds that boil below 50°C are gases, and those that boil above 250°C are either semi-volatile or nonvolatile

In the end, there are no hard and fast rules—only the preponderance of evidence.

Industrial exposures to VOCs are generally 10–100 times that of nonindustrial environments. Home and office environments are typically 2–100 times higher than that found outside. A reasonable line of logic would dictate that industrial exposures would result in more health complaints than home and office exposures. Yet, this is not the case! Many environmental professionals ascribe the complaints to exposures to a medley of chemicals and to the lack of adequate dilution of indoor air, and indoor nonindustrial air may consist of up to 300 different chemicals. The chemicals are trapped and recycled, added to, trapped and recycled. It can be a vicious cycle! Reducing building material emissions of volatile organic chemicals through awareness and proper choice of products will go a long way toward reducing VOC exposures in indoor air environments (Hess-Kosa 2011, p. 126). See examples of building materials and VOC emissions in Table 10.1.

Having reviewed VOC emissions from building materials, we may be well served to briefly review semi-volatile organic compounds (SVOCs). They are organic chemicals which have a slower rate of release than VOCs and require elevated ambient temperatures to increase the evaporation rate. They are often overlooked!

The EPA lists more than a 1000 SVOCs as high-production-volume chemicals that are produced or used at the rate of over 1 million pounds/year. In building materials, they are used as plasticizers in thermoplastic polymers such as vinyl flooring and in a variety of other building materials. See Table 10.2.

Because of their slow rate of release from sources and their propensity to remain contained until the proper conditions arise, SVOCs can persist for

TABLE 10.1

Building Materials and Possible VOC Emissions

Building Material	Possible VOC Emissions
Oil-based stains and paints[a]	BTEX, hexane, cyclohexane, 1,2,4-trimethylbenzene
PVC primer/glue and adhesives[a]	Tetrahydrofuran, cyclohexane, methyl ethyl ketone, toluene, acetone, hexane, 1,1,1-trichloroethane, methyl-iso-butyl ketone
Refrigerant from air conditioners[a]	Freons (trichlorofluoromethane and dichlorodifluoromethane)
Carpets[b]	Acetaldehyde, benzene, 2-ethylhexanoic acid, formaldehyde, 1-methyl-2-pyrrolidinone, naphthalene, nonanal, octanal, 4-PCH, styrene, toluene, vinyl acetate
Carpet adhesives[b]	Acetaldehyde, benzothiazole, 2-ethyl-1-hexanol, formaldehyde, isooctylacrylate, methylbiphenyl, phenol, 4-PCH, styrene, toluene, vinyl acetate, vinyl cyclohexene, xylenes (*n-, o-, p-*)
Foam insulation[b]	Isoprene, acetone, pentane, and carbon dioxide

[a] New York State Department of Health, VOC Compounds in Commonly Used Products. 2013. https://www.health.ny.gov/environmental/indoors/voc.htm.
[b] Hess-Kosa, Kathleen. *Indoor Air Quality: The Latest Sampling and Analytical Methods.* CRC Press, Boca Raton, Florida. pp. 209–211, 2011.

TABLE 10.2

Abbreviated List of Semi-Volatile Organic Compounds, Functions, and Applications

Chemical or Chemical Classification	Function	Applications/Sources
Phthalates	Plasticizer	PVC flooring and other plastic building materials
Brominated flame retardants	Fire retardant	Plastic building materials
Bisphenol A	Polymer	Hard plastic building materials
Pentachlorophenol	Wood preservative	Treated lumber

years within indoor air environments. Proper conditions may include, but not be limited to (1) proximity to a heat source, and (2) solar heating.

Toxic Metals and Metalloids

The more toxic, most frequently encountered toxic metals and metalloids used in building materials and some of the products in which they are used include the following:

- Cadmium is used in paint pigments and as a stabilizer in rubber and plastic materials
- Lead is used in paint pigments, water supply lines, roofing, and flashing

- Arsenic is used in paint pigments and preservatives for treating lumber
- Mercury is used as a fungicide and/or pigment in paints, and it is used in fluorescent lights
- Hexavalent chromium is used in some paints (e.g., lead chromate) mostly in the aerospace industry and chromium-containing stainless steel or painted steel—when subjected to high temperatures such as occurs in welding—converts to the more toxic hexavalent chromium

During on-site, indoor mixing, powdered pigments comprised of toxic metals/metalloids could potentially contaminate indoor surfaces and air. In older building demolition and renovation projects, disturbance of previously painted building materials could result in paint chip and dust contamination of the environment.

All toxic metal/metalloid building materials are predominantly environmental and occupational exposure hazards. As they may contribute to indoor health hazards, toxic metal/metalloid containing building materials are included within the discussion of building materials.

Thermal Decomposition and Combustion

Where there's heat, there's material degradation. Where there's fire, there's smoke. Where there's thermal degradation and smoke, there are irritating and toxic gases.

While technology and plastics have improved our lot in life, thermal decomposition and combustion products of polymer-containing building materials are becoming increasingly hazardous. Thermal decomposition is the chemical decomposition of a material as it heats up either by an increase in ambient temperatures (e.g., solar heat) or in a fire. And due to the nature of today's polymer-containing building materials (e.g., particularly the thermoplastic polymers), fires are self-perpetuating, reaching hotter temperatures faster. Flashovers occur more rapidly, and the resulting smoke is more toxic than ever before.

Prior to combustion, as the material heats up, multiple irritating and/or toxic gases are created. As combustion goes to completion, as the smoke increases, carbon monoxide and hydrogen cyanide have become the fireman's "silent killers." Other rarely discussed paradigms are that of carcinogens in fires and an increased incidence of cancer amongst firefighters.

Heating Up: Abundance of Irritants/Toxins

Many of the thermal decomposition products are respiratory tract irritants, some of which can cause severe damage to the lungs—at high exposure levels, greater than that which is likely to occur due to the solar heating of products. Although combustion may result in high exposure levels, carbon monoxide and hydrogen cyanide gases are more likely to prevail. Thermal decomposition products that cause respiratory tract irritation are grouped by location in the respiratory system.

Upper respiratory tract (URT) irritants are extremely irritating to moist surfaces such as in the nose, mouth, sinuses, and throat. Gases that cause URT irritation are the following:

- Ammonia—refrigerant
- Halogen acids (e.g., hydrogen chloride)—thermal decomposition of PVC
- Sulfur oxides (e.g., sulfur dioxide)—by-product of the manufacture of calcium silicate cement as well as the combustion by-product of organosulfur compounds such as petroleum products

Lower respiratory tract (LRT) irritants pass the URT with little or no irritation. After a one to four hour latency period, low exposures levels may result in irritation and coughing and progress to pulmonary edema. Gases that cause LRT irritation are the following:

- Carbonyl chloride (i.e., phosgene)—thermal decomposition of halogenated hydrocarbons such as Freon
- Nitrogen dioxide—thermal decomposition of melamine formaldehyde resin (e.g., composite wood)

Whole respiratory tract irritants not only irritate the entire respiratory tract but may irritate the eyes as well. Gases found to impact the entire respiratory tract include halogens (e.g., chlorine) and ozone. They are not generally considered thermal decomposition products.

Combustion Silent Killers

Most firefighters are aware that where there is smoke, there is carbon monoxide and hydrogen cyanide. Carbon monoxide can kill, and hydrogen cyanide can kill faster. Carbon monoxide is odorless and toxic by inhalation whereas hydrogen cyanide has a faint almond odor (also described a musty "old sneakers smell" by some) and is toxic not only through inhalation but skin absorption (NIOSH 2015).

Carbon monoxide displaces oxygen carried by red blood cells that deliver oxygen from the lungs to other parts of the body, preventing delivery of oxygen to the vital organs and brain. Without oxygen delivery, symptoms can result in headache, rapid breathing, nausea, weakness, dizziness, mental confusion, hallucinations, and death. Carbon monoxide is a product of the complete combustion of organic materials such as plastics and volatile organic compounds.

Hydrogen cyanide, on the other hand, interferes with the normal use of oxygen uptake in body organs particularly the more vital organs such as the central nervous system (e.g., brain), the cardiovascular system (e.g., heart and blood vessels), and the pulmonary system (e.g., lungs) (NIOSH 2015). Exposures may cause firefighters to become disoriented and agitated—to lose focus. Some have even fought others firefighters who have attempted to rescue them. Others have run away from their rescuers (Riley 2007). Not only is it an inhalation exposure, but hydrogen cyanide can rapidly enter the body through the skin absorption as well as injection and ingestion. Skin absorption is rapid, contributes to whole-body toxicity, and occurs more readily when ambient temperature and relative humidity are high (NIOSH 2015). Thus, it has been suggested that a firefighter's SCBA may not be sufficient protection. Several studies, however, indicate that hydrogen cyanide due to the combustion of building materials is at detection levels only as hydrogen cyanide is highly flammable and ultimately burns down to carbon and nitrogen. Many have suggested, however, that research is different from the real world, and real world firefighter exposures remain unknown.

According to a 2012 article published by Firefighter Nation, "the smoke that firefighters were exposed to 20 or 30 years ago is not the same as it is today. Wood, cellulose, cotton, silk, wool, etc., were bad decades ago, but they were nowhere near as toxic as the chemically-manufactured materials of today. When combined in a fire situation, these chemicals are often referred to as 'the breath from hell' (Shoebridge 2012)." Some of the polymers used in building materials that were identified in the article include the following:

- Acrylics
- Polyesters
- Polypropylene
- Polyurethane foam
- Polyvinyl chlorides (e.g., PVC)
- Thermoset polymers (e.g., phenol formaldehyde)

Synthetic rubbers (e.g., ABS) have also been implicated.

Firefighters and first responders should be aware of building materials and associated health hazards due to burning building materials. The greater the amount of polymers present in a building, the greater the risk of toxic materials—mostly during the heating process in the initial stages of a fire.

Carcinogenic Combustion Products Associated with Fires

Cancer-causing components of fires are a looming health hazard for firefighters. Studies have demonstrated statistically higher rates of multiple types of cancers in firefighters compared to the general American population. In a 2013 "Firefighter Cancer White Paper," NIOSH findings were as follows (Firefighter Cancer Support Network 2015):

- Testicular cancer (2.02 times greater risk)
- Multiple myeloma (1.53 times greater risk)
- Non-Hodgkin's lymphoma (1.51 times greater risk)
- Skin cancer (1.39 times greater risk)
- Prostate cancer (1.28 times greater risk)
- Malignant melanoma (1.31 times greater risk)
- Brain cancer (1.31 times greater risk)
- Colon cancer (1.21 times great risk)
- Leukemia (1.14 times greater risk)

During a fire, a firefighter wearing a SCBA is likely to be exposed to skin absorptions. The White Paper reported that the "skin's permeability increases with temperature and for every 5° increase in skin temperature, absorption increases 400%." Beyond a fire, before and after, both skin and inhalation exposures are likely—due largely to equipment management and hygiene. Sources of probable exposures include, but are not limited to, the following (Firefighter Cancer Support Network 2015):

- Although most firefighters wear SCBAs, their face and neck are poorly protected
- Firefighters wear dirty and contaminated turnout gear and helmets
- Many firefighters only have one set of gear which means they are continually recontaminated from previous fires
- Bunker gear goes unwashed for months at a time, even after significant fires
- Many volunteer firefighters carry their contaminated gear in their personal vehicles wherein their car is likely subjected to superheating and enhanced off-gassing of contaminants into the passenger area
- Some firefighters take their contaminated bunker pants and boots into sleeping quarters
- The interiors of apparatus (e.g., fire truck) cabs are rarely decontaminated
- Many firefighters do not take showers immediately following fires

Beyond the "Firefighters' Cancer White Paper" which claims multiple carcinogens and diesel exhaust, there is an absence in the discussion of dioxins—designated potent "known human carcinogens" by the World Health Organization (WHO).

Dioxins, Not to Be Confused with Dioxane

Dioxins are a group of compounds, the most toxic of which is 2,3,7,8-tetrachlorodibenzo-p-dioxin (2,3,7,8-TCDD). 2,3,7,8-TCDD is one of the most carcinogenic compounds known to man. It has a toxic equivalency factor (TEF) or relative toxicity of 1.0. The other dioxins range in relative toxicity from 0.1 to 0.0001 (as determined by the WHO).[*] The greater the number of chlorine atoms, the less toxic a dioxin. For instance, 2,3,7,8-TCDD with a TEF of 1.0 has four chlorine atoms, and octachlorodibenzo-p-dioxin (OCDD) with a TEF of 0.0001 has eight chlorine atoms. See Figure 10.1.

Dioxins are produced by the combustion of PCBs, a manmade dielectric fluid which is used in old fluorescent light ballasts and old transformers, and of chlorine containing plastics such as PVC plastic products.

> During July 1997, approximately 400 tons of PVC were burned in a fire at Plastimet Inc., Hamilton, Ontario, Canada. The facility was storing bales of trimmings from automobile interiors. Analysis of soot and ash samples after the fire revealed levels of dioxin sixty-six times higher than permitted in Canada. Allegedly, this fire single-handedly increased the annual dioxin emissions for the whole of Canada by 4 percent in 1997 (PVC Information 2016).

There has been considerable speculation regarding the increase in the amount of PVC plastic in buildings and increasing levels of dioxins created in this age of plastics. Subsequently, firefighters are highly likely to be exposed to dioxins—particularly by skin absorption. After a fire, uncleaned protective equipment may also provide other sources of exposure such as inhalation of airborne dioxins from dirty personal equipment stored in hot areas (e.g., the trunk of a car) and ingestion due to handling the same personal equipment and eating afterward.

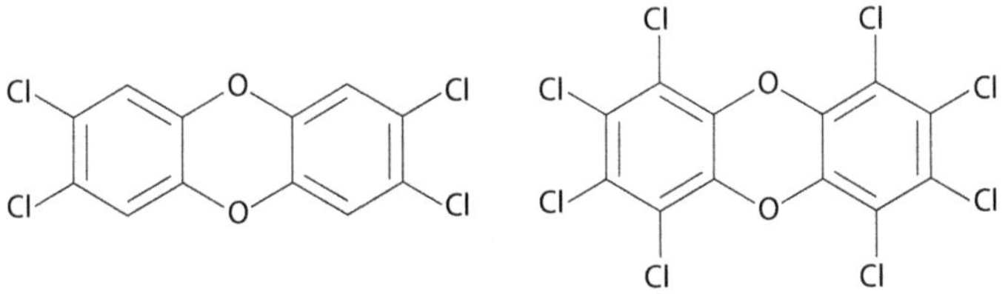

FIGURE 10.1
2,3,7,8-TCDD (left); OCDD (right).

---

[*] A TEF of 0.1 is 1/10th the toxicity of the most carcinogenic dioxin (i.e., 2,3,7,8-TCDD) which has a TEF of 1.0. Likewise, a TEF of 0.0001 is 1/10,000th the toxicity of 2,3,7,8-TCDD.

Other building materials that potentially contribute to levels of dioxins in a fire include, but are not limited to, pentachlorophenol treated wood and lumber. The latter is implicated by the EPA as "residential wood combustion" (Chlorine Chemistry Division of the American Chemistry Council 2005).

The Ever Present Asbestos

Many older buildings (e.g., built prior to 1978) and some new buildings have asbestos-containing building materials. In a fire, the building materials that contain asbestos will degrade and release cancer causing asbestos fibers. About one third of the dust and debris samples taken at Ground Zero, after the fall of the Twin Towers in September 11, 2001, contained measureable amounts of asbestos. Asbestos is a known human carcinogen and exposures are another potential carcinogenic hazard to firefighters (World Trade Center Indoor Air Assessment: Selecting Contaminants of Potential Concern and Setting Health-Based Benchmarks 2002).

Location of Building Material Emitters and the Effect of Environmental Conditions

Building material emitters are everywhere! They are exterior components. They are within interior walls, flooring, and ceiling cavities. They are in living spaces. They are in the plumbing. They are in HVAC air systems. They are simultaneously indoors and outdoors. And they are in the ever-present plastic components.

The exterior enclosure components are the first line of defense against the foibles of nature. They are comprised of (1) windows and doors, (2) roof materials, and (3) exterior wall cladding. Deterioration and irritant/toxic emissions may be due to any of a number of conditions such as any of a number of the following:

- Rainwater damage and high humidity can result in deterioration of building materials which is likely to increase product emissions.
- Excessive solar heat may cause exterior plastic building materials (e.g., vinyl siding) to deteriorate and emit toxic and irritating gases.
- Hail and high wind damage can compromise the integrity of exterior components, provide an avenue for rainwater damage, and emissions of some of the building materials within the exterior walls and roof.
- Wildlife (e.g., rodents and raccoons) and pests (e.g., carpenter ants and termites) can damage the exterior walls, a pathway to the interior—an avenue for inclement weather to breach all exterior wall barriers.

Functional framing building materials, exterior sheathing, and vapor barriers are the second line of defense, used within the exterior wall cladding and roof materials. Framing components, thermal/acoustical insulation, plumbing, and electrical wiring are also contained therein.

Damage to and improper installation of sheathing and/or vapor barriers can result in a catastrophic failure of other components, especially framing lumber. The space between the exterior components, sheathing, and vapor barrier and the interior walls/ceilings, framing materials and insulation are subject to any of a number of component failures which may result in water damage–deterioration of untreated wood, composite wood, and insulation. Emissions from products within the wall and ceiling cavities are likely to occur in the following circumstances:

- Excessive heat, especially within poorly insulated walls
- Newly manufactured building materials
- Water damaged building materials
- High humidity and heat

As most exterior wall building materials are not accessible, unseen damage and deterioration may continue for months, sometimes years—resulting in irritant/toxic emissions. The source is often difficult to track down. Generally, the search for such discrepancies starts with protestations from building occupants as they declare, "I smell mold, but I don't see it!"

Within exterior wall cavities, plumbing pipes carrying water may pose an environmental health hazard and as a source of water leaks—which can result in mold damage and deterioration of the building materials contained within. Leaks may occur due to corrosion of dissimilar metals (connectors, hangers, and joints), leaking connections, and punctures. All can occur within new construction—another source of water damage to the exterior and interior wall cavities. Water, once again, will increase many building material emissions of irritant/toxic gases and vapors.

Mechanical HVAC units have multiple emissions components. All too often, the components are not acknowledged. The components are subjected to considerable air movement and high humidity during hot ambient temperatures. Irritant/toxic emissions are mixed and conveyed to the occupied indoor air. Some HVAC components emit more than others. Other sources of indoor air pollution can also be conveyed through an HVAC system. For instance, sewage gases may be drawn into a leaking HVAC component, and refrigeration gases may leak into the system. The HVAC unit is the life's blood of the building!

Indoor walls, ceilings, doors, paint, and surface finish product emissions are delivered directly into occupied indoor spaces. Irritant/toxic emission levels are affected by the quantity of emitters, the amount emitted, fresh air exchange rates, temperature, humidity, and age of the product. If the emissions indoors become entrained in a building, they also become part of the air that is taken up in the HVAC and redistributed throughout the building.

Interior finish-out components emissions are not only indoor emitters, but occupants are also generally more routinely in direct contact with the materials. They include (1) cabinets, (2) countertops, (3) flooring materials, and (4) trim. Direct contact and a higher potential for water damage are conditions that raise the bar as to the likely impact on the occupants, and skin contact with irritant and toxic emissions in the finish-out components is more likely. Not only is there direct contact, but the emissions are also taken up by the HVAC system.

The final finishing touches are adhesives and sealants. They are everywhere, and although in small quantities, they can contribute to the overall burden of indoor irritants and toxins.

Summary

Composition and location are the key factors in considering the state of building materials. A summary of composition follows:

- The toxins of old are asbestos, lead, and sand/quartz. Asbestos and lead are all environmental and occupational health hazards whereas sand/quartz is an occupational hazard. They may not be emission hazards, but the toxins of old may pose a dust exposure hazard to not only workers but building occupants and firefighters.

- Polymers pose emission health hazards. Most plastic materials are recyclable thermoplastic polymers. Many resins are thermoset polymers and are generally formaldehyde-based resins which off-gas formaldehyde. All plastic materials and many resins also contain unbound plasticizers and polymer additives. Polymers are increasingly staking out the lion's share of building materials. They are big contributors to today's indoor air pollution. In a fire, carcinogenic formaldehyde may pose a risk to firefighters.

- Organic chemical emissions pose health hazards, albeit not as great a concern as some of the polymers. Although they represent a wide range of health effects, organic chemicals in low levels that are generally encountered indoors are irritating. Some are known carcinogens (e.g., benzene). Building materials comprised of organic chemicals (typically paints) are under tighter restriction today than they were in years past. Nevertheless, organic chemicals do contribute to the indoor "stew of toxins."

- Toxic metals/metalloids (e.g., cadmium and hexavalent chromium) are generally occupational hazards. However, they can contribute to indoor air pollution and may pose cancer risks to firefighters during a fire.
- Dioxins exposure through skin absorption, inhalation, and ingestion are potential carcinogens that contribute to the high incidence of cancer amongst firefighters.

Demolition and renovation project managers need to be aware of the possibilities of worker and environmental health hazards and, by extension, contributions of these hazardous materials to the indoor air and impact on the health of the occupants. A pre-assessment of the building materials—preexisting and new—is vital to securing the health of all concerned.

Thermal decomposition and combustion products of building materials have been discussed as well. The extreme heat such as the notoriously high temperatures in Arizona summers (e.g., exceeding 110°F) may impact building materials and off-gassing irritant gases, in turn, may impact the indoor air quality of a building. A fire, involving increasingly higher temperatures, will cause off-gassing of irritant and toxic gases. Some plastic materials will create hydrogen cyanide, and some will generate hydrochloric acid. The combustion of many building materials is likely to result in the creation of several carcinogens (e.g., formaldehyde, hexavalent chromium, cadmium, and dioxins). Awareness of building materials likely to decompose and form toxins is important to first responders and firefighters.

Understanding the composition of building materials is essential to protecting the health of future occupants as is a grasp of the impact of the function (e.g., location) of the building materials on each function material's overall emissions, contribution to indoor air pollution, and conditions to avoid. An extended discussion of building materials by function is contained within this section.

References

CAS. 2016. "CAS REGISTRY—The Gold Standard for Chemical Substance Information." *CAS*. Accessed March 30, 2016. http://www.cas.org/content/chemical-substances.

Chlorine Chemistry Division of the American Chemistry Council. 2005. "Dioxine Facts." *lForest Fires: A Major Source of Dioxins*. July 8. Accessed April 26, 2016. http://www.dioxinfacts.org/sources_trends/forest_fires2.htm.

Firefighter Cancer Support Network. 2015. "Firefighter Cancer White Paper." *Firefighter Cancer Support Network*. Accessed April 26, 2016. http://media.cygnus.com/files/base/FHC/document/2015/07/FCSN_Firefighter_Cancer_White_Paper.pdf.

Hess-Kosa, Kathleen 2011. *Indoor Air Quality: The Latest Sampling and Analytical Methods*. Boca Raton, FL: CRC Press.

NIOSH. 2015. "Hydrogen Cyanide (AC): Systemic Agent." *Center for Disease Control*. June 1. Accessed April 2, 2016. http://www.cdc.gov/niosh/ershdb/EmergencyResponseCard_29750038.html.

PVC Information. 2016. "PVC and Fire." *PVC Information*. Accessed April 26, 2016. http://www.pvcinformation.org/links/index.php?catid=3.

Riley, Carlin and Steve Young. 2007. "Hydrogen Cyanide: Deadly HCN Gas in Fire Smoke Threatens Responders." *Fire Fighting in Canada*. June 5. Accessed April 2, 2016. http://www.firefightingincanada.com/health-and-safety/hydrogen-cyanide-june-2007-1254.

Shoebridge, Todd. 2012. "Carbon Monoxide & Hydrogen Cyanide Make Today's Fires More Dangerous." *Firefighter Nation*. February 14. Accessed April 2, 2016. http://www.firefighternation.com/article/firefighter-safety-and-health/carbon-monoxide-hydrogen-cyanide-make-today-s-fires-more-dangerous.

Test America. 2011. "Volatile versus a Semi-Volatile Compounds." *Test America*. February 11. Accessed April 2, 2016. http://testamericalabs.blogspot.com/2011/02/volatile-versus-semi-volatile-compounds.html.

2002. "World Trade Center Indoor Air Assessment: Selecting Contaminants of Potential Concern and Setting Health-Based Benchmarks." *911 Digital Archive*. September. Accessed April 26, 2016. http://911digitalarchive.org/files/original/88ac5d225159e029c7d12070e19d93fd.pdf.

11

Foundations, Framing, Sheathing, and Vapor Barriers

Building foundations to framing are truly that which hold it all together. According to country singer David A. Coe, "It is not the beauty of a building you should look at; it's the construction of the foundation that will stand the test of time." Without a solid, enduring foundation, a building will crumble—the house of cards will fall!

After the foundation, framing deserves special attention! According to a forensic building engineer, structural framing reigns as the basic skeleton to progress.

> Advanced framing, as the name implies, is the intelligent use of lumber in wood framing. This unique application saves on lumber, supports better insulation and reduces the occurrence of drywall cracking—giving you a stronger, more energy—efficient home. (Joseph Lstiburek, PhD, PEng, Building Science, Inc.)

As an extension to the "world of framing," wall/roof sheathing and structural subfloors as well as the proper placement and application of vapor barriers are an integral part of framing. The whole package from foundation to framing to wall/roof sheathing and structural subflooring to vapor barriers—if not installed properly—can lead to severe structural failures. Furthermore, misapplication of sheathing and vapor barriers is a common cause of water damage and mold growth as well as ground-sourced toxic gases gaining entry into the occupied building spaces.

In addition to product emissions, water damage and ground gases can contribute to poor indoor air quality. Therefore, the more blatant structural failure modes, which are due to commonly overlooked, little known construction faults, are discussed herein with building material emissions and thermal decomposition products.

Foundation Components

Between 100 BC and 400 AD, the city of Venice was built on a foundation of wood. And the structures still stand to this day. Wood rot? No, wood pilings submerged in salt water and mud do not rot. Actually, over time, the wood petrified to stone. Yet, most structures were built on above ground pilings, and, prior to 1960, wood foundations succumbed to wood rot and pest damage. However, since the 1960s, shortly after the development of preservatives for wood that prevented rotting and insect damage, wood foundation structures have become readily accepted for low-cost lightweight stick framed residences. There are allegations that the treated wood will endure for up to 100 years.

Still standing today, over 5000 years after construction, the stone foundation—on which the Giza Pyramids of Egypt were built—has endured the test of time. During biblical times, there were references to a house built on a rock.

> A wise man built his house on a rock. The rains came down, and the streams rose. The winds blew and beat against that house. Yet, it did not fall, because its foundation was on a rock. A foolish man built his house on the sand. The rains came down, and the streams rose. The winds blew and beat against that house. The house on sand fell. (Matthew 7:24–29)

Dry stack stone foundations predated mortared stone structures, and cement in the form of lime and sand has been recorded from the building of the Pantheon. One would believe that the mortared stone structures would be longer lasting, yet mortar eventually cracks due to frost damage and erodes due to rain and water excesses. Well-built, properly stacked dry stone, however, is not subject to climate change and weather.

The first concrete-like foundations date back to 700 BC when the desert Bedouins discovered that mixing lime with water gave them cement. This was the first recorded application of cement floors—inside rubble structures. In 125 AD, the Pantheon exterior wall foundation was built with pozzolana cement (i.e., lime, reactive volcanic sand, and water) tamped down over dense stone aggregate to a depth of 15 feet. All the basic components for concrete were in play! Yet, modern concrete did not evolve until the late 1800s when reinforcing steel rods were added. Some of the original reinforced concrete structures still stand to this day—cracked but intact. In the 1900s, concrete formulation became a science. Yet, it wasn't until mid-1950s that "concrete slab foundations" came into common use.

Foundations may be full slab, pier-and-beam, or a combination thereof. They may be shallow (e.g., 1½ feet from ground level); they may be deep (e.g., 20 feet from ground level); or they may be floating slab-on-grade (e.g., molded formed concrete placed on ground). Choices of foundation type based on structure design and weight (e.g., load bearing requirements), soil types (e.g., clay, rock, or bayou mud), available materials, life expectancy, and cost. Building materials for foundations discussed herein include wood, stone, and concrete.

Timber Piles

When not chemically treated with a preservative, wood will rot and deteriorate. Older wood foundations predating treated lumber are not likely to have survived the test of time.

In mid-1930s, they began treating railroad ties with creosote. This was the first real attempt at treating lumber, but it was used strictly for treating railroad ties and later for telephone/electric poles. Let us look at the past!

Marketed and sold in the early 1940s, chromated copper arsenate (CCA) was the first wood preservative for pressure treating wood posts/timbers that were used, among other building products, in building foundations. The copper and arsenic are fungicides (e.g., prevent mold rot); the arsenic is an insecticide (e.g., prevents termite damage); and trivalent chromium provides UV protection. It was a great product! Yet, by the turn of the millennium, arsenic, a highly toxic component of CCA, was found to be leaching into the soil. CCA was durable with a life expectancy in harsh environmental conditions in excess of 46 years. Yet, as all good things come and go, and CCA's Achilles' heel was found.

In January 2004, the U.S. EPA and "industry" agreed to restrict the use of CCA treated lumber in residential and commercial construction with a few exceptions. See Table 11.1. In 2003, Europe and Australia developed similar, but enforceable, restrictions, and, in 2012, the United Kingdom classified CCA treated timbers as "hazardous waste." Yet, to this day, CCA treated posts/timbers are still marketed and sold in America. CCA remains the most trusted, most reliable treatment for use in building foundations.

As CCA environmental alarm bells were tolling, researchers launched into the development of alternative arsenic free wood preservatives. Although available around 2002, CCA replacements (e.g., alkaline copper quaternary and copper azole [CA]) have yet to be widely accepted. To this day, "CCA foundation timbers are exempt" from restrictions in the United States and have withstood the test of time!

TABLE 11.1

Abbreviated list of exceptions to U.S. EPA restrictions for CCA

Lumber and plywood for permanent wood foundations
Lumber and timbers immersed in seawater
Piles on land, in fresh and marine water
Plywood
Round poles and posts used in building construction
Sawn timbers, at least 5-inches thick, used to support residential and commercial structures
Shakes and shingles
Structural composite lumber (i.e., parallel strand and LVL)
Structural glued-laminated wood products
Utility poles

CCA treated timbers and lumber are more of a worker handling hazard and environmental contamination hazard than a toxic/irritant product emissions concern. Thermal decomposition products include carbon dioxide, carbon monoxide, copper, arsenic, and zinc.

Be forewarned! As old timbers and lumber should be considered suspect of having been treated with CCA, demolition and burning of treated wood products could result in airborne toxic levels of arsenic!

Stone Foundation Building Materials

Many of the old stone foundations of yesteryear still remain to this day. While they captivate the log cabin and back-to-nature groups in industrialized nations, today's stone foundations are still the predominant foundation of choice, particularly in Third World countries. Stone foundations may have a long life (e.g., in excess of 100 years)—assuming the soil cooperates. They are not as costly as concrete foundations in time and materials, and stone is certainly the way-to-go in remote wilderness areas.

Stone foundations may be dry stacked stone or mortared. Dry stacked stone relies on a craftsman's skills, while mortared stone allows for gaps and voids. They may be simple padstone; they may be exterior wall support; or they may be exterior wall, rubble filled.

In terms of emissions, stone could contain naturally occurring radioactive materials (e.g., uranium) that decay and emit radon gas—a silent killer. The U.S. EPA speculates that radon may be responsible for about one-third of all lung cancer deaths to nonsmokers in the United States. While they have been dammed as major sources of radon by some, granite and quartz are defended by others. All minerals are not alike. Granite may or may not contain uranium. Quartz, which is in most mineral formations, may or may not contain uranium. Local governments, however, are likely to have assessed radon emissions by area. A little research can go a long way!

Concrete Foundations

Simply stated, "Concrete foundations are concrete and rebar." Yet, simple is an understatement. Beyond the basics, concrete foundations consist of (1) concrete with additives; (2) fill material; (3) a vapor barrier that resists water intrusion through the foundation; (4) rebar that rusts and rebar that resists rust; and (5) pesticides.

Concrete is Portland cement, sand, aggregates, and additives. The Portland cement is typically powdered limestone. The sand and crushed stone aggregates are comprised of silica, various amounts of the respirable health hazard crystalline silica. Additives, which alter the concrete properties such as workability, curing temperature range, set time, and color, "may contain lead." Plasticizers and nonhazardous fibers (e.g., polypropylene fibers) are generally added as well. Once cured, concrete components should not pose an emissions hazard—unless the concrete is cut, chiseled, and/or ground. In the latter scenario, smaller, respirable dust "could" pose a crystalline silica health hazard.

Fill, sometimes referred to as hard pack, is typically nonhazardous road base or compacted local soil. Yet, with the rise in the popularity of recycling, demolished roads, foundations, and buildings are an option. A word of caution! Recycled concrete may contain unidentified health hazards (e.g., asbestos, pesticides, and/or toxic waste). Whereas the solids, such as asbestos, do not pose a threat while contained within a slab, the more volatile toxins may enter a building through porous concrete.

A vapor barrier is applied over the fill and footing prior to placement of the rebar and concrete. Its purpose is to prevent moisture from rising up through the porous concrete foundation. Vapor barriers were not used in concrete foundations until the 1950s. The original 6-mil polyethylene plastic sheeting was a lightweight, inexpensive vapor retarder. It got damaged during the concrete pour and was minimally effective. Today, 15-mil plastic sheeting is an ideal vapor barrier, and 10-mil plastic sheeting is considered minimal. Yet, 6-mil plastic sheeting has been used in the past and is still used to cut corners to this very day. If inadequate or improperly installed, moisture "will" rise through a breach in the barrier, through the concrete and into the building—resulting in damage to the slab, excessive moisture in the building, and potential mold growth within occupied spaces. Product emissions from a slab vapor barrier (e.g., plasticizers and additives), if any, are unlikely to pass through the concrete.

Moisture protection of underground concrete walls to be used as a living space must be held to a higher standard than above ground walls. Moisture can seep through porous concrete or cracks in the subterranean walls and flooring. Besides microbial and algae growth, signs of penetration include concrete efflorescence (i.e., white crystalline powder), mineral deposits, and rust stains. Vapor barriers should be located on the external wall of a basement or other in-ground structures (e.g., subterranean "earth homes"). This should allow inward drying to the basement where moisture can be removed by ventilation or dehumidification. Exterior moisture barriers are a good start while interior moisture barriers should be avoided at all costs (Lstiburek 2006b).

Rebar may be carbon-hardened steel or rust-resistant stainless steel and fiberglass. In foundations, glass fiber-reinforced polymer (GFRP) has a life expectancy of over 100 years, far exceeding that of steel (alleged to be about 10 years). Yet, the cost is considerably higher and it must be sized and shaped at the factory. Nevertheless, GFRP is being used in civil engineering projects (e.g., bridges) and occasionally in building foundations. Post tension steel cables, an alternative to reinforced steel bar, are generally coated (e.g., grease), covered (e.g., polypropylene), or GFRP. There are no perceived and/or reported toxic/irritant emissions or health hazards associated with any form of in-foundation rebar.

Concrete and Steel Support Piles

Although treated timber piles have been the residential choice-of-the-day, steel, concrete-filled pipes, or precast reinforced or prestressed concrete support piles are requisite for larger, high load bearing buildings. There are no perceived and/or reported toxic/irritant emissions associated with any of these materials.

Framing Components

The frame of a building is the skeletal system upon which all other building components must rely. Without a strong backbone, a structure will fall!

Dating back to 500 AD, man built timber framed homes similar to today's hand crafted log cabins. By the twelfth century, Europe had refined timber construction. Some timber framed churches remain to the very day (e.g., Winchester Cathedral). It wasn't until mid-1800s that sawmills were able to cut dimensionally smaller, lightweight "lumber." By the late 1800s, steel framing had its start in Chicago where high-rise buildings were in large-scale demand. The Empire State Building was built in 1930, prior to the Great Depression and World War II. About the same time as steel was introduced, iron-reinforced concrete and precast concrete blocks saw the light of day. Engineered wood framing products were developed in the 1960s. Structural insulated panels were developed in the 1970s, and insulated concrete forms were created in the 1990s. Other nonconventional framing methods, deserving mention only, include straw bale, rammed earth, and cob house—all of which may require some degree of stick framing.

Stick, or wood, framing is the most common residential and small commercial building method. Building materials are conventional lumber, treated lumber, and engineered wood. Steel framing is commonly used in large commercial, high-rise structures, and industrial complexes. Concrete form and precast concrete blocks are universally used in a variety of structures. And the energy efficient, more costly structural insulated panels and insulated concrete forms are a niche market, used mostly in the more expensive, energy conscious homes and small commercial buildings. Framing material types may also include dissimilar building materials—steel load-bearing I beams with lumber framing, steel exterior framing with lumber interior framing, and concrete exterior framing with steel interior framing. The combinations are unlimited!

Wood Framing

Wood framing, also referred to as "stick framing," has evolved over time to accommodate practical and environmentally sensitive concerns regarding diminishing reserves of lumber. As good wood began to rot and large lumber became scarce, technology found its way. "A solution for every problem!"

Conventional lumber is from fast growing, smaller trees. Rot is managed with wood treatment. Engineered wood accommodated the need for larger, structurally sound lumber. So, rags to riches, wood framing becomes sustainable and cost effective for all.

Conventional Lumber

Lumber is by far the most popular construction framing material, because it is readily available, easy to work with, and comparatively less expensive than other framing materials. Choices for lumber include Douglas fir, Western/Eastern S–P–F, Hem-fir, Southern yellow pine, and fir and larch.

Douglas fir (*Pseudotsuga menziesii*) is neither a true fir nor a pine or spruce. Also known as Oregon pine or Douglas spruce, it is an evergreen conifer species native to Western North America, America's most abundant softwood. It is reputed to have dimensional stability and superior strength. Douglas fir has been reported to cause skin irritation and an increased likelihood of splinters getting infected which is likely due to the resin while handling the lumber. Fine dust may cause worker nausea and giddiness. However, once installed, irritating/toxic emissions are highly unlikely.

Western/Eastern S–P–F is a composite of spruce, pine, and Douglas fir. All are considered "white" lumber and are batched as one, indistinguishable from one another. All can cause worker skin irritation, and fine dust can exasperate asthma. Irritating/toxic emissions are highly unlikely.

Hem-fir combination is a reference to a group of timber found in the Western United States—Western hemlock (*Tsuga heterophylla*) and five of the true firs (genus *Abies*). This group is second only to Douglas fir in terms of abundance, production volumes, strength, and versatility (Western Wood Products Association 1997). Do not confuse hemlock lumber with poison hemlock. The Western hemlock is a tree, and poison hemlock is a parsley-like herb. Western hemlock and the true firs are skin irritants only.

Southern yellow pine is native to the Southern United States. It is hard, dense, and warm in color. The resin can cause worker skin irritation, and the fine dust can exasperate asthma. Irritating/toxic emissions are highly unlikely.

Fir-larch is a mixture of Douglas fir and Western larch (*Larix occidentalis*), sold as either individual species or as one species grouping. The two species are intermixed because they have similar characteristics and are used in similar ways. Western larch can cause worker skin irritation and may cause hives and lesions. Irritating emissions are highly unlikely.

In conclusion, toxic/irritant emissions from conventional lumber are highly unlikely. Thermal decomposition products are carbon dioxide and carbon monoxide.

Treated Lumber

Most wood species are attacked by wood-destroying organisms (e.g., insects, fungi, and bacteria) when exposed to environmental stresses (e.g., moisture, soil contact, and water submersion). Termites and carpenter ants will eat all wood products. Fungi, likewise, will consume moisture damaged wood products. Yeast and bacteria will grow in wet areas. So, in defense of nature's bounty, man has developed wood preservatives—insecticides and fungicides.

Wood preservatives are controlled by the U.S. Federal Insecticide, Fungicide, and Rodenticide Act which is under the U.S. EPA. Studies and the advancement of new technology fall under the purview of technical committees within the American Wood Protection Association (AWPA) and ASTM International (Lebow 2010, pp. 15-1 to 15-2).

The U.S. EPA is responsible for regulating and registering all such wood preservatives, and federal law requires that prior to "selling or distributing preservatives" in the United States—a company must ensure, with reasonable certainty, that a preservative will cause no harm to human health or to the environment. To accomplish this, applicants must produce over one hundred different scientific studies and tests. Registration is not an overnight wonder, but it only applies to the selling and distribution of wood preservatives, not the products. This is an important point in that the treated wood products are not federally regulated. There are no restrictions on the sale of treated wood products. Whereas the application of a wood preservative is restricted, the sale of and final use of the treated wood is not restricted.

Before a wood preservative is approved for pressure treatment of structural members, it is evaluated by the AWPA. Thus, as each preservative has its own unique characteristics, the AWPA developed a set of Use Category System (UCS) standards to simplify identification and appropriateness of various wood preservatives. They range from UC1 for interior construction, above ground, dry conditions for insect treatments only, UC2 for interior construction, above ground, damp conditions for fungal and insect treatments, to UC5C for salt or brackish water and adjacent mud zones for saltwater organisms. See Table 11.2. With an exception of restricted use CCA, all wood preservatives are UC1, and most are UC2—non-inclusive of borates (AWPA 2015). The UC1 and UC2 standards apply to framing lumber.

In the United States, treated lumber is generally affixed with information regarding the UCS approved end use of the lumber along with the type of wood preservative, retention levels, warranty information, warnings, and much more. All this information is presented in small print on a 2¾ by 7/8-inch, front and back tab which often goes ignored—particularly the environmental, health and safety warning information on back. See Figure 11.1.

In construction, pressure treated lumber and timber are generally used for (1) sill plates that are in direct contact with foundations, (2) floor joists and flooring on pier-and-beam foundations, and (3) other areas where moisture,

TABLE 11.2

AWPA Use Category Designations

Use Category[a]	Brief Description
UC1	Interior dry
UC2	Interior damp
UC3A	Exterior above ground, coated with rapid water runoff
UC3B	Exterior above ground, uncoated or poor water runoff
UC4A	Ground contact, general use
UC4B	Ground contact, heavy duty
UC4C	Ground contact, extreme duty
UC5A	Marine use, northern waters (salt or brackish water)
UC5B	Marine use, central waters (salt or brackish water)
UC5C	Marine use, southern waters (salt or brackish water)
UCFA	Interior above ground fire protection
UCFB	Exterior above ground fire protection

Source: American Wood Protection Association. 2016. "Information for Homeowners." *AWPA.* Accessed November 8, 2016. http://www. awpa.com/references/homeowner.asp.

[a] Wood preservatives are categorized in accordance with the minimum average retention level of the preservative. For example, Alkaline Copper Quaternary (Type A or D) with a retention level of 0.17 pounds per square foot (psf) is assigned a category of UC1,2, and a retention level of 0.40 psf is assigned a UC4A. The higher the retention level the more durable a preservative with 0.60 psf being the highest for ACQ Type A or D which is assigned a use category of UC4B.

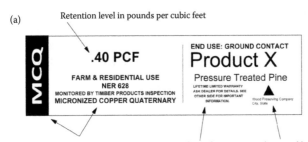

(a) Retention level in pounds per cubic feet

Treatment description   Warranty and other information   Producer and location

(b)

Important information
• For interior and exterior application. Use fasteners and hardware that are in compliance with the manufacturer's recommendations and the building codes for their intended use.
• Do not burn preservative wood. Wear dust mask & goggles when cutting and sanding wood.
• Wear gloves when working with wood. Do not use preserved wood as mulch.
• Some preservatives may migrate from the treated wood into soil/water or may dislodge from the treated wood surface. Upon contact with skin, wash exposed skin areas thoroughly.
• Dispose in accordance with federal, state, and local regulations.
For additional infonnation, visit www.producer

FIGURE 11.1

Sample of affixed information tab on all tread lumber and plywood with front of tab (a) and back of tab (b).

fungal, and insect damage might reasonably be anticipated (e.g., exterior door jambs and headers). In some areas of the world, especially where there are extreme conditions (e.g., constant rain; invasive, difficult to control Formosa termites), "all framing lumber" may require treatment.

Poorly designed exterior cladding can result in water damage and wet rot fungal growth and damage to "untreated sheathing and lumber." Wet rot fungal growth can cause structural damage to a building. See Figure 11.2.

On the other hand, treated lumber with some commonly used treated lumber may have visible mold growth on the surface of newly purchased treated lumber. For example, the author discovered after purchasing several boards of pressure treated plywood and leaving them in the back of an open pickup truck for a couple of days—staining which appeared to be mold growth. The suspect stain was confirmed by a laboratory to be a type of mold—*Paecilomyces variotti*. This finding only demonstrates that once we think we have things under control, someone throws a curve ball. See Figure 11.3. The type of wood preservative was micronized copper quartenary.

Wood preservatives are generally categorized: (1) waterborne or (2) oilborne. There is a laundry list of chemicals applicable to each category, but only a handful are EPA approved. Even fewer are readily available commercially for building construction purposes at a competitive price. Some are used in niche, difficult-to-find applications. Some are extremely expensive. Some are phased out due to lack of interest. And many manufacturers "create their own formulations" out of the EPA approved chemicals. As all things are possible, be vigilant when purchasing non-EPA preservative treated lumber and/or lumber that has not been assigned an AWPA use category.

The preservatives disclosed in this section are AWPA-listed chemicals. Each chemical has a specified preservative retention requirement to allow assignment to each category. The greater environmental stresses require

FIGURE 11.2
Fungal damage to untreated structural wood studs—wet rot (a Basidiomycetes fungus).

FIGURE 11.3
Surface mold growth on pressure treated plywood—full sheet of plywood (a); close-up of section of full sheet (b).

greater product retention, and the greater amount of product, the greater the weight of the treated lumber. And the reliability of lumber that does not have an AWPA Use Category should undergo some level of scrutiny. The possibilities are endless! Thus, some unproven, yet potentially exceptional preservatives and wood preservative systems may not be listed herein.

Waterborne Preservatives

Alkaline copper quaternary (ACQ) preservative is copper (e.g., 66.7%; copper oxide; micronized copper) with a quaternary ammonium compound (e.g., 33.3%; 15-odecyl dimethyl ammonium chloride). The copper component (up to 70% by weight) is the primary fungicide, and the quaternary ammonium compound is a secondary fungicide as well as an insecticide. A drawback to ACQ, however, is that its corrosive to steel and requires more expensive double-galvanized or stainless steel fasteners. There are four different ACQ formulations—Types A, B, C, and D. All contain copper with different combinations of quaternary ammonium compounds and carrier solutions (i.e., ammonia; ethanolamine). See Table 11.3. Carrier solutions are regulated worker exposure toxins, irritants of the eyes, skin, and respiratory tract, and quaternary ammonium compounds are also irritants. ACQ treated lumber gives off a slight "ammonia-like odor." There have been no reported odors, emissions, and/or irritation to building occupants associated with ACQ treated lumber. However, building material emissions are possible at

TABLE 11.3
Summary of Wood Preservatives

	Initial Use Year	Special Characteristics	Appearance/Odor	Special Fasteners Required[a]	Handling Hazards
Restricted use preservatives		*EPA restricted use in 2004*			
CCA; CCA-C[b]	1940	47.5% chromium trioxide, 18.5% copper oxide, and 34.0% arsenic pentoxide; carrier: water; restricted use (see Table 11.1)	Green-brown weathers to gray/no odor	—	Irritant, sensitizer Toxic arsenic pentoxide
Waterborne preservatives[d]		*Interior, exterior above ground, and ground contact*[c]			
ACQ[d]					
ACQ-A	1992	Discontinued in 2000	—	—	
ACQ-B	1992	66.7% copper oxide and 33.3% didecyl dimethyl ammonium chloride (DDAC); carrier: ammonia; primarily used to treat Western species such as Douglas fir	Dark green-brown/slight ammonia odor	X	Severe irritant
ACQ-C	2002	66.7% copper oxide; 33.3% alkyl dimethyl benzyl ammonium chloride (ADBAC); carrier: ammonia-ethanolamine	Variable green-brown/ammonia odor	X	Severe irritant
ACQ-D	2002	66.7% copper oxide and 33.3% didecyl dimethyl ammonium chloride (DDAC); carrier: ethanolamine; used to treat most lumber with the exception of Western species	Light green-brown/ammonia-like odor	X	Severe irritant
MCQ	2012	Micronized copper and quaternary ammonium chloride	Light green/slightly ammonia odor	X	Severe irritant

(Continued)

TABLE 11.3 (*Continued*)
Summary of Wood Preservatives

	Initial Use Year	Special Characteristics	Appearance/Odor	Special Fasteners Required[a]	Handling Hazards
CA (also referred to as Womanized®)[e]		*Interior, exterior above ground, and ground contact[c]*			
CBA-A	1995	49% copper oxide, 49% boron (i.e., boric acid), and 2% azole (i.e., tebuconazole); has been discontinued	Green-brown/no odor	×	Slight irritant
CA-B	2002	96.1% copper oxide and 3.9% azole (i.e., tebuconazole); carrier: water	Green-brown/no odor	×	Slight irritant
CA-C/MCA	2012	96.1% micronized copper and 3.9% azole (i.e., tebuconazole and propiconazole); carrier: water; red and yellow iron oxide colorants	Red and yellow colorants/no odor	×	Slight irritant
CX-A	2005	50% Cu-HDO, 16.3% copper carbonate micronized elemental copper, and 5.0% boric acid; restricted in aquatic areas: toxic to aquatic organisms	Blue violet colorant/no odor	×	Slight irritant
ACZA	1940	50% copper oxide, 25% zinc oxide, and 25% arsenic pentoxide; carrier: ammonia in water high-temperature stable: performs well with Douglas fir	Olive to blue-green/slight ammonia odor	×	Slight irritant
CuN-W	1940	Copper naphthenate waterborne is less common than solvent-suspended	Dark green-brown/no odor	Recommended	Slight irritant
Organic preservatives		*Interior and exterior above ground only[c]*			
PTI		American Wood Protection Association (AWPA) approved for above ground use only; can paint and/or stain; wax stabilizer added to prevent warping and splitting when drying new on the market; not widely available; generally special order required	Color additive to indicate treatment/solvent odor	—	Irritant; slightly to moderately Toxic components in organic solvents

(Continued)

TABLE 11.3 (*Continued*)

Summary of Wood Preservatives

	Initial Use Year	Special Characteristics	Appearance/Odor	Special Fasteners Required[a]	Handling Hazards
CuN	1940	Copper naphthenate is most commonly suspended in solvents (e.g., mineral spirits)	Bright green to light brown/solvent odor	–	Irritant; slightly to moderately Toxic components in organic solvents
Borates		*Interior only[c]*			
SBX	1980s	Borate compounds (e.g., boric acid; sodium octaborate) nontoxic but borate compounds do not become fixed in the wood; when exposed to moisture may lose effectiveness; floor beams and internal structure members only	No color/no odor	–	–
Other		*Unlisted by the AWPA*			
Sodium silicate-based preservatives	1800s	"Glass Wood"; rot and decay resistant; surrounds wood fibers with protective, nontoxic, amorphous glass matrix	No color/no odor		

a Corrosion resistant fasteners: hot dipped galvanized steel, stainless steel, silicon bronze, and copper.

b Federal Insecticide, Fungicide, and Rodenticide Act—basis for regulation, sale, distribution and use of pesticides in the U.S. EPA-authority to suspend or cancel the registration of pesticide if subsequent information shows that continued use would pose unreasonable risks.

c As per the American Wood Protection Association (AWPA).

d EPA identified ACQ as an alternative to CCA in 2014.

e Ammonia may be included during pressure treatment for better penetration to Douglas fir.

elevated temperature, but not likely (EPA 2014a). Thermal decomposition products include carbon dioxide, carbon monoxide, nitrogen oxides, organic chlorides, aldehydes, amines, ammonia, and copper compounds.

Alkaline copper DCOI (ACD) is ACQ-C with the addition of a co-biocide to protect against copper tolerant fungi (i.e., brown rot). This co-biocide is 4,5-dichloro-2-N-coty-4-isothiazolin-3-one (DCOI). DCOI is chemically similar to methylisothiazolinone which can cause allergic contact dermatitis, and DCOI is alleged to cause skin allergies and chemical burns with constant contact such as worker handling of treated products. As DCOI is a semi-volatile, building material emissions are possible at elevated temperature, but not likely. Thermal decomposition products include carbon dioxide, carbon monoxide, nitrogen oxides, organic chlorides, aldehydes, amines, ammonia, and copper compounds.

Ammoniacal copper zinc arsenate (ACZA) is similar to CCA with a few improvements that make long-term retention of the active ingredients (copper, zinc, and arsenic) a big plus as regards environmental concerns. ACZA is also highly resistant to leaching in water, more so than CCA, and at a retention rate of 1.25 lb. ft^{-3}, ACA treated lumber has no failures after 60 years as opposed to 46 years for a CCA-B retention rate of 1.04 lb. ft^{-3} or 25 years for a CCA-C retention rate of 0.79 lb. ft^{-3} (Lebow 2010, pp. 15–7). Ammonium bicarbonate dissolves the metals, and ammonium hydroxide is the carrier which evaporates from the treated wood, leaving the active ingredients behind. Due to the presence of arsenic, ACZA is subject to the same health hazards and restrictions as CCA. Yet, ACZA and CCA components are solids. ACZA treated lumber emissions are not likely. Thermal decomposition products include carbon dioxide, carbon monoxide, copper, arsenic, and zinc. Burning ACZA treated lumber could result in airborne toxic levels of arsenic!

Borate preservatives (SBX) are nature's powerhouse for defense against fungi, termites, beetles, carpenter ants, and scorpions. AWPA acceptable borate compounds include disodium octaborate tetrahydrate (DOT), sodium octaborate, sodium tetraborate, sodium pentaborate, and boric acid. As per the AWPA, SBX with a retention of 0.28 lb. ft^{-3} are excellent for treating typically preservative-resistant Formosa termites as opposed to SBX with a retention of 0.17 lb. ft^{-3} that does not protect from Formosa termites (AWPA 2015). The main drawback to SBX is that—in the presence of water—it will leach out of treated lumber, dissipate, and lose its effectiveness. Thus, it is applicable only on dry surfaces such as framing lumber. If the lumber should somehow get wet (e.g., a water leak in the plumbing), its effectiveness will dissipate and be rendered ineffective. Beyond pre-pressure treated lumber, SBX can be and is often applied to the surface of sill plate lumber prior to wall placement. Zinc borate has a lower water solubility than other borates and is typically added to the finish or wax coat that is applied to wood composites. Borates are odorless and not toxic to humans. Borate treated lumber does not pose a toxic emissions hazard. As borates are not combustible, the thermal decomposition products of treated lumber would be carbon dioxide and carbon monoxide.

CA is one of the most commonly encountered pressure treated wood preservatives in lumber at home improvement stores. An all-around low toxic preservative, CAs are designated Type A, Type B, and Type C. Copper boron azole Type A (CBA-A) consists of copper (e.g., copper oxide; micronized copper), boron (i.e., boric acid), and azole. It was discontinued shortly after production—for lack of interest. CA-Type B (CA-B) contains copper oxide and azole, no boron. Copper oxide is a fungicide, and the azole (e.g., propiconazole or tebuconazole) is a secondary fungicide and insecticide (EPA 2014). CA-Type C, also referred to as micronized copper azole (MCA), is a variation of CA whereby suspended nanoscale copper is the fungicide, the azole a secondary fungicide and insecticide. Azoles may cause slight eye, skin, and respiratory irritation. CA treated lumber emissions are not likely. Thermal decomposition products include carbon monoxide, nitrogen oxides, formaldehyde, and copper compounds.

CCA is a restricted pressure treated wood preservative. See discussion in Timber Pile Foundation Materials. CCA treated timbers and lumber are more of a worker handling hazard and environmental contamination hazard than of a toxic/irritant product emissions concern. Thermal decomposition products include carbon dioxide, carbon monoxide, copper, arsenic, and zinc. Burning CCA treated lumber could result in airborne toxic levels of arsenic!

Copper HDO, bis-(N-cyclohexyldiazeniumdioxy) copper (Cu-HDO), is a single component preservative that has limited human toxicity and irritant studies. However, Cu-HDO treated lumber is restricted from use in aquatic environments and should not be used in the construction of beehives. It should be noted that Cu-HDO may not be a stand-alone preservative, practically speaking, but a component of other formulations and/or trade names. For example, Wolmanit® CX-10 contains 3.5% bis-(N-cyclohexyldiazeniumdioxy) copper, 16.3% copper carbonate hydroxide, and 5% boric acid. Under another name and formulation, Cu-HDO may not fall under the AWPA guidelines for standardized wood preservatives. Cu-HDO treated lumber emissions are not likely. Thermal decomposition products include carbon dioxide, carbon monoxide, and "irritants."

Copper naphthenate (CuN-W) is an organometallic compound. CuN is also available in a fuel oil carrier which is more restrictive in use in occupied structures but is not as corrosive as CuN-W to fasteners. Worker exposures may result in severe eye, skin, and respiratory irritation. CuN-W is not considered toxic to humans, but it is very toxic to aquatic life. For this reason, wood products treated with CuN-W should not be used in and/or around surface water but is acceptable in occupied structures. CuN-W treated lumber emissions are possible, but not likely. According to manufacturers, thermal decomposition products of CuN-W and carbon dioxide, carbon monoxide, and "unknowns."

EL2 is DCOI (98%; main fungicide component of ACD), imidacloprid (2%; termite resistant component of PTI), and a moisture control stabilizer. As the lion's share of EL2 is DCOI, anticipate similar worker skin allergies and chemical burns. EL2 treated lumber emissions are possible, but not likely.

KDS has copper oxide (41%), polymeric betaine (33%; 19odecyl polyoxyethyl ammonium borate), and boric acid (26%), and KDS-B does not have boric acid. The copper oxide is a primary fungicide, polymeric betaine is a secondary fungicide and an insecticide; and boric acid is an insecticide. According to a U.S. EPA approved pesticide label, contact exposures to polymeric betaine may cause irreversible eye damage and skin burn, and it can be fatal if inhaled (U.S. EPA 2009). Polymeric betaine is a quaternary ammonium compound similar to that found in ACQ, KDS has a much higher content of polymeric betaine which elevates the irritation levels of the KDS product. Based on the U.S. EPA approved label and the higher content of polymeric betaine, worker exposures are likely, and handling KDS treated lumber may result in severe eye irritation and skin burn. People in the vicinity when the lumber is cut and sawed may also experience symptoms. As for KDS treated lumber emissions, the components are semi-volatiles and not likely to pose an emission hazard to building occupants. KDS treated lumber is processed and sold mainly in Germany, and product information was unattainable at the time of the writing of this book. The treated lumber is a rare find on the U.S. marketplace. Building emissions are not likely.

Propiconazole–tebuconazole–imidacloprid (PTI) is an organic, copper and arsenic-free wood preservative. Combined together as one, propiconazole and tebuconazole are "synergistic" azole fungicides. The synergistic azole partnership makes up a primary fungicide as opposed to the single azole that is a secondary fungicide in CA. Imidacloprid, the third component, is an insecticide which resists, not kills termites. Azoles and imidacloprid may cause slight eye, skin, and respiratory irritation. Yet, all are semi-volatiles and not likely emitted from treated lumber. Thermal decomposition products include carbon monoxide, nitrogen oxides, hydrogen chloride, and hydrogen cyanide.

Ammonia copper citrate is no longer commercially available—due to lack of interest, not due to toxicity. Other AWPA preservatives not mentioned herein are typically not used in building materials. For example, creosote is generally applied to utility poles and railroad ties, not framing lumber.

Oil-Based Preservatives

Oxine copper, copper-8-quinolinolate (Cu8), is an oil-borne organometallic compound consisting of 10% copper-8-quinolinolate, 10% nickel-2-ethylhexanoate, and 80% inert ingredients. Beyond pressure treated wood, Cu8 is used as a lumber/timber spray to control sapstain fungi ("bluestain"), surface mold, and insects. Cu8 is the only EPA-registered preservative that is permitted by the U.S. Food and Drug Administration for treated-wood contact with food (e.g., produce containers). Cu8 has also been used to treat webbing, cordage, cloth, leather, and plastics. All things considered, Cu8 treated wood is neither an environmental nor a human health hazard. However, the manufacturers claim that contact with the Cu8 chemical may cause severe eye irritation as well as skin and respiratory tract irritation. Thus, Cu8 can be a worker handling concern. It is, however, a semi-volatile organic, not likely to pose a building material emissions hazard. According to manufacturers, thermal decomposition of the Cu8 may be carbon dioxide, carbon monoxide, nitrogen oxides, and other "potentially toxic and irritating fumes."

Copper naphthenate (CuN) is an organometallic compound. Beyond its use as a pressure treatment preservative, CuN (oilborne) is available as a stand-alone wood treatment. In other words, it may be superficially sprayed or brushed on wood products. The AWPA does not, however, recommended its use within inhabited structures where occupants may potentially be exposed—exposures resulting in severe eye, skin, and respiratory irritation. Although it is not considered toxic to humans, CuN is very toxic to aquatic life. Wood products treated with CuN should thus not be used in and/or around surface water. According to manufacturers, thermal decomposition products of CuN are carbon dioxide, carbon monoxide, and other "unknowns."

In brief, toxic/irritant emissions from treated lumber are not likely. There is only consensus on the thermal decomposition of treated wood products being carbon dioxide and carbon monoxide. Each preservative listed herein has other toxic by-product including (1) organic chlorides, aldehydes, amines, ammonia, and copper (e.g., ACQ and ACD); (2) copper, arsenic, and zinc (e.g., ACZA); (3) nitrogen oxides, formaldehyde, and copper compounds (e.g., CA); (4) copper, arsenic, and zinc (CCA); (5) irritants (Cu-HDO); (6) hydrogen chloride and hydrogen cyanide (PTI); (7) undisclosed (8) potentially toxic and irritating fumes (e.g., oxine copper); and (9) unknowns (e.g., CuN).

Be forewarned! Burning CCA and ACZA treated lumber could result in airborne toxic levels of arsenic!

Engineered Lumber Products

Engineered lumber, also referred to as composite wood, is manufactured by bonding dimensional lumber, wood strands, and veneers to give strength and stability to lumber that would otherwise be poorly suited for structural framing. Bonded composite wood products are made of fast growing, immature trees that when bound together attain a high structurally integrity—meeting or exceeding the qualities of slower growing, mature trees. Composite woods are cost effective, and composite lumber eliminates waste associated with warped, twisted, and otherwise usable wood. Engineered framing wood is made by different wood types, configurations, and additives—each serving a different purpose.

Differentiating Products

Glued-laminated timber, often referred to as glulam, is a timber product manufactured from dimensional lumber—typically 2-inches thick, custom widths up to 14½ inches. The lumber stock is joined together end-to-end and glued together to almost unlimited depths and lengths—straight and curved, treated and untreated. According to one manufacturer:

> Glulam is available in several species of wood including southern pine, Douglas fir, Port Orford cedar, and Alaskan yellow cedar. Southern pine species can be preservative pressure treated *prior to gluing*, which unlike treatment *after gluing,* allows the treatment to reach the center of the beam for additional weather resistance. Port Orford cedar and Alaskan cedar are naturally decay resistant and therefore don't require preservative treatment. Douglas fir is the exception in glulam. Doug fir cannot be preservative treated without incising (a process that scars the beam with thousands of slit-like holes in the surface), nor is it naturally decay resistant. Topical treatments are available, but not nearly as effective as pressure treatments. (Timber Systems 2015)

Glulam is an exceptional structural support used to create beams capable of spanning large distances and to provide load bearing column support. Visible surfaces can be stained, painted, or veneered—an artist's cornucopia of designs.

Cross laminated timber (CLT) is similar to glulam timber except that the stock lumber is alternately layered crosswise. CLT is made of lumber stacked together at right angles and "glued over the entire surface." Exceedingly strong and resilient, CLT lumber/panels are usually custom-made, prefabricated off-site, and assembled on-site as load bearing walls, floors, and roofs. CLT can be a stand-alone wood frame—unconventional to say the least. The Science Section of the New York Times observed the odd dichotomy:

> A tall wooden structure would seem to be a collapse waiting to happen, but a building made from cross-laminated timber is stronger than a conventional wood-frame structure, in which two-by-fours and other relatively small components are tied together by materials like plywood and plasterboard. (Fountain 2012)

Laminated veneer lumber (LVL) is manufactured by bonding multiple layers of "thin-sliced lumber" to create thicker, strong, resilient lumber. It is similar to plywood. Whereas plywood has cross-grain layering, LVL has wood grain in the same direction. LVL is stronger, straighter and more uniform than conventional lumber, and it is used for headers, beams, rim board, joists, and wood I beam flange material. LVL beams are manufactured to lengths up to 80 feet.

Parallel strand lumber (PSL) consists of long, thin, "parallel strands" (i.e., strips of wood) bonded together with adhesive. It is similar to LVL. The only difference is that PSL is strips of wood as opposed to wide, thin-sliced lumber. Processing and use are the same. Beams can be manufactured up to 60 foot lengths.

Oriented strand board (OSB) is manufactured by bonding short, thin, "cross-oriented strands" of lumber. OSB is used to make I beam webbing. The OSB I beams are used mostly in floor and ceiling joists, and they can be manufactured up to 60 foot lengths.

Components, Emissions, and Heat Decomposition By-Products of Lumber

The wood components are generally a combination of hardwoods and/ or softwoods—species generally not specified by the manufacturer. They may disclose a "wood dust" component. Some manufacturers affirm that "allergenic Western red cedar" is not in the mix, and others include a more expanded list of excluded wood species. Most manufacturers' disclose health effects of exposures to wood dust as being mechanical irritation and/ or respiratory sensitization to workers only, not to building occupants.

> When ... machined (sawn, sanded, drilled, routed, planed, etc.) wood dust is produced. Wood dust and splinters may cause irritation of the nose and throat, eyes and skin. Some woods may also be sensitizers, and some people may develop allergic dermatitis or asthma. (Wesbeam 2005)

As for the adhesives, most "engineered wood products" used in framing are typically, not always, bonded with a PF adhesive and paraffin wax mixture. The finished product is then coated—to protect the wood from moisture penetration and damage. The coating is frequently a polyurethane film (e.g., paint-on, spray-on clear coat), and/or it may be a "proprietary secret." The PF adhesive and adhesive/wax mixture are moisture resistant, not water proof, and the coating provides additional protection from moisture damage.

Although PF is the most commonly used adhesive in framing timbers and lumber, other glues are possible. MF, a less water resistant glue, is cheaper, and phenol–resorcinol–formaldehyde, a more water resistant glue, is more expensive. UF, the least expensive, least moisture resistant adhesive, is not "normally" used in timber and/or lumber. Each poses a potential formaldehyde emission hazard and the amount of off-gassing is based on the resin type, manufacturing method(s), temperature, and relative humidity. UF is the most likely to emit formaldehyde and deteriorate when wet, and phenol–resorcinol–formaldehyde is the least likely to emit formaldehyde and does not readily deteriorate when wet.

As contractors are becoming more environmentally aware, some manufacturers are providing "health hazard information" in their spec sheet and/ or MSDSs regarding formaldehyde emissions testing on engineered wood products typically used in framing. For example, one manufacturer disclosed chamber testing performance for LVL processed with phenol formaldehyde:

> In newly manufactured LVL, which is the worst case scenario, formaldehyde emission has been measured in the range 0.03–0.05 ppm using the large scale chamber test method. As LVL products have emission levels of 0.03–0.05 ppm, well below the WHO recommended level of 0.1 ppm, under reasonably foreseeable circumstances it is unlikely that the presence of traces of formaldehyde in the product poses a health risk. Formaldehyde gas is irritating to the nose and throat, eyes and skin. It is

recommended that storage areas be well ventilated to avoid any irritating effects of a build-up of formaldehyde. (Wesbeam 2005)

The thermal decomposition products for wood are carbon dioxide and carbon monoxide. For further details, see Chapter 5: Thermoset Plastics and Chapter 13: Paint Pigments. Each of the composite lumber types may also be treated in the same fashion as that of "Treated Lumber."

Steel Framing

You may ask, "Steel, what's the big deal?" Steel emissions are an improbability—unless the welding/cutting of steel is in enclosed areas in and/or around public areas, in buildings undergoing renovation with the HVAC system still operating.

Steel frame buildings are "structural steel" (e.g., I-beams) with interior "stud walls" of either wood or steel. The structural and stud steel are not one and the same.

Most structural steel, used in all forms of construction, is comprised of carbon steel, painted with a rust protective coating. While welded carbon steel fumes are minimally toxic, the protective coatings are often overlooked by many contractors. Protective coatings for steel may be comprised not only of lead, but hexavalent chromium—both highly toxic components that can become problematic during welding/torch cutting processes. Do not fall into the trap that lead-based paint has been banned from use in all instances within the United States and other countries! There are no bans on lead-based paint on structural steel within the United States. Although they have strayed from its use, many manufacturers are substituting nontoxic pigments (e.g., iron oxide) instead of lead and/or lead chromate. This is not to say that "all" U.S. manufactures have opted not to use the more durable, yet more toxic, coatings. Foreign manufacturers—China, India, and other countries—are most likely to use lead coatings. And there is a high likelihood that most of the older structural steel coatings contain lead and/or lead chromate. Caution is advised when welding and torch cutting all structural steel—old and new. You would be well advised to research protective coating components of new and/or used structural steel.

Stud steel (e.g., studs, tracks, and headers) used in building construction is generally corrosion resistant galvanized steel—base metal and metallic coating. The base steel is typically carbon steel with about 1% manganese (a toxic metal), and the metallic coating contains mostly zinc with small amounts of aluminum and antimony (a slightly toxic metal). The stud steel is generally saw cut and joined with hardware. As welding is not likely, toxic fume exposures are not likely as well.

In some instances, a niche market will require stainless steel for environmentally and/or aesthetically exposed structural members. Stainless steel is a steel alloy of chromium and a few other metals (e.g., molybdenum and nickel). The greater the chromium content, the greater the corrosion resistance. The chromium in stainless steel is, however, not the toxic hexavalent form of chromium. Within the architectural community, all things are possible—there are no limits to the coatings and therefore components. The niche market is a world unto itself. So, once again, further research may be indicated!

Concrete Formed Walls

Concrete formed walls can be painted with a very broad brush from quick, easy, and cost effective to labor intensive, complex, and expensive. The more common ones are tilt wall, insulated concrete formwork, and cement blocks.

In a tilt wall construction, the building's walls are poured on location into forms with a rebar grid for structural integrity and with cutouts for window, doors, and other wall penetration. After curing, the formed slabs, referred to as "tilt wall panels" or "tilt-up panels," are raised by crane into position around the building's perimeter. The panels are attached (i.e., bolted or welded) to panels and roof joists. An alternative to on-site construction is precast which is formed and poured at the manufacturing facility, transported to the job site. The construction process for tilt wall panels is similar to that of concrete foundations without the vapor barrier.

An insulated concrete formwork (ICF) is a stay-in-place concrete forming system for a steel reinforced, solid poured, monolithic concrete wall. A rigid polystyrene form is held in place with plastic, snap-in-place building blocks. Rebar is strategically positioned within the form and concrete is poured within the wall cavity. Now, here is where ICF differs from tilt-wall—the form is not removed and the concrete wall remains in place. ICF is used for exterior, interior, load-bearing, and non-load bearing walls in commercial, industrial, and residential construction. The wall systems provide a solid framework, thermal insulation (up to R-30), sound insulation, and vapor barrier all-in-one. On the other hand, the ICF system is complex, requiring special labor skill sets for installation of walls, plumbing, and electricity, and the price tag is high.

Concrete block walls are none other than concrete-formed building blocks with cavities into which rebar is imbedded prior to stacking. Steel mesh is in place for joint control during the stacking. Blocks are manually stagger-stacked and mortared. Concrete is poured into the cavities, and the process repeats itself until the wall is completed. Whereas concrete block structures are sturdy and durable over time, cement block framed walls are more expensive than wood framing.

Reality check! Cinder blocks and concrete blocks are not one and the same. Concrete block aggregate is finely crushed stone or sand, and cinder block aggregate is coal cinders or ash. Concrete block is heavier and stronger than cinder block, and some building codes expressly prohibit the use of cinder blocks in construction projects.

Components of aforementioned concrete formed wall (e.g., polystyrene foam, plastic blocks, rebar, and concrete) have been discussed at length previously (see "Polymeric Foams" and "Concrete Foundations"). Concrete formed wall component product emissions are unlikely. Yet, cutting, sanding, and abrading concrete products are likely to result in the release of crystalline silica exposures to workers. Thermal decomposition products of styrene are carbon dioxide and carbon monoxide, and concrete is not combustible.

Exterior Wall/Roof Sheathing and Structural Subflooring

Sheathing is a protective covering with a purpose that has been defined and redefined over the passage of time, and structural subflooring, the base upon which underlayment and finished flooring is placed, is a new concept. Until 1909, sheathing and flooring was basically tongue-and-groove lumber which had poor thermal insulating qualities and was flammable.

So, naturally, the construction industry gravitated toward newly emerging products that were not only structurally sound (e.g., brace and stiffen the walls from wind shear) and did not require additional bracing, and sought that which was thermally insulating, fire resistant, moisture resistant, and less expensive—the best of all worlds. Fiberboard, plywood, and gypsum board were introduced at about the same time. Although they had some of these qualities, the initial sheathing and subflooring building materials were far from perfect. Then, OSB was introduced in the late 1970s. It was the end-all-be-all, close to the perfect. OSB has since taken the market by storm. Other more recent advancements such as cement board, glass mat faced gypsum board, and zip board approach the perfect world—albeit they are expensive.

Asphalt-Impregnated Fiberboard Sheathing

Fiberboard, also referred to as black board, grayboard, softboard, and buffaloboard, is recycled "waste!" As an exterior wall and roof sheathing, fiberboard is a composite building material made of cellulose products which include bagasse (i.e., sugar cane fibers), bark, flax, grass, hemp, jute, peanut shells, reeds, sawdust, straw, wood pulp tailings, and shredded paper. Most fiberboard is comprised of cellulose fibers held together with petroleum pitch/asphalt and/or starch and wax, coated with asphalt and/or wax (e.g., Georgia-Pacific Fiberboard and Knight-Celotex Fiberboard), some with starch, carbon black, clay, and wax (e.g., StructoDek®-low slope, high density roof sheathing), and others with a copper pesticide additive (e.g., tricupric borate; Homasote® 440).

When exposed to excessive moisture, fiberboard swells and rots. In 1996, a homeowner's class action suit sued International Paper, the manufacturers of Masonite™ products, for water damage to and deterioration of fiberboard wall sheathing. The jury concluded that the siding was defective. Yet, despite being a pariah, fiberboard is still sold and encountered in buildings today! Deterioration was not exclusively a wall sheathing problem.

Many of the older structures that had fiberboard roof sheathing, without the benefit of a solid backing, have since been compromised and exposed to the rain and harsh environments—"sturdy" turned to mush. Tread lightly! As for the more recent fiberboard roofing, it is usually supported by a corrugated steel roof.

Of the fiberboard components, as disclosed by some of the manufacturers, asphalt (or bitumen) and petroleum pitch have potential building material emissions.

> According to the ACGIH, asphalt (as a benzene soluble aerosol) is toxic, but manufacturers consistently disclose asphalt is "non-hazardous per OSHA 1910.1200." Well, this is true! In accordance with OSHA, asphalt is not listed. NIOSH associates asphalt fumes with eye and upper respiratory tract irritation; they conclude that there is insufficient evidence to indicate benzene soluble aerosol. (NIOSH 2000)
>
> As disclosed on the "asphalt" MSDSs, petroleum asphalt also contains hydrogen sulfide—a naturally occurring component of petroleum products (e.g., sour gas). Hydrogen sulfide has a rotten egg odor threshold (0.0094 ppm) at levels well below the strictest exposure limit (ACGIH: 1 ppm). It is an upper respiratory tract and eye irritant. Elevated temperatures may result in asphalt component emissions from exterior wall and roof fiberboard—tar-like odor. The exterior sheathing of a building is subject to solar heating and elevated temperatures, but there have not been any reports of a "tar-like odor" within occupied building spaces where asphalt-like fiberboard has been installed as an exterior sheathing.
>
> Petroleum pitch is similar to petroleum asphalt. Both are processed petroleum products, the chemical composition of which is dependent upon the crude oil—unknown variables. The main difference between the two is that asphalt is a heavy, viscous liquid while petroleum pitch is a brittle solid. And where "asphalt" is listed by the ACGIH as a toxin, petroleum pitch is not. However, producers of petroleum pitch identify trace amounts of polynuclear aromatic (PNA) hydrocarbons such as naphthalene and benzo(a) pyrene. However, PNAs are high molecular weight compounds that are not likely to pose an emissions problem at elevated ambient temperatures.

Thermal decomposition of cellulose and asphalt and petroleum pitch containing fiberboard include carbon dioxide, carbon monoxide, sulfur oxides, and hydrogen sulfide.

Plywood Sheathing and Interior Subflooring

Plywood is manufactured by bonding multiple cross-grain layers of thinly sliced wood sheeting to create thicker, strong, resilient panels. The glue used for bonding "exterior plywood"—engineered to resist water damage—is generally phenol–formaldehyde but may be MF, a less water resistant glue, or phenol–resorcinol–formaldehyde, a more water resistant glue. Although it is a poor exterior material, UF bonded plywood may be used as interior deck sheeting. Yet, formaldehyde emissions are more likely with the less expensive urea formaldehyde plywood, and if the plywood were to get wet, it will not only swell, cup, and delaminate but release formaldehyde—even after new product off-gassing has long since run its course.

OSB Sheathing and Interior Subflooring

OSB is manufactured by bonding short, thin, "cross-oriented strands" of lumber. The adhesives used in OSB are typically a mixture of PF and equal or lesser amounts of polymeric diphenylmethane diisocyanate (PMDI; also referred to as polyurethane). Paraffin wax emulsion is often added to the resins and/or as a coating to extend water resistance. Borate preservatives (e.g., zinc borate) are added as a preservative in treated OSB.

Higher quality, more water-resistant OSB is formulated with greater quantities of polyurethane glue than PF—creating a strong bond with greater water resistance (e.g., AdvanTech®). The same formulation can be extended to waterproof sheathing by overlaying the polyurethane glued OSB with PF impregnated Kraft paper (e.g., Zip System™ Roof and Wall Sheathing). The latter "system" requires a specialty tape to seal the edges and around windows and doors and serves as a stand-alone building moisture system—building moisture wrap is unnecessary!

Formaldehyde and isocyanate (e.g., MDI) emissions are possible but not likely.

Gypsum Board Exterior Sheathing

Gypsum board, also referred to as drywall, wallboard, Sheetrock®, and plasterboard, is a processed gypsum product. The gypsum is pulverized to a fine powder, heated to about 350°F to drive three-fourths of the water component—making a dry gypsum plaster (i.e., Plaster of Paris). See Figure 11.4.

Gypsum plaster, reinforcing fibers (e.g., cellulose and/or glass fibers), binder (e.g., starch), additives (e.g., vermiculite in DensArmor Plus® Fireguard), and

$$CaSO_4 \cdot 2H_2O \xrightarrow{\triangle} CaSO_4 \cdot 0.5\,H_2O + 1.5\,H_2O \uparrow$$

FIGURE 11.4
Chemical formulation for converting gypsum into gypsum plaster (calcium sulfate dehydrate).

water are mixed to make a slurry. The slurry is poured onto and covered with heavy sheets of paper or glass fiber mat, and it is kiln dried. The end product is a rigid board that is then cut to size—typically 4-foot by 8-foot boards. Gypsum board, used mostly as an interior wall covering, has become increasingly popular, in recent years, as exterior sheathing—especially the inorganic glass fiber mat (e.g., DensGlass Gold®) which does not support mold growth when it gets wet.

The paper backing on gypsum board, however, will support mold growth and significantly deteriorate when wet. Thus, paper-covered gypsum board and oil-coated paper-covered gypsum are not good candidates for exterior sheathing. These products have, however, been discovered in buildings built between 1950 and 1965—in North American structures. It was not only used as sheathing in low cost buildings but as a component in roofing systems as well.

Gypsum board has no identified toxins/irritants which could contribute to product emissions—notwithstanding undisclosed additives. Thermal decomposition of gypsum board is likely to be carbon dioxide and carbon monoxide from the paper-backed gypsum board and steam from the water content in fire resistant gypsum.

Be forewarned! Mined gypsum contains crystalline silica—a potential worker exposure hazard.

Cement Board Sheathing and Interior Subflooring

Cement board, also referred by trade name Hardibacker®, Durarock®, Wonderboard®, and Versarock®, is a Portland cement product. Cement board is made of a slurry of Portland cement, cellulose/glass fibers, filler (e.g., sand), additives, and water. Furthermore, some manufacturers identify sand in the ingredients listing on their MSDS—at levels as high as 50% by weight (e.g., Hardiebacker® cement board). The thin slurry is then poured into a form and pressed into and covered with a fiberglass mesh.

Although the component of sand may have crystalline silica, there are no identified toxins/irritants emissions associated with cement board. Cement board is not combustible!

Be forewarned! When sawing, breaking, and rasping cement board, workers are potentially exposed crystalline silica hazards. During renovation projects, cement board (e.g., Transite™, Eternit™, HardieFlex™, etc.), prior to its phase out in the late 1980s, had 15%–25% asbestos fibers.

> Over 95% of the chrysotile asbestos fiber mined worldwide is mixed with cement to form roofing, siding, pipes and many more products. These products have been used for decades throughout the world, but due to bans on their use in many countries they are now marketed primarily in developing countries (e.g., China, India, Indonesia, etc.). (Oberta 2015)

Vapor Barriers

The shell of a structure is the first line of defense between the building occupants and outdoor environment—wind and rain. Rain can be and is driven through leaking and porous exterior walls (e.g., brick mortar) and roof materials by the force of wind, gravity, and capillary action. This is the liquid phase of water, not to be confused with vapor. "The liquid phase as rain and ground water has driven everyone crazy for hundreds of years but can be readily understood—drain everything and remember the humble flashing" (Lstiburek 2006a).

In an effort to subvert further movement of water into the interstitial spaces, migration by water vapor, proper placement of a vapor barrier can help avert—(1) wood rot; (2) mold growth; (3) steel corrosion; and (4) insulation R-value reduction. And a vapor barrier is meant to do that very thing—retard the migration of water vapor.

Water vapor migrates from a high vapor pressure (e.g., warmer air) to a low vapor pressure (e.g., cooler air). You ask, what does this have to do with anything? In warmer climates, water vapor will migrate from outdoors to the cooler interior conditioned air—if not blocked by a vapor barrier. In colder climates, water vapor will migrate from the warmer interior conditioned air to the outside, if not blocked by a vapor barrier. Location matters!

Traditionally, vapor barriers, also referred to as vapor retarders, reduce the migration of water vapor. Yet, vapor barriers and vapor retarders are not one and the same. Traveling into the land of correctness—terms matter! In the building sciences, vapor barrier is a term to describe "vapor impermeable" retarders. All vapor barriers are vapor retarders, but not all vapor retarders are vapor barriers. Barriers and retarders are defined by vapor permeance (perm) which is the rate of transfer of water vapor through a material. See Table 11.4.

According to IRC definition, the only true vapor barrier is polyethylene sheeting (e.g., Tyvek™), and 30-pound felt is a vapor retarder. The plastic vapor barrier is impermeable. Other vapor retarders are semi-impermeable. What does this tell us? The plastic sheeting is more likely stop the flow of liquid phase water than the other vapor retarders.

The cardinal rule for building enclosures that has bedeviled building engineers and architects is quite simple (Lstiburek 2006a).

- Keep water out
- Let water out if it gets in

According to a forensic building engineer who investigates building failures, the key to prevention is avoidance (Lstiburek 2006a).

- Avoid using vapor barriers where vapor retarders will provide satisfactory performance

TABLE 11.4

Vapor Diffusion Retarders

Class	Diffusion Rate	Permeability	Example of Wall Vapor Retarder Materials
I	≤0.1 perm	Vapor impermeable "barrier(s)"	Polyethylene sheet (porous and nonporous) Rubber membranes
II	0.1–1.0 perm	Vapor semi-impermeable	30-Pound asphalt-coated paper Felt and bitumen-coated Kraft paper Polystyrene rigid insulation with no facing (1-inch thick) Polystyrene rigid insulation with perforated facing (1/2-inch thick) Non-perforated foil Polypropylene-faced rigid insulation Kraft faced fiberglass batt insulation
III	1.0 perm–10 perms	Vapor semi-permeable	15-pound asphalt-coated paper/felt Latex paint Micro-perforated foil Gypsum board painted on one side

[a] Class Definitions Excerpted from International Residential Code 2002 (IRC).

- Avoid using vapor retarders where vapor permeable materials will provide satisfactory performance and encourage drying instead of moisture prevention
- Avoid the installation of vapor barriers on both sides of assemblies (e.g., double vapor barriers)
- Avoid the installation of vapor barriers—such as polyethylene vapor barriers, foil faced batt insulation, and reflective radiant barrier foil insulation—on the interior of air-conditioned spaces
- Avoid the installation of vinyl wall coverings inside exterior walls of air-conditioned spaces

The list of vapor barriers/retarders is somewhat limited. The longest standing, tried and true—of present day wall and roof vapor retarders—is the ever present asphalt-saturated felt. However, polyethylene sheeting is running a close second and gaining widespread popularity. Other less popular materials that have been considered and occasionally used are Kraft paper adhered to fiberglass and mineral wool insulation, polystyrene board, foamed-in-place insulation (e.g., polyurethane spray insulation), aluminum covered insulation, vinyl wallpaper, and paint. True moisture blocking materials, used in some roofing, retaining walls, and basements. Placement of vapor barriers, dependent on temperature and humidity, can be exterior to the wall studs in warmer climates and interior to the wall studs in colder climates.

Asphalt-Saturated Felt

Asphalt-saturated felt, also referred to as asphalt-impregnated felt, asphalt felt paper, and felt paper, comprises pressure treated unwoven fibers impregnated with water resistant asphalt. Back in the 1800s, paper was coated with asphalt, but the tar-impregnated paper tore easily and rotted when it got wet. As a stand-alone product, asphalt impregnated paper, or cellulose fiber, was not reliable. Thus, the paper was replaced with more durable, but still susceptible to water damage, asphalt-impregnated heavy Kraft paper and organic rags (e.g., jute, burlap, etc.) which may still be found in houses dating back to the late 1800s. In 1910, asbestos came into common use as a component of roofing felt (U.S. EPA 1990). Then, in 1935, glass fiber was developed, later to be used as a sole component in the production of a completely inorganic asphalt felt which is still in use today—as a wall vapor retarder and as roofing underlayment. The cellulose felt paper is also used on flat roofs where hot asphalt is applied to the surface.

A couple of the types of asphalt-saturated felt available today include, but are not limited to

- Grade D felt paper—0.6 lb.–6 lb.
- Roofing felt—15 lb. and 30 lb. (roof and wall vapor retarder)

Regulated potential emissions, associated with the present day asphalt-saturated felt, include asphalt fumes and hydrogen sulfide (in trace amounts). Although saturation asphalt has a softening (potential off-gassing) temperature of 120–140°F (50–60°C) (NIOSH 2003), temperatures easily reached within a solar heated exterior wall or roof, toxic/irritant emissions have not been reported. Yet, however unlikely asphalt off-gassing may be, a tar-like odor, emanating from an exterior wall into the occupied building, should raise suspicion.

Be forewarned! During wall and roof renovation projects, workers are potentially exposed to asbestos when removing asphalt-saturated paper and/or felt—in buildings constructed prior to 1990.

Plastic Vapor Barriers

Plastic vapor barriers, also referred to as polyethylene sheeting, plastic housewrap, and Tyvek-like sheeting, are thin sheets of impermeable plastic—not only a strong, stretchable, tear-resistant "vapor barrier" but an air barrier as well—unless the plastic sheeting is purposely micro-perforated, rendering it a semipermeable vapor retarder. Other specialty plastic vapor barriers have evolved in order to accommodate different needs. Specialty and nonspecialty plastic vapor barriers include, but are not limited to

- Spun-bonded polyethylene (e.g., Tyvek® HouseWrap)
- Plastic roof underlayment (e.g., Roof Top Guard II™)

- Self-adhered air and vapor barriers (3 M™ Air and Vapor Barrier)
- Grooved spun-bonded polyethylene (Tyvek® StuccoWrap™)
- Self-adhesive high-performance membrane (Visqueen Self Adhesive Membrane)
- High-performance membrane (e.g., Stego Wrap)

HouseWrap®, 175 µm thick, is an above grade vertical wall vapor barrier. Roof Top Guard II™, five layers of woven and spun-bonded polypropylene/polyethylene fabric and film, is a 50-year warranty roof vapor barrier. Stego Wrap, 15 mils thick, is a heavy under slab vapor barrier, and Visqueen Self Adhesive Membrane, is a polymer modified asphalt adhesive on a 1.5 mm thick polyethylene membrane used for subsurface (e.g., basement) vapor barriers. StuccoWrap is grooved to allow for downward water movement. The examples herein of different plastic vapor barriers are only the tip of the iceberg, and some plastic vapor barriers are used in consort with asphalt saturated vapor barriers—a bond break (i.e., secondary vapor barrier) is recommended by some contractors and forensic building engineers in most instances where a building exterior is stucco (Lstiburek 2006a).

Thus, plastic vapor barriers come in all manner of specialty formats, and competition is fierce—in terms of pricing and/or innovations. That which is today's plastic vapor barrier may not be tomorrow's! So, that said, let's look at today's product disclosures.

Most plastic vapor barriers are predominantly high-density polyethylene (HDPE) with plasticizers and additives (e.g., UV protection). The spunbonded is a continuous nonwoven, non-perforated sheeting that is made by spinning fine HDPE fibers that are fused together to form a strong, uniform web, and most vapor barriers are strong and resilient as opposed to hard and brittle. Without plasticizers, HDPE is hard and brittle. With plasticizers, HDPE is strong and resilient. Yet, most manufacturers disclose their vapor barriers are strong and resilient, while they do not/are not required to disclose unregulated product components. Some manufacturers do, however, disclose the presence of "additives." Some disclose polypropylene instead of, or in combination with, polyethylene. Some use adhesives (e.g., 3 M™ Air and Vapor Barrier—acrylic adhesive listed as a trade secret).

Emissions from component plasticizers and some plastic additives (e.g., UV light stabilizers) may cause eye, skin, and upper respiratory irritation. Off-gassing of plasticizers at elevated temperatures encountered within wall cavities is possible. See "Plasticizers" and "Plastic Additives."

During a fire, polyethylene/polypropylene thermoplastic sheeting, if not treated with a fire retardant, will create its own fuel (e.g., ethylene or propylene) and "feed the fire." Thermal decomposition products of complete combustion are carbon dioxide and carbon monoxide.

Be forewarned! Polypropylene plastic that has not been specially treated with fire retardant additive readily burns. Once set on fire, the flame will be self-sustaining.

Summary

Although external to a building, foundations, framing, sheathing, and vapor barriers are not without their challenges not only in terms of product emissions and combustible by-products but also environmental impact and worker exposures to toxins during construction. Each building material—old and new—harbors its own surprises!

Concrete foundations have crystalline silica which when cut or sanded will release toxic dust. Treated timber foundations may release arsenic into the environment. Stone (e.g., granite) under and/or as a component of foundations may potentially emit radon gases.

Some conventional framing lumber are skin irritants. Treated lumber may pose environmental health hazards. Engineered lumber may emit formaldehyde. Painted steel beams may emit lead and/or hexavalent chromium when welded. Concrete formed walls may release toxic crystalline silica dust when cut or sanded.

Fiberboard sheathing and asphalt-saturated felt emit asphalt fumes. Plywood and OSB sheathing may emit formaldehyde particularly when wet. Old cement board and asphalt-saturated roofing felt potentially release asbestos fibers.

Plastic vapor barriers potentially emit plasticizers and additives—eye, skin, and eye irritants. The World of Plastics is unavoidable. It is here to stay!

Additionally, the installation of vapor barriers is rife with complexities, and misapplication frequently culminates with water intrusion which may lead to mold growth in the exterior framing members and, ultimately, to litigation.

> As I studied the various suggested approaches to vapor barrier applications, it became readily apparent that some well-intended contractors and product salespersons recommend solutions that are frowned upon by forensic building engineers and building sciences experts. Best to "get it right" during the construction process!

Simple guidelines are provided within the section on "Vapor Barriers."

References

American Wood Protection Association. 2016. "Information for Homeowners." *AWPA*. Accessed November 8, 2016. http://www.awpa.com/references/homeowner.asp.

AWPA. 2015. "Information for Homeowners." *References and Technical Information.* Accessed May 14, 2015. www.awpa.com/references/homeowner.asp.

EPA. 2014a. "Chromated Copper Arsenate (CCA): ACQ-An Alternative to CCA." *Pesticides: Regulating Pesticides.* February 7. Accessed May 9, 2015. http://www.epa.gov/oppad001/reregistration/cca/acq.htm.

———. 2014b. "Chromated Copper Arsenate (CCA): Copper Azole-An Alternative to CCA." *Pesticides: Regulating Pesticides.* February 7. Accessed May 9, 2015. http://www.epa.gov/oppad001/reregistration/cca/copperazole.htm.

Fountain, Henry. 2012. "Wood That Reaches New Heights." *The New York Times.* June 12. Accessed May 26, 2015. http://www.nytimes.com/2012/06/05/science/lofty-ambitions-for-cross-laminated-timber-panels.html?_r=0.

Lebow, Stan T. 2010. "Wood Preservation." Chap. 15 in *Wood Handbook*, eds. Richard Bergman et al. Madison, WI: Forest Products Laboratory.

Lstiburek, Joseph. 2006a. "BSD-106: Understanding Vapor Barriers." *Building Science Corporation.* November 24. Accessed June 8, 2015. http://www.buildingscience.com/documents/digests/bsd-106-understanding-vapor-barriers/.

———. 2006b. "BSD-103: Understanding Basements." *Building Science Corporation.* November 27. Accessed June 15, 2015. http://www.buildingscience.com/documents/digests/bsd-103-understanding-basements/.

NIOSH. 2000. *Health Effects of Occupational Exposure to Asphalt.* Atlanta: CDC.

NIOSH. 2003. *Asphalt Fume Exposures During The Application of Hot Asphalt to Roofs.* Document, Cincinnati, OH: NIOSH. https://www.cdc.gov/niosh/docs/2003-112/default.html.

Oberta, Andy. 2015. "Myths and Facts about Asbestos-Cement." *The Environmental Consultancy.* Accessed June 8, 2015. http://www.asbestosguru-oberta.com/ACMyths&Facts.html.

Timber Systems. 2015. *Glued-Laminated Timber.* Accessed June 26, 2015. http://www.timbersystems.com/products/glulam-timber.

U.S. EPA. 1990. *Asbestos/NESHAP Adequately Wet Guidance.* Bulletin, Office of Pollution Prevention and Toxics, Washington, DC: U.S. EPA.

U.S. EPA. 2009. "Pesticides." *EPA.* November 5. Accessed May 15, 2015. http://www.epa.gov/pesticides/chem_search/ppls/075269-00003-20091105.pdf.

Wesbeam. 2005. "MSDS: Laminated Veneer Lumber-Untreated." *GPEMS.* August. Accessed May 19, 2015. http://www.gpems.com.au/Brochures%20in%20pdf/MDS%20wes-untreated-safety.pdf.

Western Wood Products Association. 1997. "Hem-Fir." *Western Wood Products Association.* March. Accessed May 7, 2015. http://www2.wwpa.org/SPECIESPRODUCTS/HemFir/tabid/299/Default.aspx.

12

Exterior Enclosure Components

The visible exterior of a building is its first line of defense. It is the façade. It is the aesthetics. It is the keeper of the castle!

The exterior enclosure is comprised of the non-load bearing visible exterior. The enclosure, also referred to as the building envelop, is the first line of defense against inclement weather, thermal discomfort, wind, and structural damage. It covers the load bearing support and vapor barriers contained within. As the first line of defense, the exterior enclosure includes the roof, the walls, and the windows and doors. Each is a part of the whole!

The roof is the primary line of defense. The walls are the secondary line of defense. The windows and doors are the weakest link—least efficient in protecting from outdoor environmental conditions, most susceptible to leaks. Each component type has an extensive pallet of building materials choices—some natural materials, some manmade, and some synthetic.

Windows and Doors

From the belly of dark-and-dank open caves to the magic of natural light and climate controlled cocoons, modern civilization has come a long way. We have come to expect functional value in windows and doors—lots of natural light, temperature/humidity management, and security.

Although glass sheets and/or panels are installed into windows for the primary purpose of allowing natural light in, they must be supported and framed. A window frame supports the glazing (i.e., glass). A door, on the other hand, does not normally require a support frame—unless the door is a sliding or folding track glass or glass panel door. In the latter, the glass is supported by a frame. Sliding glass doors are framed, and partial glass and decorative glass inserts (e.g., leaded or beveled glass) are a component of the door. Thus, the component window/door frame and door materials are similar, if not the same.

Window frames and doors may be comprised of wood, metal, resin-impregnated fiberglass, and rigid vinyl. They may be one or a combination of these materials such as exterior vinyl clad wood window frames and decorative copper exterior solid wood doors. They may have insulation (e.g., polyurethane [PU] foam). They may have stiffeners (e.g., honeycombing).

They may have reinforcing components (e.g., steel bar) for security. Windows have a gaskets/seals and a drain plane with weep holes, and doors may have rubber spacer balls (to prevent damage to wood due to swelling and shrinking).

Wood

Wood doors and window frames provide a rich aesthetic appeal with a few surprises. Yet, exterior wood window frames and doors are high maintenance, particularly in climates where there are temperature extremes.

Solid wood window frames and doors are cut, joined, glued, sanded, and sometimes finished at a millworks factory or by a specialty wood artisan. For exterior window frames and doors, a special waterproof adhesive (e.g., Gorilla Glue or Elmer's E7050 Carpenter's Wood Glue) is generally, not always, used to bond the components. Non-waterproof adhesives used in exterior wood products will separate and interior wood products used outdoors may deteriorate.

Interior hollow-core wood doors are typically made of thin panels of medium density fiberboard (MDF), also referred to as hardboard and chipboard. MDF is thermoset plastic resin-impregnated wood fibers, generally saw dust. The thermoset plastic resin is generally melamine– or urea–formaldehyde. The thin panels are framed with lumber, and the core consists of an inner honeycombed panel which is also coated with a thermoset plastic resin. The MDF panels and top/side strips are laminated, stained/sealed, and/or painted.

In newly manufactured interior hollow-core doors, formaldehyde emissions are possible from the MDF panels/strips and from the honeycombed core. Laminating adhesives, depending on glue type, may potentially emit volatile organic compounds (VOCs) (e.g., solvent-based contact cement) and formaldehyde (e.g., urea formaldehyde [UF]).

Interior hollow-core wood doors have occasionally been used as exterior doors. In this case, excessive radiant heat and rain will have a profound effect not only on door deterioration but also on formaldehyde emissions. Mold and formaldehyde emissions should be anticipated wherein hollow-core wood doors have been installed as exterior doors. At first blush, one would think, "How is it that someone would use an interior door as an exterior door?" Well, it happens!

Used in either hollow- or solid-core doors, "high density fiberboard (HDF) door skins" are the new door component on the block. HDF is made in a similar process to that of MDF, but it has a higher density. HDF is thermoset resin-impregnated, steam-exploded wood fibers. During the molding process, a wood veneer may be stamped into the surface of the HDF. The end product "appears to be a solid wood door." The manufacturer of the veneered HDF—as depicted in Figure 12.1—alleges the product is highly resistant to mold and infestation, and not nearly as moisture sensitive as MDF (Cosmos 2015).

FIGURE 12.1
Veneered HDF door skin. (Courtesy of Consmos, Shandong, China.)

For this to happen, the thermoset resin must be moisture-resistant phenol–formaldehyde or resorcinol–formaldehyde. This reduces the likelihood of formaldehyde emissions in elevated temperatures and moisture.

Solid-core is a hollow core interior/exterior door with a laminated solid core component. The solid-core component is typically engineered wood (e.g., particleboard and OSB board) which is thermoset plastic resin-impregnated wood. The most common exterior solid-core doors are hardwood veneer glued on the surface of particleboard. Due to their weight, veneered particleboard doors are often perceived to be solid wood. They look good on the showroom floor!

A particleboard core is a potential formaldehyde emitter, and—the hotter it gets, the wetter it gets—it will off-gas formaldehyde with greater frequency. Out-of-sight, out-of-mind. Yet, another potential source of formaldehyde exposure is hiding in the morass of building materials.

All wood products may be stained and finished with a protective coat, or they may be painted—either at the factory or on-site. On-site coatings pose the potential for worker and area occupant exposures to toxic/irritating chemicals. After installation, emissions of formaldehyde from veneered particleboard doors are possible, and emissions from laminate/veneer glues, stains/sealers, and paints are a considerable risk as well. For more in-depth information regarding protective coating and adhesives emissions, see Chapter 16: Adhesives.

Be forewarned! Interior doors used as exterior doors are subject to deterioration and water damage.

Looking beyond the visible skin, toxins reside within "hollow" core wood doors. Potential polystyrene and PU insulation emissions include styrene and isocyanates, respectively, and honeycombed Kraft paper stiffener potentially off-gases formaldehyde. Thermal decomposition products of isocyanates include hydrogen cyanide.

Aluminum/Steel and More

Metal window frames and doors are generally hollow core or core filled. All metals readily conduct heat and transmit noise. Metals corrode (e.g., aluminum); rust (e.g., galvannealed steel); change color (e.g., copper); darken and lose their luster (e.g., silver). Some are light weight (e.g., aluminum); some moderately heavy (e.g., stainless steel); and some very heavy (e.g., copper and silver). Alloyed steel has two to four times the tensile strength of alloyed aluminum—the higher the tensile strength, the greater the security. All things considered, aluminum is lightweight, durable, and has low tensile strength; steel is moderately heavy, moisture sensitive, and has high tensile strength; and both conduct heat and transmit noise.

Steel door thickness and steel stiffeners bestow security. The lower the number, the thicker the steel. For instance, 20-gauge steel is 0.359 inches thick and 12-gauge steel is 0.1046 inches thick. The thicker the door, the more difficult it is to damage. Yet, thickness alone does not impart security. Security doors generally have 20- to 12-gauge steel stiffeners welded top-and-bottom within a door core. Stiffeners are spaced 4–8 inches, and insulation is installed in the cavities.

Most, if not all, metal doors are "hollow" core metal doors. Potential polystyrene, PU, and resin-coated mineral wool insulation emissions include styrene, isocyanates, and formaldehyde, respectively.

Fiberglass

Fiberglass window frames and doors were first introduced in the early 1980s—the latest-and-greatest. They are light weight, dent and scratch resistant, and strong. High-quality fiberglass structures will last for years without fear of mold and rust. Poor quality, inexpensive fiberglass structures will, however, crack and deteriorate.

The hollow core in fiberglass window frames and doors are treated in a similar fashion to that of metal window frames and doors. Insulation treatments are typically polystyrene or PU foam. Rigidity may be imparted by light weight, moisture-resistant balsa or polypropylene honeycomb core panels. Security stiffener is typically aluminum, and fire rated fiberglass doors may contain gypsum.

Thermal decomposition can be a concern wherein PU insulation is used within the door core. Hydrogen cyanide is a toxic gas associated with all PU foam.

Rigid Vinyl

Rigid vinyl window frames and doors were introduced in the middle 1900s and became popular in the 1990s. They are durable, scratch resistant, easily customized, energy efficient, and inexpensive. What is there not to like about vinyl? Let us have this discussion.

Rigid vinyl is "unplasticized" polyvinyl chloride (PVC-U) which is a thermoplastic resin. Over half of its weight is chlorine gas, and the PVC plastic decomposes at low temperatures, resulting in product breakdown. When unstabilized PVC is exposed to elevated temperatures (such as solar heating), product emissions include regulated, toxic/irritating hydrochloric acid and a variety of minor organic compounds. Warpage and deterioration due to heat degradation particularly of exterior vinyl doors and window frames is highly likely, especially in areas where there is a lot of reflective heat and/or extreme temperatures such as temperatures encountered in Arizona during the hot summer months.

Be forewarned! Vinyl does expand and contract in extreme climates. Where there is expansion and contraction to the extreme, vinyl is going to weaken and break down according to heat load (Lee 2013). PVC begins to soften at 149°F (65°C) at which point it begins to lose its integrity, and as temperature increases, distortion becomes more evident, see Figure 12.2.

Andersen Window Corporation recently released a vinyl window frame (i.e., Fibrex®) which is a composite of 40% wood and 60% PVC polymer. Fibrex is twice the strength of non-composited vinyl. It is energy efficient. It requires little or no maintenance. And it is warranted "not to warp."

The hollow core in vinyl window frames and doors is treated in a similar fashion to that of metal window frames and doors. Insulation treatments are typically polystyrene or PU foam. Rigidity may be imparted by light weight, moisture-resistant balsa and/or "polypropylene honeycomb" core panels.

FIGURE 12.2
Vinyl siding deterioration—solar heat (left) and fire damage (right). (Courtesy of InspectAPedia at www.inspectapedia.com)

Thermal decomposition can be a concern wherein PU insulation is used within the door core. Hydrogen cyanide is a toxic gas associated with all PU foam.

Hollow Core Door and Window Frame Treatments

Core insulation forestalls heat and noise transmission, and it helps minimize moisture buildup which causes corrosion and rust. Most exterior doors and many window frames are insulated, and some interior doors are insulated. Polystyrene and PU are common foam insulation materials. Potential regulated emissions include styrene (i.e., monomer of polystyrene) and isocyanates (i.e., TDI, MDI, and HDI components of PU).

The styrene monomer of polystyrene has a sharp, sweet odor detectable at 0.34 ppm (U.S. EPA 2013), below the ACGIH exposure limit of 20 ppm. Emissions from polystyrene at temperatures below 158°F (70°C) have not been reported. In a fire, polystyrene will feed flammable styrene to the fire as long as the ignition source remains. Otherwise, it will melt and snuff itself out.

The isocyanate components of PU are eye, mucous membrane, skin, and respiratory tract irritants and sensitizers. They do not have an odor! The thermal decomposition products are carbon dioxide, carbon monoxide, oxides of nitrogen, and hydrogen cyanide.

Beyond the more common foam materials, specialty types of insulation have been introduced to the marketplace. For instance, Thermofiber® Safing™ is a fire-resistant insulation. It resists temperatures up to 2000°F (1093°C) and attenuates sound. According to the manufacturer, Thermofiber Safing is a mineral wool fiberboard impregnated with a phenolic resin (e.g., phenol formaldehyde).

Phenolic resin-impregnated material may emit formaldehyde when the product has recently been manufactured, and formaldehyde may potentially off-gas from the treated product when exposed to the extreme heat and humidity. The potential for water-resistant phenol formaldehyde to off-gas formaldehyde under these conditions is, however, "considerably less" than that of water-deteriorating UF.

Hollow core doors require a rigid component so they won't collapse. Honeycombed paper provides compression rigidity, and steel provides both compression rigidity and security. A honeycombed paper core is the cheapest, most commonly encountered core component. Honeycombed core material is a heavy Kraft cardboard-like paper formed into hexagonal cells, bound and impregnated with a phenolic resin (i.e., phenol formaldehyde)— partially or completely. Formaldehyde emissions are possible.

Be forewarned! Honeycombed core may be referred to as such but not be resin-impregnated Kraft cardboard-like paper. For instance, the Chinese Sing Honeycomb Panels are alleged rigid insulation, a composite of expanded polystyrene foam and wood fibers.

Thermal decomposition products of PU foam insulation are hydrogen cyanide. Generally, the hydrogen cyanide levels are only detectable at low levels wherein there is one source. Yet, in some indoor environments, multiple PU foam building materials and furnishings may contribute to the whole.

Glazing Treatments

Window and sometimes door glass (e.g., sliding glass doors) may be breakable glass, tempered safety glass, polycarbonate bulletproof glass, fire-rated glass, leaded stained glass, and/or mirrors. Window installation, or glazing, is traditionally done with putty/clay-like pastes, synthetic sealants, and more recently with glazing gaskets.

Glazing pastes are generally a mixture of that which is referred to as whiting (i.e., finely powdered calcium carbonate), linseed oil, and other proprietary products. The mix is beaten, or kneaded, to the consistency of dough. Calcium carbonate and linseed oil are neither health hazards. When, however, subjected to solar heat, the linseed oil may soften and give an optical appearance to occupants that the windows are weeping which has been perceived to be the source of a health hazard. This can be a distraction. If anything, linseed oil is beneficial to human health!

Synthetic sealants most commonly used are silicone. Once installed, silicone does not pose an emission health hazard. For more information regarding different sealants, see "Sealants" in Chapter 16.

The glazing gaskets are a newly minted approach to window glazing and hold great promise as a durable, long lasting approach to window installations. One such glazing gasket is NuGlaze® which is a rigid PVC gasket with extremely high anti-UV properties. The manufacturer alleges that if treated properly, NuGlass gasket window glazing "should last many, many happy years ... keeping water, air, dust, smoke, pollution and smells out" (Nu-Glaze 2014). Once again, rigid PVC may emit hydrogen chloride. Although the manufacturer may claim that the product has high anti-UV properties, it is unlikely that UV protection will also protect from PVC deterioration due to excessive infrared (i.e., radiant heat). Hydrogen chloride emissions from PVC glazing are possible.

Roofing Materials

Good roof, bad roof! The roof can be either a structure's "crowning glory" or the bane of the buildings' occupants. Material choices range from inexpensive to expensive, from natural to synthetic. And with natural resources becoming scarcer and more expensive, there is a trend toward recycled and/ or engineered, petroleum-based roofing materials.

Asphalt Shingles

Asphalt shingles are the most commonly encountered roofing material in the United States. The tiles are inexpensive and typically have a 20–30 year warranty. There are the standard three-tab strip shingles and the more three-dimensional "architectural shingles" which are also referred to as laminated shingles. Architectural shingles are two shingles in one. By laminating two shingles, the shingles are stronger and more durable than the traditional shingles while mimicking the look of wood or slate.

Beyond single as opposed to double thickness, asphalt shingles are organic or fiberglass. Organic asphalt shingles are comprised of asphalt-impregnated organic wood fibers, coated with colored inorganic mineral granules, whereas fiberglass asphalt shingles consist of asphalt-impregnated inorganic fiberglass mats, coated with colored inorganic granules. Organic asphalt shingles are slightly more expensive than inorganic; organic shingles are thicker, heavier, and more durable than inorganic; organic shingles require more asphalt than inorganic; organic shingles have a poor fire rating as compared to inorganic; and organic shingles perform better in cold climates than inorganic.

Be forewarned! Fiberglass asphalt shingles tend to crack and split more than the organic wood fiber asphalt shingles, see Figure 12.3.

On a hot roof, asphalt emissions from asphalt-impregnated organic and fiberglass shingles are possible, but not likely to impact building occupants. The asphalt "fumes" are more likely to dissipate into the outdoor environment—not withstanding a heating, ventilation, and air conditioning (HVAC) fresh air intake in the vicinity of an asphalt shingle roof.

Wood Shakes and Shingles

Wood shakes and shingles are made predominantly of cedar and occasionally of redwood and southern pine. Cedar naturally resists insects

FIGURE 12.3
Cracking fiberglass asphalt shingles. (Courtesy of InspectAPedia at www.inspectapedia.com)

and ultraviolet light damage, and it withstands hail and heavy storms. Wood is an excellent insulator—when installed properly; it has a 30–40 year lifespan; it is vulnerable to moisture; and it requires routine maintenance.

The Achilles' heel for wood shakes/shingles is that wood will incur moisture damage and is flammable—neither of which has a solution. Wood can be pressure treated to prevent moisture damage, and it can be impregnated with a fire retardant to reduce its flammability. Emissions from pressure treated and fire-retardant wood shakes and shingles are not likely. However, handling and management of treated wood shakes/shingles may pose an occupational health hazard.

Clay and Concrete Tiles

Hand formed and dried clay tiles originated over 5000 years—in China and Babylon. By the eighteenth century, molded, ground baked clay tile roofs were standard in Europe, and today oven baked clay tiles are used throughout the world, predominantly in warm climates. Clay roof tiles have an average life span of 100 years, and they are not prone to color fading.

The most commonly encountered clay and concrete tiles are the Spanish-style barrel design or Mediterranean-style. Yet, they can mimic slate and wood shakes. Clay tiles are baked molded clay. The finished product is either unglazed natural clay color or glazed to any color desired. The array of glazing pigments may comprise toxic metals—cobalt (e.g., blue), copper (e.g., reds, greens, and black), lead (e.g., red and yellow), chromium (e.g., yellow), and manganese (e.g., purple).

Concrete roof tiles originated in Bavaria in the nineteenth century. To this day, they still stand as a testament to their durability. They were introduced in the early 1900s. Recent estimates show that concrete tiles account for 90% of all roofs in Europe and the South Pacific (Boral 2002). Concrete roof tiles are about half the price of clay roof tiles; they have an average life span of 30–50 years; and the color does fade over time.

Concrete roof tiles mimic all designs that are available for clay tiles. Concrete tiles are made by mixing Portland cement, sand, and water. The mixture is poured into molds and cured. Thus, it does not require baking— an energy conservation process—and the finished product only gets harder with age. Color may be added to the surface of the finished product, or it may be an added component in the mix. All color pigments potentially encountered in the clay tile glaze also apply to the concrete roof tiles.

Finished clay and concrete products are not likely to emit toxin/irritant vapors or gases. However, rainwater runoff from a roof may potentially carry toxic metals from the pigmented glazes and paints and impact the quality of water collection systems. Clay and concrete both contain quartz. The presence of quartz means, "Cutting or demolition of the finished product(s) can result in a release of toxic/carcinogenic crystalline silica."

Slate Tiles

Slate roofs date back to the era of kings and castles. As slate was extremely expensive and time consuming to install, only the wealthy could afford slate roofs. In the 1800s, slate became more readily available and was installed more frequently in European structures. Worldwide slate reached its highest production in the late 1800s and early 1900s. Then came the less costly asphalt shingles. Slate roofing is, however, to this day highly desirable and long lasting with a 100–200 year life expectancy, and the craft is highly specialized. Due to this life expectancy, the roof flashing must also have a long life expectancy. For this reason, expensive copper flashing is a complement to slate roofs.

Slate is a shale rock formed from compressed sediment that over time become thin, flat, layered clay-like stone. Shales are highly variable in content, mostly a mix of clay, quartz, calcite, and other minerals. Some of the other minerals may lend color to the shale. Mostly a dark gray, shales with carbon compounds are black. Shales with ferric oxide are red. Shales with iron hydroxide are brown. Shales with metal silicates minerals are green. Of all the various components of shale (or slate), quartz contains amorphous and crystalline silica.

Although toxic/irritant emissions are not likely from either the slate or copper flashing, workers may potentially be exposed to toxic/carcinogenic crystalline silica dust when working with slate.

Metal Sheets and Panels

During the eighteenth century, copper was used on domes and cupolas, and lead was used to fill in and/or around clay tiles and wood shakes/shingles voids, on flat/low sloped roofs, on domes and cupolas, and on awkward shaped roofs., Sheet iron roofing had its origin in the late 1700s. Lead was, and still is, used in plumbing roof vents and flashing. By the early 1800s, corrugated iron and zinc galvanized metal, and tin-plated steel roofing were included in the mix of early metal roofing materials. By the twenty-first century, the arsenal of metal roofing materials has grown to include rust-resistant Terne-coated steel, stainless steel, factory-coated roof panels (e.g., baked on enamel and Teflon®), and Galvalume®. They can be rolled into panels and stamped into various forms to simulate shakes, shingles, and tiles.

Be forewarned! Terne is another term for "lead," a lead alloy which usually contains 80% lead, about 20% tin, and 1%–2% antimony.

Although toxic/irritant emissions are not from likely metal roofs, lead and Terne-coated roofs may pose toxic worker exposure when handling the material without precautions (e.g., gloves). Beyond emissions, lead component and coated roofing materials may pose an environmental health hazard through rainwater runoff.

Built-Up Roof

Built-up roofing (BUR) systems, also referred to as tar and gravel, were first introduced in the 1840s. A BUR system consists of "multiple plies" of rein-forcing material (e.g., organic felts, fiberglass mats, and polyester), layered with tar, and coated with gravel or decorative stone. A ply of reinforcing material is laid down, and hot or liquid tar is mopped on the reinforcing material. Another ply is laid down and tar mopped on the reinforcing mate-rial once again—layer upon layer. This redundancy provides flexibility and durability for low-pitched or flat roofs. The tar, also referred to as bitumen, is typically petroleum asphalt or coal tar pitch.

Petroleum asphalt is a product of the nondestructive distillation of crude oil. It is a brown to black semisolid or solid substance. It is applied by ambi-ent temperature deposition or by heat softened asphalt application.

Ambient temperature deposition of asphalt is performed by solvent pre-cipitation or emulsion deposition. Solvent precipitation, also referred to as "cutback," is where the semisolid/solid asphalt is dissolved in a solvent (e.g., petroleum distillates) which after evaporation leaves an asphalt deposit. As much as 50% of liquid asphalt (or cutback) content is solvent. After "ambient temperature application" of cutback, solvent emissions are highly likely and may take several weeks to completely off-gas. Solvent vapors may cause eye and respiratory tract irritation, and some of the components may be regu-lated (e.g., benzene) (U.S. Department of Health and Human Services 2000).

Emulsified deposition, also called asphalt emulsion, is a mixture in immisci-ble water with asphalt and an emulsion such as soap. Asphalt deposition relies on water evaporation, and there are no known irritant/toxic emission hazards.

Heat softened asphalt application, or hot processing, is the most com-monly encountered BUR process. Oxidized-asphalt is heated in a kettle to a temperature where the solid asphalt is soft enough to either hand mop or mechanically spread onto a roof surface. Different types of asphalt are pro-cessed at the refinery, and each type has a different use and melting point. In the United States, mopping-grade roofing asphalts are classified by roof slope and application temperatures (e.g., melting point). Yet, to maintain its application temperature, it must be heated to higher temperatures (e.g., up to 100°F) than the melting point in the kettle. The types of asphalt are characterized by roof slope and melting point (U.S. Department of Health and Human Services 2000, p. 9).

- Type I: "Dead level," no roof slope; asphalt melting point of 330–355°F
- Type II: 0.5–1.5:12 roof slope; asphalt melting point of 365–390°F
- Type II: 1–3:12 roof slope; asphalt melting point of 395–420°F
- Type IV: 2–6:12 roof slope; asphalt melting point of 430–445°F

The word on the street is that asphalt fumes are highly flammable—a fire hazard and, in some cases, an explosion hazard when contained within a closed kettle (U.S. Department of Health and Human Services 2000, p. 10).

Asphalt fumes have repeatedly been found to cause serious eye irritation, skin irritation, and photosensitization, and upper respiratory tract irritation. In addition, several nonspecific symptoms (e.g., nausea, headaches, fatigue, decreased appetite, and stomach pain) have been reported by some roofing workers. Where does this lead us in terms of building occupant exposure?

Asphalt fumes tend to drift to occupied spaces through HVAC open makeup air intake vents located within the vicinity of roofing activities. In addition, the fumes may drift into wall and roof penetrations (e.g., doors and windows). Once cooled, asphalt ceases to produce irritating fumes. Some asphalt also has as much as 2% of hydrogen sulfide in the petroleum product. Hydrogen sulfide is highly flammable, highly toxic, and highly irritating, and the rotten egg odor of hydrogen sulfide is five ten thousandth its exposure limits (i.e., 0.005 ppm/10 ppm). Yet, it is not clear as to hydrogen sulfide's contribution to the odor of asphalt and/or to the eye and upper respiratory tract irritation attributed to asphalt. Asphalt fume odors are particularly noticeable during hot roofing activities. After, however, "hot application" and cool down, asphalt fume emissions are unlikely. Coal tar fumes are a different story!

Coal tar pitch is a tar that contains polycyclic aromatic hydrocarbons (PAH) and is produced by the destructive distillation of a relatively soft, bituminous coal. It is similar in function and use to petroleum asphalt—with slightly different toxic components. Asphalt may contain hydrogen sulfide, and coal tar does not. Coal tar pitch volatiles contain carcinogenic PAH at levels of 1%–2% (Alcoa 2011), and asphalt has only detectable levels (e.g., <0.1%) of PAHs (Valero 2011). An NIOSH published study demonstrates the elevated coal tar pitch PAH fume condensates as opposed to asphalt. See Table 12.1 for toxins/carcinogens associated with asphalt and coal tar pitch fumes.

Both coal tar and asphalt fumes can cause eye and respiratory tract irritation. The melting point for coal tar 86–356°F and is used on "dead level" to 0.25:12 roof slope, but due to its low melting point, coal tar pitch tends to soften on a hot roof. This tendency is a double-edged sword. As it softens, coal tar can "heal" BUR cracks and tears on hot days, and coal tar pitch volatiles can also pose an emission hazard on hot days.

Coal tar pitch volatiles are regulated as benzene soluble fraction (e.g., PAHs) and can cause eye and respiratory tract irritation. On the other hand, the flash point for coal tar pitch is greater than its melting point. Thus, coal tar does not pose a fire hazard as asphalt does.

In BUR systems, petroleum asphalt fumes and coal tar pitch volatiles pose irritant/toxic emission hazards during hot application to workers and potentially to building occupants. Cutback ambient temperature spreading of asphalt and coal tar pose irritant/toxic emission hazards during application

TABLE 12.1

Comparative mean concentrations of various PAHs in Type I roofing asphalt and coal-tar pitch fume condensates, µg/mL

PAH	Petroleum Asphalt, Fume Condensates		Coal Tar Pitch, Fume Condensates	
	232°C	316°C	232°C	316°C
Naphthalene[c,d]	22	4	>1800	1700
Anthracene/ phenanthrene	180	53	>960	2960
Pyrene	70	9	>2070	1790
Benz(a)anthracene[b]	11	10	570	330
Chrysene[c,d]/ triphenylene	25	19	460	300
Benzofluoranthenes[c]	2	4	230	230
Benzo(e)pyrene	6	8	42	51
Benzo(a)pyrene[a,d]	2	2	96	85
Dibenzanthracenes[b]	2	ND	12	ND

Excerpted: NIOSH: Health Effects of Occupational Exposure to Asphalt. CDC, Atlanta, Georgia (2000) p. 15.

[a] IARC Cancer Review Group 1. (Carcinogenic to humans.)
[b] IARC Cancer Review Group 2A. (Probably carcinogenic to humans.)
[c] IARC Cancer Review Group 2B. (Possibly carcinogenic to humans.)
[d] Has an OSHA Permissible Exposure Limit.

and for an extended period of time after application. During kettle heating, asphalt is a likely fire and potential explosion hazard. The thermal decomposition products are carbon dioxide and carbon monoxide.

Many of BUR potential exposure hazards are rectified with modified bitumen (MB) membranes. They are a transition between BUR and synthetic roof membranes.

MB Roof Membranes

MB roof membranes were developed in Europe in the 1960s and have been in use in the United States since the middle 1970s. MB roof membranes are typically composed of two layers of reinforcing material (e.g., polyester fabric) that are coated with hot polymer-MB and sometimes, not always, a layer of mineral granules (e.g., limestone). The membranes are factory manufactured and applied on site by one of several methods—torch down, self-adhering, hot mopped, or liquid applied.

The most common "polymer-modifiers" are atactic polypropylene (APP) and styrene-butadiene-styrene (SBS)—factory manufactured membranes, and the application process is highly variable, dependent upon the polymer modifier in the membrane.

- AAP, a thermoplastic polymer, is added to bitumen to extend longevity and resilience. APP-MB is applied to polyester fabric, fiberglass mat, or polyester/fiberglass mat—two ply layering which is considerably less than BUR systems. At the factory, the sheet treated fabric and/or mat are cut to manageable size for handling. Most membranes are pretreated with bitumen on the underside to allow the manufactured APP-MB membranes to be heat-welded or torch-applied, circumventing the need for a messy hot kettle and mopping. Yet, torch application does not get a pass in terms of fire hazards. Torching asphalt to its melting point may cause fire spread. It has been alleged by some that torch down has caused building fires.
- SBS, a synthetic rubber, is added to bitumen to provide more longevity, resilience, and flexibility. SBS-MB is applied to two layers of fabric and/or mat—cut and rolled for ease of handling. SBS-MB membranes are self-adhering, hot mopped, or liquid applied.

With a couple of exceptions, application techniques of the APP and SBS membranes are similar to BUR. There is only one coat, not several coats of bitumen, and bitumen adhesives can be pre-applied at the factory or applied on site. So, all the irritant/toxic emission hazards are similar, but to a much lesser degree, to that of BUR systems. The polymers may contain plasticizers and/or additives some of which may be irritants. See Section II, "Polymers". Thermal decomposition products are carbon dioxide and carbon monoxide.

Synthetic Roof Membranes

All synthetic roof membranes are either thermoset rubber (e.g., ethylene propylene diene monomer [EPDM]) or thermoplastic polymers (e.g., PVC). EPDM was first manufactured in the United States in the early 1960s. About the same time, single ply PVC roof membranes were first introduced in Europe in the 1960s. Ten years later, PVC membranes found their way into the United States. Other synthetic membranes have since been introduced, and in an effort to attain a competitive edge, manufacturers continue to develop different formulations of the tried-and-true favorites (e.g., EPDM and PVC) while researching newer and better polymers.

Rubber Membranes

Although single ply rubber membranes are cheaper than thermoplastic polymers, rubber membranes are not living up to their alleged life expectancy of 30 years. The membrane is not the problem. The installers are allegedly the problem. Rubber membranes can fail shortly after a new roof installation, failures due to poor seals around roof penetrations, flashing, and seams.

Today, the most commonly encountered rubber membrane is EPDM with chlorosulfonated polyethylene (CSPE), and neoprene nipping at the heels of the entrenched front runner.

EPDM, and other rubber membranes, are installed by one of three methods: (1) ballasted; (2) mechanically attached; and (3) fully adhered. Ballasted membranes are held in place by round stones or flat slabs. Mechanically attached roof membranes—in low wind areas—are held in place with nails. The fully adhered installation membranes are more costly and have a greater performance than the other methods. Rubber membranes are adhered by any of a number of acrylic water-based, solvent-based, and/or low-VOC adhesives. See Chapter 16: Adhesives.

Possible exposure details regarding the main rubber roof membranes, as discussed in Chapter 7, "Elastomeric Polymers," are summarized as follows:

- EPDM emissions of unregulated diene component(s) (e.g., 1,4-hexadiene) and some of the additives (e.g., stabilizers and antioxidants) may cause eye, skin, and respiratory tract irritation. Combustion products include carbon oxides and hydrocarbons.

- CSPE emissions of unregulated components (i.e., unknown, but reported), plasticizers, and some additives may cause eye, skin, and respiratory tract irritation. Furthermore, toxic and regulated carbon tetrachloride emissions are likely. Combustion products include carbon oxides, nitrogen oxides, and hydrogen cyanide.

- Neoprene emissions of unregulated components (i.e., unknown, but reported) and additives may cause eye and respiratory tract irritation. In addition, toxic and regulated chloroprene emissions are possible, but unlikely.

Thermoplastic Polymer Membranes

The most commonly encountered thermoplastic roof membrane is PVC, also referred to as vinyl; and the second most common is thermoplastic polyolefin (TPO). PVC roof membranes were first manufactured in Europe in the 1960s, later to enter the United States market in the early 1970s. As the rumor mill has it, PVC roof membranes had a rough start in the United States but have since surpassed all other synthetic roof membranes in terms of quality, durability, and reliability. The rationale for it not taking the marketplace by storm is that PVC membranes, requiring hot-air welded installation, are more costly than rubber membranes.

In the manufacture of PVC membranes, biocides, plasticizers, ultraviolet light inhibitors, heat-stabilizers, color pigments, and polyester/fiberglass reinforcement are added during the polymerization process—without which there would be diminished flexibility and stability. PVC product emissions are likely at elevated roof temperatures.

- PVC polymers have a low melting point of 167–194°F (75–90°C), temperatures that are possible on roofs in hot climates. According to one study performed in Central Texas on a clear 90°F day, black single ply "roofing materials absorb and retain solar energy so that the surface temperatures reach 170°F–200°F" (National Coatings Corporation 2014). White acrylic coating roofing material retains less than 110°F. Thus, a white single ply PVC membrane is not likely to reach the melting point of PVC, and a black single ply PVC membrane is likely to exceed the melting point. Most, not all, PVC roof membranes are white!

- Plasticizers are not matrix bound and when released into the ambient air may cause eye and respiratory tract irritation—especially with the elevated temperatures generally encountered on roof tops. The more flexible a PVC membrane, the more plasticizer in the membrane. As a plasticizer dissipates, a thermoplastic roof membrane with begin to lose its flexibility.

- Polymer additives, not matrix bound, when released into the ambient air may pose a toxic and/or irritant hazard. Elevated temperatures may be contributory to chemical emissions from additives. Furthermore, in some cases, water runoff of some water-soluble, toxic additives can pose an environmental hazard.

Hydrogen chloride is a product of PVC thermal decomposition in the earlier stages of a fire, and the chlorine acts as a fire retardant. When, however, PVC does burn, it burns slowly, and the combustion products include carbon oxides and hydrogen chloride.

TPO roof membranes, as compared to PVC membranes, were a late bloomer. Initially manufactured and installed in Europe, TPO found its way into the United States market in the early 1990s. Initially, it was thought to have the same characteristics as PVC membranes, but it was cheaper—in the United States—to manufacture. Failures occurred allegedly due to its thinner (e.g., 50% thinner), poor formulation, and U.S. climatic extremes (TPO Roofing 2014). Today, however, quality is up, cost is up—at or exceeding that of PVC.

A "polyolefin," also referred to as a polyalkene, is a class of organic thermoplastic polymers that includes, but is not limited to, polyethylene and polypropylene. In the manufacturing process, plasticizers and polymer additives (including fire retardants) are mixed with the alkene monomers prior to polymerization. The end product is less temperature sensitive than PVC, and there is no halogen (e.g., chlorine) content—unless a halogenated fire retardant (e.g., brominated compounds) was added to the formulation. TPO product emissions are not as much a concern as that of PVC membranes.

- TPO polymers have a considerably higher melting point than PVC products. The melting point for polypropylene is of 226–338°F (139–170°C), and the melting point for polyethylene is of 248–356°F

(120–180°C). These temperatures are not likely to be met or exceeded on even the darker colored roofs.

- As with PVC products, plasticizers are not matrix bound and when released into the ambient air may cause eye and respiratory tract irritation—especially with the elevated temperatures generally encountered on roof tops. Plasticizer emissions are less likely, but still possible, from TPO roof membranes.
- As with PVC products, polymer additives, not matrix bound, when released into the ambient air may pose a toxic and/or irritant hazard. Elevated temperatures may be contributory to chemical emissions from additives. Furthermore, in some cases, water runoff of some water-soluble, toxic additives can pose an environmental hazard.

Polyolefin fires are self-perpetuating releasing flammable alkenes as they burn. The combustion by-products are carbon oxides.

In review of the thermoplastic roof membranes, plasticizers and additives are possible emission hazards associated with PVC and TPO roof membranes. In addition, PVC may emit hydrogen chloride in the early stages of a fire. Beyond carbon oxides, the combustion byproducts of PVC include hydrogen chloride.

Exterior Wall Cladding

Face appeal, face function! Exterior wall cladding is the functional character that defines a building. The face may bespeak great beauty. It may display creativity. It may herald a vision. It may foreshadow a frugal reserve. Or it may simply be functional—to protect and weatherproof. All exterior wall cladding serves as a non-load bearing exterior skin—a "protective façade."

Exterior wall cladding is the outermost covering (or skin) on a building, exclusive of windows, doors, soffits, and trim. Mankind has progressed from mud huts which are still the choice of wall coverings in third world countries to synthetic rubber walls at the other extreme. The most commonly encountered building covering in the United States is siding—wood, cement, metal, and vinyl. The types of the outermost coverings are grouped as follows: (1) siding and panels; (2) masonry; (3) adobe and stucco; (4) poured concrete; and (5) synthetic/faux products.

Siding and Panels

Siding is a term of many meanings! It is, in the strictest sense, exterior wall cladding. Yet, to the layperson, sales persons, many builders, and herein, siding refers to long, thin panels—the most commonly recognized of which is "vinyl siding." Other siding building materials include, but are not limited to, wood, metal, fiber cement, and fiberboard. Each comes in rectangular panels as well, providing easier, quicker coverage, and panels may be formed into different formats (e.g., faux stacked stone vinyl panels). Siding/panel materials discussed here include (1) metal; (2) wood; (3) fiber cement; (4) hardboard; and (5) vinyl.

Metal

Metal siding/panels, an extension of metal roofing, may include, but not be limited to, baked-on painted steel, unpainted zinc galvanized metal, tin-plated steel, stainless steel, and copper. Metal siding is not likely to produce toxic/irritant emissions, and combustion products are unremarkable. For more details, see "Roofing Materials" section.

Be forewarned! Air tight metal panels on the exterior of a building may cause moisture to accumulate, result in mold growth and rotting within the exterior wall cavity.

Wood

Wood siding/panels, also used in roofing and sheathing, may be untreated (e.g., cedar), treated with preservatives (e.g., ACQ treated pine), and/or engineered (e.g., plywood).

> In dust form only, some natural woods (e.g., western red cedar) may cause eye and respiratory sensitization and irritation. Wood siding, however, is not likely to produce toxic/irritant emissions. Combustion products are unremarkable.
>
> With a few exceptions, most wood preservatives are relatively nontoxic. The exceptions-to-the-rule are arsenic-containing chemicals (e.g., ACZA) which could pose an environmental concern, not an emissions hazard. Including the arsenic-containing preservatives, all treated wood siding materials are unlikely to produce toxic/irritant emissions. Combustion products of arsenic treated wood can result in toxic levels of arsenic, and most wood preservative combustion products also include exposures to copper.
>
> Engineered wood siding/panels are a comprised of thermoset adhesive(s) and wood—wood fibers (e.g., engineered wood siding) and sheet wood (e.g., marine plywood). Many are also pressure treated with preservatives to prevent rot, pest and fungi damage of the siding/panel wood component. Exterior engineered wood adhesives are generally phenol formaldehyde but may be melamine or resorcinol formaldehyde. In order of greatest to least probable, formaldehyde off-gassing is possible from melamine-, phenol-, and resorcinol-formaldehyde resin components—especially when the products are newly manufactured or when exposed to extreme heat and moisture. However, formaldehyde

emissions out-of-doors are unlikely to impact the air quality of building occupants. Combustion products are unremarkable.

Fiber Cement

Fiber cement siding, also referred to as concrete siding, is a composite material made of sand, cement, and cellulose (i.e., wood) fibers. Although the sand component can pose a crystalline silica occupational exposure hazard, emissions are irritant/toxic but emissions are highly unlikely—unless old cement siding (comprised of asbestos fibers instead of cellulose fibers) is damaged. Cement siding is not combustible.

Hardboard

Hardboard siding, also referred to as pressboard and "masonite," is a composite of wood fibers (e.g., sawdust), wax, and a thermoset resin (e.g., phenol formaldehyde). The mixture is heated and pressed into sheets. Then, wood-grain embossed paper is laminated to the surface of the hardboard, and the sheets are cut to size. First introduced in the 1970s, hardboard siding was installed extensively from the 1980s to the 1990s. And then, the house of cards came tumbling down!

> Within short order, newly installed hardboard siding started showing the ravages of rain and other water sources such as gutter overflow, sprinkler systems, and power washing. The fibers soak up water. Water is conveyed up from panel to panel and into the wall cavity. As it does not deteriorate, the resin-impregnated laminate on the surface of the hardboard obscures the swelling, bowing, and rotting that occurs within. Damage often goes unnoticed for extended periods of time (e.g., up to two years). This inexpensive siding became the homeowner's curse, and they sought compensation. In 1998, a class action lawsuit against Masonite Corporation was settled for one billion dollars. All class members with failing Masonite™ hardboard siding installed and incorporated on their property between 1 January 1980 and 15 January 1998 were entitled to make claims (Lieff Cabraser Heimann & Bernstein, LLP 2015).

After the lawsuit, Manonite Corporation ceased production. By 2002, other manufacturers allegedly followed suit. Yet, "hardboard" is still being marketed and sold today.

Perhaps, hardboard siding undergoes a slightly different manufacturing process than that of years past. Perhaps, the earlier hardboard siding was compressed wax and wood fibers with a UF binder (which deteriorates when wet) and a phenol formaldehyde laminating glue (which resists deterioration when wet). Perhaps, the name has changed—engineered wood siding.

Clarity is illusive! But most hardboard siding, today, is formulated with a mixture of hardwood fibers, wax, and phenol and/or resorcinol formaldehyde binder. Irritant/toxic formaldehyde emissions are possible from hardboard and engineered wood siding; yet, occupants are not likely to be exposed. Combustion products are carbon oxides and aldehydes.

Vinyl Siding

Vinyl siding, also referred to as PVC siding, is a thermoplastic polymer that is comprised of ethylene and chlorine. When unstabilized and exposed to intense radiant heat, PVC emissions of "regulated, toxic/irritating" hydrogen chloride gas and detectable organic compounds occur. The chlorine component renders PVC fire resistant, but vinyl will burn—slowly, releasing irritating/toxic chlorine gas.

Be forewarned! Vinyl siding has been reported to deform as it softens when exposed to extreme heat such as glass-amplified solar energy or heat from reflective surfaces.

Masonry Products

Masonry dates back to times of old when pyramids arose in the Sahara desert landscape and castles were fortified with stone. It is the art of building structures using individual units laid in and bound together with mortar. These individual units may be stone, brick, cement blocks, and/or decorative tiles. And all masonry building materials are comprised of natural stone.

Natural Stone

First, let's address the difference between rock and stone. This question will come up. Simply stated, rock is that which is in the ground whereas stone connotes human manipulation of nature's mineral bounty. Rocks are sporadic. All granite is not alike. All slate is not alike. All travertine is not alike. Components, colors, and patterns are regional. Nature is fickle indeed. Where is this going? Why focus on variability? In assessing toxic components, we are subject to nature's recipe. So, let's look at the rock pile!

Rocks are categorized according to their origin. Igneous rocks originate in the depths of Hades. Sedimentary originate on the surface, and metamorphic rocks resides in limbo between igneous and sedimentary rock formations.

Igneous is Latin for fire. Igneous rock is produced by the crystallization and solidification of molten magma (i.e., rock, volatiles, and solids) formed deep within the earth's core. It reaches the surface by volcanic activity, lava flow, or rise to up to a few kilometers below the surface. As the molten magma cools, igneous rocks are formed. A common igneous stone used as exterior wall cladding is "granite" which has up to 70% quartz content. Quartz

(7 Mohs hardness) is silicon dioxide, also referred to as silica. Crystalline silica is a regulated worker health hazard!

Whereas igneous rocks are born within the bowels of the earth, sedimentary rocks are born on the surface of the earth. Sediment (i.e., mud, sand, gravel, and/or clay) formed into soft rocks overtime. Color is created by a diversity of Earth's elements (e.g., red iron rust), and many sediments show impressions of prehistoric/recent life forms and surface activity (i.e., fossils, tracks, and ripples). A couple of the sedimentary stones used in exterior wall cladding are limestone (3 Mohs hardness) which is comprised of calcite and travertine (2.5–3.5 Mohs hardness) which is comprised of calcium carbonate. "Sandstone," a common masonry product, has a considerable amount of quartz. The quartz content of sandstone makes it denser, harder than many of the other sedimentary rocks, and its quartz content is region dependent.

Metamorphic rocks are morphed rocks, changed forms of igneous and sedimentary rocks—shaped by heat, pressure, chemicals, and stress (e.g., stretching, squeezing, and shearing). A couple of metamorphic stones used as exterior wall cladding are "marble" which is recrystallized limestone (e.g., predominantly calcite) and slate (2–4 Mohs hardness). Marble has as much as 60% silica content.

Beyond the basic composition that allows for identification of the various types of rock, Earth delivers a hodgepodge of other chemicals and/or mineral substances into the mix. The take away is, nature delivers unknowns. Without a site specific assessment of quarried materials, a supplier has limited information regarding the stone. Move along, no problem here! In most cases, the "no hazardous components" statement is likely to apply. One area that rarely gets attention is the element of radium and the off-gassing of radon.

> All natural stone contains detectable amounts of radioactive (e.g., uranium) and/or radon. In the 1990s, the news media reported that granite countertops can release dangerous levels of radon. Subsequent U.S.A. studies indicated that the fears were unfounded. Studies had been done in homes and in laboratories, with and without air ventilation. They were conducted by university graduate student, state health departments, and Consumer Reports. All concluded "low detectable levels of radon" (Consumer Reports News 2008). However, the region of stone origin was not included in the study. A considerable amount of granite is mined in Kazakhstan, Canada, and Australia. These countries are, in respective order, the largest producers of uranium in the world (two-thirds of the world's production in 2014) (World Nuclear Association 2015). Should not stone mined in these regions be suspect? Just a point to ponder!

Due the natural air movement around a building, it is unlikely that stone will off-gas radon at sufficient levels to impact occupants within a building. However, the negative perception of wary occupants, particularly toward granite, may warrant a pre-installation evaluation.

Exclusive of sedimentary rocks, most natural stone products contain silica—amorphous and/or crystalline silica. Crystalline silica dust is a regulated occupational toxin, not a building material emissions hazard.

Brick

From ancient times until today, brick has evolved from air-dried mudbrick to oven or kiln-dried brick to chemically processed masonry units. Amazingly, old mudbrick, dating back to 7500 BC, is still manufactured to this day. Even the most commonly used kiln-dried brick dates back to 3000 BC. In 1927, calcium silicate brick, a German creation, became popular and has become a major masonry building material in Europe. Each form of brick is progressively a slight improvement over the preceding one.

In "dry climates" such as in the Middle East, mudbrick, also referred to as adobe brick, is extremely durable, and accounts for some of the oldest existing structures in the world. Mudbrick is a composite of earth mixed with water and organic material such as straw or dung. The earth component is typically local and is comprised of clay, silica sand (55%–75% by weight), and silt. Furthermore, modern adobe brick is stabilized with up to 10% of either emulsified asphalt or Portland cement to protect against water damage. The brick is shaped by hand or packed into forms, and air dried in the heat until hard. The final product is either installed with mortar or mud and may be coated with lime-based mud (old style) plaster, whitewash (e.g., slaked lime and calcium carbonate). Although occupational exposures to silica dust are possible, product emissions are highly unlikely.

Kiln-fired bricks are similar in composition to mudbricks, but they are more durable and long lasting—made possible by extreme heat (900°C [1652°F] to 1300°C [2372°F]). Kiln-fired brick has clay, sand (50%–60%), slaked lime (i.e., calcium hydroxide), and iron oxide. The more iron oxide in the mix, the stronger and the redder the brick. Once again, as in mudbrick, product emissions are highly unlikely.

Calcium silicate bricks and other masonry units are distinct from clay bricks in that they do not contain clay. The masonry units are manufactured by mixing calcium lime (e.g., calcium oxide) with a fine sand and water mix. The mix is molded under high pressure, followed by high-pressure steam curing. It is a high-density, brick that is typically yellowish in color. Manufacturers claim that a wide variety of colors can be produced ranging from white to earth tones similar to adobe and kiln fired bricks, and masonry units can be formed to appear as nature stone. Although occupational exposures to silica dust are possible, product emissions are highly unlikely.

Be forewarned! Manufactured masonry products are generally intended for above grade installations. Regardless of their composition, masonry products are inherently absorptive, and as such, are not intended for use below grade or in contact with wet soil or mulch. These conditions will create a "rising damp"—a slow upward movement of water in the wall and moisture penetration into the lower wall cavities.

Stucco

Although stucco has been traced back 5000 years to what is now Iraq, it was not until the revival period of the Renaissance that stucco became widely accepted in Italy and other parts of the old world. Stucco Veneziano was conceived for long lasting, partially submerged Venetian structures. At this time, stucco was made of lime, sand, and water with animal and plant fibers added for strength. By the 1800s, Portland cement replaced lime, and the basic composition of stucco, also referred to as render, became Portland cement, sand, and water.

The "lime stucco" of old had been hard, easily broken, or chipped by hand. It also had a medley of additives that appears to be more like a witches' brew than a stucco mix. The concoction included blood, urine, eggs, varnish, wheat paste, salt, sodium silicate, alum, tallow, beeswax, and whiskey.

The "Portland cement stucco" is hard and brittle, easily cracked if the base on which it was applied was unstable. Today's Portland cement stucco has additives such as lime, fiberglass, and acrylics. The lime provides self-healing properties in the presence of water. The fiberglass acts as a binder for greater strength. The acrylic additives are used in the finish coat for color retention, reduction in cracking, reduced water retention, and increased bond/compressive strength.

Stucco is a process! It is either applied directly to masonry products (e.g., brick, stone, and concrete) or adhered to a lath surface. Wood lath, in the form of furring strips, was replaced by metal lath in the late 1800s. After a surface has been prepared, two to three coats of stucco are applied—less for surfaced masonry products (e.g., scored, gouged, and weathered stone) and more for a lath substrate. The multiple coats consist of (1) a scratch coat; (2) a floating brown coat; and (3) a finishing coat. The base scratch coat and brown coat contain binder (e.g., fiberglass) and one or more additives whereas the finishing coat is only plaster (e.g., lime or Portland cement) or plaster with sand, pigments, and additives (e.g., acrylics). The greater the amount of texture desired, the greater the amount of sand required.

As with many of the stones, worker exposure to silica is possible, but stucco product emissions are unlikely.

Synthetic/Faux Walls (e.g., EIFS)

Synthetic exterior walls are impersonators of traditional masonry products. They mimic the beauty of traditional building materials without the expense. They enable builders access to a wide variety of forms and shapes that were previously not attainable without special stone masons or wood artisans (e.g., elaborate stone columns/cornices and wood corbels). They often provide increased thermal insulation and water resistance. Yet, they are typically an extension of the "World of Plastics"—faux building materials—many of which were born around the 1970s.

Some of the more commonly encountered synthetic exterior wall building materials are faux brick veneer, faux stone panels, faux wood beams, faux wood siding, and synthetic stucco systems. Exclusive of vinyl, most synthetic "faux" building materials are comprised of pigment colorant(s), a high-impact PU surface, and a PU foam backing. In recently manufactured PU products, highly irritating isocyanate emissions are likely. However, without a means to become entrained within the occupied spaces in a building, isocyanate vapors are likely to dissipate into the outdoor environmental, and are not likely to pose an indoor air quality health hazard. Combustion products are likely to be carbon oxides, oxides of nitrogen, and hydrogen cyanide.

Synthetic stucco systems are not a faux finish. It is an "exterior insulation and finishing system" (EIFS). The exterior insulation is covered with a fiberglass reinforcing-mesh that is imbedded in the scratch coat and finished with traditional stucco. It was used in Europe after World War II to repair buildings damaged during the war, and it found its way into North America in the 1980s for use as a "wall system." The insulation, which is glued or mechanically fastened to the moisture barrier, is typically polystyrene. The polystyrene may occasionally be substituted by other polymeric foams such as polyisocyanurate.

> Polystyrene foam is the most commonly used foam used in an EIFS system. Styrene emissions are possible, but not likely, yet a polystyrene fire is self-supporting. Combustion products are carbon oxides.
>
> Polyisocyanurate foam emissions and combustion products are similar to that of polyurethane. Highly irritating isocyanate emissions is possible, but not likely, to be emitted but not likely to impact building occupants. The combustion products are carbon oxides, oxides of nitrogen, and hydrogen cyanide.

In the late 1980s, environmental consultants and forensic engineers began to sound the alarm over moisture buildup within EIFS systems. Then, in the late 1990s, the house of cards began a downward spiral.

> In 1998, a North Carolina homeowner described his encounter with structural damage. As he was putting a new shutter on his garage, the homeowner missed the shutter, and the nail disappeared through the stucco wall. He pondered, "Where did the nail go?" He probed deeper into the wall, looked within, and he was "horrified" to find "Styrofoam" and a pervasive "rot." An EIFS lawsuit was filed, and the case was settled out of court. Subsequently, there have been hundreds of lawsuits against EIFS manufacturers. Lawyers dubbed EIFS the "ultimate roach motel." Water got in, but it couldn't get out! As builders began to shy away from the product, EIFS manufacturers redesigned the system to include a drainage plane, a weather-resistive barrier between the sheathing and

exterior insulation, as well as some improvements in the reinforcing mesh to increase strength—impact resistance. Some of the original critics and expert witnesses began to change their tune. Today, EIFS appears to be making a comeback!

Be forewarned! Although EIFS has evolved and is allegedly considerably improved, experienced professionals should be consulted and warranties secured prior to installation.

Finishing Touches

A hodgepodge of building components—ever-present and/or decorative—complete the exterior. These include, but are not limited to, soffits, roof vents, fascia, drains, columns, and corbels. Not to belabor a point and not to be redundant, building materials encountered in the finishing touches are discussed within the preceding topics. For example, soffits may be wood, metal, cement board, or vinyl. Columns may be wood, concrete, natural stone, and synthetic stucco. The emission from these components may be contributory to the whole.

Summary

The culmination of all things on review—the exterior enclosure—is a medley of many possibilities. Most are from the bowels of the earth, posing a potential worker exposure hazard, not a material emissions hazard to building occupants. Whereas some building materials pose a potential environmental toxics exposure hazard, very few exterior building materials pose a potential for an irritant/toxic emissions that may impact building occupants. The synthetic products (e.g., plastics, foams, and rubber) can be the bane of the green movement or the salvation of natural resources.

Whereas most exterior building components emissions impact outdoor air, window and door emissions may impact outdoor air and indoor air.

- The exterior skin of wood veneered window frames and doors and of solid wood core doors may potentially emit formaldehyde—particularly newly manufactured materials. Elevated radiant heat on vinyl window frames and doors may result in vinyl degradation and emissions of hydrogen chloride.

- Out-of-sight, out-of-mind, the core of doors—all hollow core doors (e.g., wood, metal, vinyl, and fiberglass)—is an often overlooked source of toxic/irritant emissions. Possible insulation emissions comprise isocyanates (e.g., PU foam), styrene (e.g., polystyrene foam), and formaldehyde (e.g., Thermofiber Safing), and rigid core material emissions are formaldehyde (e.g., resin-impregnated honeycombed paper). Newly manufactured products are more likely to emit toxic/irritant substances than older products.
- Glass sealant/gasket emissions of hydrogen chloride are possible.

Roofing materials are strictly external, and only through the infiltration of the outdoor air to indoor spaces can roof-top emissions impact indoor air quality. And very few roofing materials are likely to emit irritant/toxic chemicals.

- Asphalt roof shingles may potentially emit detectable levels of asphalt
- BUR systems of petroleum asphalt and sometimes coal tar pitch may potentially emit irritant/toxic chemicals (e.g., asphalt fumes, hydrogen sulfide, and PAH) particularly during the application stage
- Pre-manufactured MB roof membrane emissions are similar to BUR systems—to a much lesser extent
- Synthetic rubber roof membranes emissions may be irritating (e.g., diene components, plasticizers, and additives) or toxic (e.g., carbon tetrachloride)
- Plastic membrane emissions may be irritating (e.g., plasticizers and additives)

Exterior wall cladding materials are similar in some respects to roofing materials—in terms of materials and the impact of exterior wall emissions. Wall cladding emissions, however, cover more surface area, and even fewer exterior wall cladding materials are likely to emit irritant/toxic chemicals.

- Vinyl siding that is exposed to intense radiant heat may deteriorate and emit hydrogen chloride

All things considered, exterior building materials may be the least of one's concerns in terms of product emissions and indoor air quality. Yet, as Murphy's Law is alive and well, all things are possible. Off-gassing of formaldehyde from doors and windows may contribute significantly to other sources of formaldehyde within a structure. Off-gassing of hydrogen chloride from one or several building materials (e.g., vinyl windows/doors, roof membranes, and siding) may cause building occupant eye irritation. Are not multiple sources of irritant/toxic emissions likely?

References

Alcoa. 2011. *MSDS: Coal Tar Pitch*. Pittsburgh, PA, July 19.

Boral. 2002. "History of Concrete Roof Tiles." *Boralna*. Accessed July 7, 2015. http://www.boralna.com/rooftiles/history-of-concrete-roof-tiles.asp.

Consumer Reports News. 2008. "Buzzword: Radon." *Consumer Reports News*. June 30. Accessed July 31, 2015. http://www.consumerreports.org/cro/news/2008/06/buzzword-radon/index.htm.

Cosmos. 2015. "Door Skin HDF Molded with Design, Veneered HDF Door Skin, Door Skin HDF, Molded Door Skin, Natural Veneer Door Skin." *Cosmos*. Accessed June 29, 2015. http://www.consmos.com/Door_Skin_HDF_molded.html#OSBpictures.

Lee, Jenica. 2013. "Problems with Vinyl Windows Every Homeowner Should Know About." *Southwest Exteriors*. April 16. Accessed June 29, 2015. http://www.southwestexteriors.com/problems-vinyl-windows-every-homeowner-should-know-about.

Lieff Cabraser Heimann & Bernstein, LLP. 2015. "Masonite Defective Product Class Action Lawsuit." *Lieff Cabraser Heimann & Bernstein, LLP Case Center*. Accessed July 30, 2015. http://www.lieffcabraser.com/Case-Center/Masonite-Defective-Product-Class-Action-Lawsuit.shtml.

National Coatings Corporation. 2014. "Cool Roof Overview." *National Coatings*. Accessed July 20, 2015. http://www.nationalcoatings.com/cool-roof-overview.

Nu-Glaze. 2014. *Nu-Glaze – The Alternative to Putty*. Accessed June 30, 2015. www.nu-glaze.co.za.

TPO Roofing. 2014. "Comparative History of TPO vs PVC in the US vs Europe." *TPO Roofing*. Accessed July 20, 2015. http://www.tporoofing.org/guide-for-roofing-contractors-comparative-history-of-tpo-vs-pvc-in-the-us-vs-europe.

U.S. Department of Health and Human Services. 2000. *Health Effects of Occupational Exposure to Asphalt*. Atlanta, Georgia: CDC.

U.S. EPA. 2013. "Styrene." *Technology Transfer Network—Air Toxics Web Site*. October 18. Accessed February 25, 2015. http://www.epa.gov/ttn/atw/hlthef/styrene.html.

Valero. 2011. *MSDS: Asphalt-Oxidized Roofing*. San Antonio, Texas, April 27.

World Nuclear Association. 2015. "World Uranium Mining Production." *World Nuclear Association*. May 22. Accessed July 31, 2015. http://www.world-nuclear.org/info/Nuclear-Fuel-Cycle/Mining-of-Uranium/World-Uranium-Mining-Production.

13

Plumbing, Electrical, and Mechanical Systems

After the skeleton, environment enclosure, and exterior adornments, a building begins to function, take on a life of its own in the form of electrical, plumbing, and mechanical systems without which a building would be a mere shell. Comfort is the venue!

Historically, plumbing dates back to the lead aqueducts of Rome in 30 BC and 220 AD.

> An empire brought down by one of its signature innovations, the aqueduct—it is a theory that has stuck with the public, although experts have long been skeptical of its merits. It turns out that the theory was half-right: In a new study in the *Proceedings of the National Academy of Sciences* ... a group of French and British researchers report that the tap water in ancient Rome was indeed contaminated with lead, with levels up to 100 times higher than those found in local spring water at the time. But while Roman tap water might not have passed modern-day standards, it's almost certain that the contamination wasn't extensive enough to be responsible for the collapse of Roman civilization. (Walsh 2014)

Although the majority of people living in larger towns and cities in the United States had electricity by 1930, rural America didn't have electricity until 1939.

> Some of the older electric cables have a black canvas-like covering, as opposed to plastic, and some has a metal jacket. Some remain functional to this day—as long as the cables haven't been damaged, and the insulation hasn't become so brittle that it has flaked off. Even knob-and-tube wiring with ceramic insulators are still functional in old houses. (Gibson 2016)

Without electricity, air conditioning (AC) would not have been possible.

> Although they were introduced in urban America in 1931, window air conditioners were enjoyed only by the wealthy. The large cooling systems cost between $10,000 and $50,000 which is equivalent to $120,000 to $600,000 today. (Green 2015)

Well, we have come a long way! Plumbing has transitioned from lead to plastic. Electricity transmission has transitioned from knob-and-tube wiring to plastic wiring. And AC has transitioned from expensive and inefficient to affordable and insulated central HVAC systems. Modern technology has given us a multitude of options. Each has an Achilles' heel.

Plumbing: Water and Sewage/Wastewater Pipes

In ancient times, primitive systems to provide clean water and to remove waste—in densely populated civilizations—were made of wood, clay, bamboo, and stone. Then, around 500 BC, the Romans gave birth to aqueducts on a grand scale that ultimately, prior to the fall of the Roman Empire, provided clean water to approximately 1,000,000 inhabitants. The Roman water/wastewater system was to become one of the greatest achievements post Christianity. Aqueducts were massive systems of above ground and underground water conveyances that were built of stone, brick, and volcanic cement. Water was conveyed to and/or through the aqueduct system by gravity and inverted pressurized siphons. At the end of the line were enormous holding tanks, or cisterns. These cisterns once filled were capable of supplying pressurized water to the populations through a vast water surface distribution system of lead pipes. As lead was used as a sweetener in the wines consumed by the upper class and the general population drank lead-containing potable water, some speculation abounds that lead poisoning may have contributed to the fall of the Roman Empire. Beyond the water supply systems, sewer mains, made of hewn stone, carried wastewater and sewage into the Tiber River. Subsequently, wastewater and sewage was universally dumped into the local waterways and continued to be a common practice for many years thereafter—well into the 1800s.

Where there were no waterways, civilization reverted to chamber pots, cesspools, containerized vase port-a-potties, street dumping, sewage ditches, conveyance to crops and rivers, and waste pools beneath castles. A point of interest, in 1598, the British Royal Court issued a proclamation:

> Let no one, whoever he may be, before, at, or after meals, early or late, shall foul the staircases, corridors, or closets with urine or other filth.

These systems, if one could call them systems, went on for over 1000 years before poor sanitation and dirty drinking water were recognized to be the cause of water-borne diseases such as cholera and typhoid. Progress was slow, but progress was a must. In Chicago, the first sewers were hollowed-out logs and they drained by gravity into the Chicago River. Due to a heavy rain storm in 1885, all the sewage that had been dumped into the Chicago River backed up, went upstream, contaminating Lake Michigan's drinking water. This resulted in a typhoid and cholera outbreak that killed 11%–13% of the city's population. Efforts were taken to prevent such backups, continuing to dump the sewage in the Chicago River. Sewage treatment facilities were yet to be developed until the 1900s. It may be surprising to many that U.S. sewage treatment facilities were limited to just a few large cities in 1940. See Table 13.1.

The British National Public Health Act of 1848 was the first plumbing code to set the stage for Europe and the rest of the world. Eighty years later, in 1928, the United States published its first plumbing code. Although the health considerations of proper water treatment, sewage and city waste treatment facility advances were not accomplished until later in time.

Plumbing comes from the Latin term "plumbum" meaning lead. The term plumber arose from the skilled lead Roman artisans who installed and repaired—roofs and gutters, water and sewage pipes. In the United States, lead was used in potable water supply systems up until the 1930s, and lead solder was used in copper/brass plumbing joints until it was banned in 1986. And to this day, lead is still found in plumbing from street mains and some of the older buildings. See Figure 13.1.

Other countries marched to the beat of their own drummer. In the United Kingdom, lead plumbing was installed in buildings up until 1970. Thus, buildings constructed in Britain before 1970 should be suspect (Drinking Water Inspectorate 2010).

Most present day modern water supply systems are comprised of copper, PVC, and cross-linked polyethylene (PEX), and sewage systems are comprised of galvanized steel, cast-iron, and PVC. Plumbing fixtures/couplings are predominantly brass and sewer gas vents are galvanized steel, cast-iron, PVC, and lead.

TABLE 13.1

Estimated Percent of Sewage That was Treated by Large U.S. Population Centers

City	Percent of City's Population's Sewage That Was Being Treated	Population in 1940
Cleveland	85	1,200,000
Chicago	70	4,400,000
New York	25	8,100,000
Philadelphia	15	2,000,000
Los Angeles	5	1,300,000
Detroit	0	1,600,000
Boston	0	2,000,000
St. Louis	0	950,000

Source: Adapted from Editorial in *Sewage Works Journal*, by F.W. Mohlman, Vol. 12, No. 1, 1940.

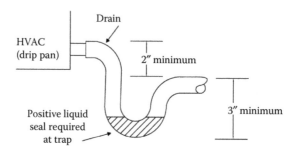

FIGURE 13.1
Lead fixtures in water supply lines.

Metal Water and Wastewater Pipes

From the 1800s until the 1960s, water and wastewater pipes have historically been metal. Although many of them are being or have been phased
out, metal water and sewer pipes are durable yet problematic—for different
reasons.

Galvanized steel pipe was commonly used in plumbing from the 1800s
to the mid- to late 1900s. Although an inexpensive plumbing component,
old zinc-coated steel corrodes, rusting from the inside out. Accumulated
rust reduces the pipe diameter and water pressure. Rust bestows a brown-
ish color to potable water. And ultimately, the pipe walls rust, and the pipe
leaks. This topic is noteworthy in that potable water and indoor air quality
may be impacted by leaking supply pipes, and noxious gases may escape
through leaking sewer pipes. This is one of those out-of-sight, out-of-mind
scenarios!

Be forewarned! During renovation/restoration projects, brown colored
potable water and sewer odors emanating from the foundation are symptoms of plumbing leaks—a topic to investigate.

It may surprise some that copper water supply pipe did not become popular until the 1960s. The more expensive, noncorrosive copper replaced the
inexpensive, corrosive galvanized steel. Yet, wherein the newer copper pipe
and the older galvanized steel are joined, the point of contact will result in
corrosion. Dissimilar metals make poor bedfellows!

Thick-walled, high carbon-content cast iron pipes date back to the early
1800s when they were initially used for water distribution mains. In the 1890s,
cast iron sewer pipes came to be the go-to choice for plumbing. Although
durable and long lasting (life expectancy: 75–100 years), cast iron was and is
expensive to purchase and install, and shifting soils, tree root damage can
break the joints and/or damage the pipes. Cast iron sewer pipes can be found
in structures as recent as the early 1970s.

Corrosion resistant stainless steel pipes are rare, used mostly in corrosive environments. They are used predominantly in salt water, marine

environments, and in industrial facilities. Stainless steel is a steel alloy with a minimum of 10% metallic chromium.

Metal pipe emissions are not likely, and metal is not flammable. The devil is within the toxins—toxins in potable water and leaking sewer pipes and air vents, not product emissions.

Plastic Pipes

With advances made in the 1960s, plastic pipes began to replace metal pipes and became commonplace within the plumbing industry in the mid-1980s. Plastics are corrosion resistant, durable, inexpensive, and light weight—the "World of Plastics" began to replace metal pipes.

In the 1930s, PVC water and wastewater pipe was introduced by the Germans. Many of the earlier PVC pipes installed in central Germany are allegedly still in use to this day. After World War II, PVC water and wastewater pipe began its rise in popularity within the United States, becoming the industry standard, replacing metal pipes. Yet, PVC is not without its problems. It becomes brittle in freezing temperatures; it will warp in hot water; and extreme temperatures may result in the release of hydrogen chloride. PVC has a maximum service temperature of 140°F (60°C) (Georg Fischer Harvel LLC 2015). Subsequently, technology developed and introduced chlorinated-PVC (CPVC). CPVC is less brittle and has a maximum service temperature of 200°F (93°C) (Georg Fischer Harvel LLC 2015). Although slightly more expensive, CPVC became "the preferred alternative" to PVC pipe by the mid-1980s. Although emissions in cool to temperate water temperatures are unlikely, elevated temperatures and above-ground PVC pipes are likely to degrade and become brittle due to solar UV light. The combustion products of PVC include carbon dioxide, carbon monoxide, and hydrogen chloride, and CPVC may produce small amounts of chloroform and carbon tetrachloride as well (Georg Fischer Harvel 2015).

An alternative to CPVC wastewater pipe, acrylonitrile butadiene styrene (ABS) pipe is less rigid than CPVC, less likely to crack, and it is more resilient to freezing temperatures. Yet, ABS pipe degrades more quickly in sunlight; it and may warp in uneven heat conditions (e.g., boiling hot water followed by cold water down the drain); and it is less resistant to chemicals than PVC/CPVC pipes (Plastics International 2015). Emissions and/or leaching from ABS are unlikely. At the onset of a fire, however, styrene is released—providing fuel for the fire. The combustion products of ABS include carbon dioxide, carbon monoxide, and hydrogen cyanide.

PEX was developed in the 1960s and introduced in the United States in the 1980s—about the same time as CPVC was recognized. PEX tubing is used for water supply only, not for wastewater. Presently, flexible PEX tubing is replacing the more rigid CPVC. PEX tubing is freeze damage resistant and can withstand temperatures up to 200°F (93°C). PEX tubing is made from high-density polyethylene (HDPE) granules that are melted, cross-linked extruded.

Cross-linking improves the chemical and temperature performance of a polymer, giving it flexibility and added strength. There are three different processes.

> PEX Type A, the "peroxide process," is alleged to have residual by-products in the final product. One of these alleged products is methyl tertiary butyl ether (MTBE). Although the levels recorded to be leaching into potable water are low, non-problematic (according to the U.S. EPA), some people claim that MTBE renders the water undrinkable due to its offensive taste and odor. Allegations have yet to be proven!
>
> PEX Type B, the "moisture cure process," involves cross-linking in the presence of water and a catalyst. There are "no reports" of associated foul drinking water.
>
> PEX Type C, the "electronic irradiation process," involves cross-linking by exposing the extruded tubing to electron beams. There are "no reports" of associated foul drinking water.

Leaching of chemicals from PEX Type A process residue into the water supply is possible, but is not likely to be a health hazard. Product emissions are unlikely in all other types of PEX, and combustion products include carbon dioxide, carbon monoxide, aldehydes, and VOCs.

Electricals

Thomas Edison exhibited electric lighting in a few select New Jersey residences on New Year's Eve—December 31, 1879. The "wonder of electricity" has since evolved. The earliest "knob-and-tube" electrical systems have been used in buildings from the 1890s to the 1970s when thermal insulation in attics became popular. Subsequently, the hot splices with glowing connections or overheated conductors could easily ignite an insulation fire. Knob-and-tube systems—hot and neutral, rubberized cloth wiring with ceramic insulators—are still found in older homes to this day.

Beginning in Edison's time, the original residential wiring systems that used wire insulation that was comprised of cotton-wrapped gum-rubber. Vulcanized rubber contains sulfur which has a corrosive effect on copper. So, the copper was tinned to protect it from the sulfur. Furthermore, rubber becomes brittle and cracked with age—losing its insulating qualities, posing a potential fire hazard. Rubber was indeed problematic! So, during the 1950s, thermoplastics (e.g., PVC) began to replace rubber.

From the mid-1960s through the 1970s, aluminum wire became popular due to the rising costs of copper. Yet, aluminum wire was not without its problems—requiring special connections and anti-oxidation treatment of exposed joints. Improper design and installation could and did result in fires. Subsequently, the more costly copper wiring remains the electric conductor of choice.

Conductor Insulation

With the introduction of plastics, most modern electric wire/cable insulation has found its home in "thermoplastics"—some with paper wrap around the individual conductors. PVC insulation is used in the lion's share of commercial and residential buildings whereas rubber-like synthetic polymer insulation is used on industrial cables and on underground cables.

On the other hand, some industrial and hot environment conductor insulation is nonorganic—compressed minerals. The compressed minerals include, but are not limited to, mica, magnesium oxide, and asbestos. Although it is commonly accepted that asbestos was used to insulate electrical equipment and wiring prior to 1980, it is not clear as to the extent to which it was used or whether it is still used in products sent from China. It has been the author's observations that pre-World War II asbestos was used mostly to insulate primary electrical distribution lines on military bases. Older buildings—particularly industrial and military structures—may still contain asbestos in the cables and wiring.

Be forewarned! Asbestos has been encountered in electrical insulation within older buildings and may potentially be encountered in newer structures—particularly in industry, military, and hot environment facilities.

Thermoplastic PVC-coated cable emissions are unlikely to impact indoor air quality in a newly constructed building—unless the size of the cable is not adequate and/or the PVC-coated is exposed to extreme temperatures such as may occur in poorly insulated attics and/or in contact with extreme temperatures (e.g., ovens). Under these conditions, hydrogen chloride emissions are possible. The combustion products of PVC are carbon dioxide, carbon monoxide, and hydrogen chloride.

Conduit/Raceways

From the grooved wood molding (late 1800s) to flexible metal conduit (1902) to gas pipe conduit (1910s) to thermoplastic Bakelite conduit (late 1920s) to thermoplastic PVC conduit (1960s) to nonmetallic flexible conduit (1981) to nonmetallic tubing (1983)—the history of conduit in a nutshell! Conduit is a system used for protecting and routing electrical wiring.

Flexible conduit is still used to this day for vibration isolation (e.g., parking garages) and for accommodating tight, closed quarter areas (e.g., can lights and attic vents). The more common type (e.g., flexible metallic conduit) is made of zinc galvanized metal or a more expensive stainless steel. Some metallic conduit is covered with a soft thermoplastic (e.g., plasticized PVC) for waterproofing; some flex conduit is covered with a thermoplastic with cross-woven galvanized steel; and some flex conduit is comprised of rolled stainless steel, galvanized steel, and copper or a combination thereof. Aluminum may be alloyed with one or more of these metals.

Rigid nonmetal conduit, however, is the most common conduit used in most buildings today. It is generally rigid PVC (not likely to contain plasticizers)—gray in color and light weight.

Rigid metal conduit is zinc galvanized steel or aluminum. It is heavier and thicker than other forms of conduit. It can be waterproofed with a thermoplastic such as PVC.

Thermoplastic PVC-coated conduit emissions are unlikely to impact indoor air quality in a newly constructed building—unless exposed to excessive heat. Examples include, but are not limited to, exposed areas in poorly insulated attics, excessive heat around water heaters, and contact with floor heater systems and radiators. Hydrogen chloride emissions from PVC products can reasonably be anticipated where exposed to excessive temperatures.

Heating, Ventilation, and Air Conditioning

From the wood burning cast iron Franklin stove (1742) to coal fired cast iron boilers with heat delivered by steam through radiators and furnaces and with convective heat delivered through ducts (1885) to gas/oil fired boilers with fan forced air furnaces (about 1935), heating preceded electricity dependent air conditioning.

Willis Carrier invented the first mechanically blown air over cooling coils (1902) which was used in industry to thermally manage product quality. Then, the first residential AC unit, 7-feet high, 6-feet wide, and 20 feet long, was installed in the Charles Gates Minneapolis mansion (1914), and it wasn't until much later that the window AC units were invented (1931)—available only to the wealthy. In 1947, a British scholar wrote, "The greatest contribution to civilization in this century may well be air-conditioning—and America leads the way" (Steinmetz 2010). Post World War II, window units became accessible. Over 1 million units were sold in 1953 alone.

It was not until the 1970s that central HVAC became available in all buildings—residential, commercial, institutional, and industrial. By 1980, the United States which had only 5% of the world's population consumed more air-conditioning than all the other worldwide countries combined (Steinmetz 2010). In the interest of energy efficiency, there have been, since the earlier units, many changes in the design and mechanical operation of HVAC systems.

Beyond the mechanical components of the various HVAC units, nonmetal, nonmechanical components and thermal/acoustical insulation is important to the efficiency of each unit. But let us not stop here! While thermal/acoustical insulation is an important component in HVAC systems and health hazards, poor design and maintenance can greatly impact the indoor air quality as well. The water condensate and evaporative spray from the cooling coils/drip pan can be sources of biological growth. A poorly desig-nated "p" trap drain and associated vent can be a source of water damage to areas in and around the HVAC unit. Sewer gases, irritating/toxic chemicals, and/or wildlife droppings (e.g., raccoons and rodents) may be sucked into the HVAC. And refrigerant gases and heating fuels can pose a separate set of health hazards. Despite our reliance on HVAC systems to supply clean air to the occupants of a building, we often overlooked likely sources of multiple health hazards lurking "within our air distribution systems."

HVAC Thermal/Acoustical Insulation

As the purpose of HVAC systems is the controlled conveyance of condi-tioned air, it behooves one to maintain air temperatures from the air han-dling unit (AHU) through the supply plenum (i.e., air distribution box) to the air supply ducts and into the occupied spaces. Yet, temperature loss is inevitable. Generally accepted thermal loss from the AHU to the air sup-ply vent(s) is up to 20°F in insulated residential units, while poor insula-tion, partial insulation, and no insulation can result in considerable loss in efficiency and deterioration of the conditioned air—well in excess of a 20°F.

After World War II, fiberglass insulation was used for HVAC thermal/sound dampening. Today, most, if not all, AHU fan housing units are internally insu-lated with semirigid fiberglass insulation. The air plenum insulation may be semirigid interior mats or rigid panels. The air supply duct may be insulated with loose fill, flexible exterior wrap, or it may be comprised of rigid duct board panels. The air return plenum may be the entire ceiling space—predominantly in large commercial buildings. In smaller HVAC units, the return air enters a plenum box into a return air duct. Returns are rarely insulated!

Manufacturer disclosures of product components are often illusive. Consensus is, "Nothing here. Move along." So, it is not unusual for an indoor air quality consultant to overlook that which is out of plain sight and off the radar. Does it not deserve a second look? Let us delve a little deeper!

All manufacturer SDSs for fiberglass plenum/duct insulation do affirm, at a minimum, up to 30% "thermoset resin." Some disclose the fiberglass is impregnated with up to 30% "polymerized" or "cured" phenol and UF resin. Some disclose a "proprietary binder." Most, if not all, claim "formaldehyde free" fiberglass plenum/duct insulation. Yet, polymerized UF and to a lesser extent phenol formaldehyde will release formaldehyde in the presence of moisture. This point was made in a Johns Manville SDS that states, "Trace amounts of formaldehyde may be released when contacted with moisture, including humidity." This release is most prevalent in conditions of high heat and humidity. Emissions of regulated irritating formaldehyde are pos-sible—particularly where interior HVAC insulation is wet!

Many manufacturers' SDSs for fiberglass plenum/duct insulation disclose up to 5% of the fire-retardant antimony oxide. Antimony and antimony com-pounds are regulated toxins that can cause eye, skin, and mucous membrane irritation. Although there are more severe symptoms associated with most antimony compounds (e.g., antimony trioxide), the toxicity of antimony oxide is unclear. The release of antimony oxide powder from damaged insulation and its health effects on building occupants are unclear.

Some manufacturers of fiberglass plenum/duct insulation that repels water disclose that their product contains up to 5% of a nonregulated skin and eye irritant—polyethylene terephthalate (PET), a woven thermoplastic polymer which is likely to contain a plasticizer. The woven polymer is, in turn, likely to be comprised of recycled PET which has been demonstrated to contain an unbound plasticizer [e.g., di(2-ethylhexyl) adipate (DEHA)].

> Recent studies have shown that recycling bottles made of PET can in fact be dangerous. Recycled PET beverage bottles have been found to break down over time and leach into the liquid. The toxin DEHA appeared in a water sample from recycled bottles. DEHA has been shown to cause liver problems, possible reproductive difficulties, and is suspected to cause cancer in humans (Kovacs 2016).

Some products contain other woven polymers (e.g., latex rubber and acrylic-based polymers) may or may not contain plasticizers which are typically not reported by the manufacturers. In conclusion, unbound plasticizers in woven polymers are potentially released within the airstream in an HVAC system, contribute to eye, skin, and respiratory tract irritation. Emissions of nonregulated irritating plasticizers from woven polymers are possible.

Plenums are often exposed to high humidity and condensate spray from the cooling coils. In many commercial units, the mixing plenum and sometimes the AHU insulation is waterproofed with an asphalt/polymer coating. Due to the elevated temperatures inside an HVAC system during the colder seasons, asphalt emissions are unlikely.

Guilty by Association: Refrigerant Gases and Heating Fuels

For builders and consultants involved in renovation projects, it behooves one to become familiar with environmental/toxic gas sources that may impact planning decisions and post construction investigations. Although not building materials, refrigerant gases and heating fuels are associated with most HVAC systems.

Refrigerant Gases

Contained within the cooling coils, refrigerant is what makes AC possible. Since inception of mechanical AC in 1902, the most commonly used AC refrigerant was R-22 Freon (i.e., chlorodifluoromethane). Then, in 1987, an international environmental agreement (i.e., the Montreal Protocol) established requirements that began a worldwide phase-out of ozone-depleting chlorinated hydrocarbons (e.g., R-22 Freon) (EPA 2014).

The transition away from ozone-depleting R-22 Freon to systems that rely on replacement refrigerants (e.g., fluorinated hydrocarbons) has required redesign of heat pump and air conditioning systems. New systems incorporate compressors and other components specifically designed for use with specific replacement refrigerants.

The Montreal Protocol requires the U.S. to reduce its consumption of HCFCs by 99.5% below the U.S. baseline. Refrigerant that has been recovered and recycled/reclaimed will be allowed beyond 2020 to service existing systems, but chemical manufacturers will no longer be able to produce R-22 to service existing air conditioners and heat pumps.

Acceptable transition/replacement refrigerants are non-ozone depleting fluorinated hydrocarbons (e.g., R-410A and 407C) and ammonia. As all chlorinated Freon and fluorinated Freon-substitutes are odorless, nonflammable, and nontoxic, leaks are not likely to be detected, and, short of displacing oxygen in the air, they are not likely to pose a health hazard to occupants. However, due to the United Nations IPCC Listing of chlorinated hydrocarbons and the European Union inclusion of fluorinated hydrocarbons as greenhouse gases, ammonia—which has not been designated a greenhouse gas—has become an acceptable alternative for use as a refrigerant in HVAC cooling systems.

Ammonia was used as a refrigerant as early as the middle 1800s for the artificial production of ice and later as a food refrigerant. It has not until recently been used in HVAC cooling systems. Its use, however, has been restricted to water chillers and large-scale air cooling systems. In Europe, ammonia refrigeration systems are used to air condition hospitals, public buildings, airports, and hotels. In North America, ammonia refrigerants are being installed in institutional, commercial, industrial, and large office buildings. Ammonia has a pungent, irritating odor with an odor threshold ranging from 0.043 to 53 ppm (AIHA 2013); it is toxic (ACGIH 15 minute limit: 35 ppm); it is highly corrosive to copper; and it is nonflammable. Those who promote ammonia as a refrigerant claim, "Ammonia's distinctive (odor) is detectable at concentrations well below those considered to be dangerous, and ammonia is lighter than air, so if any does leak, it will rise and dissipate in the atmosphere" (Goodway 2009). As they are generally located outdoors, chillers and large-scale air cooling systems are unlikely to pose an indoor exposure hazard—unless winds convey leaking ammonia gas, if there is a leak, into the occupied interior. For example, leaking ammonia can be drawn in with the makeup air on an associated HVAC system.

Heating Fuels

Natural gas, propane, and oil are alternatively used to generate heat in an HVAC furnace—as opposed to electric furnaces and heat pumps. Natural gas and propane are most common whereas fuel oil is occasionally found in older structures in the North American climates. Only about 8% of U.S. homes use oil heat today. Most are in the Northeastern United States and were built back in the day when oil was the cheapest way to keep toasty through the long winters. As the price of oil has surpassed that of natural gas by up two to three times, many communities have since put gas lines into neighborhoods that didn't have them in the past, opening the door for homeowners to switch out old inefficient oil furnaces for more efficient gas units. In today's climate of environmental and climate change, the inefficiency of electricity production is noteworthy! According to the American Gas Association, 70% of the energy used to generate electricity is lost during the process of generating, transmitting, and distributing the electricity. Yet, heating fuels can pose hidden hazards!

Incomplete combustion of heating fuels will result carbon monoxide—an odorless, colorless, toxic gas. If the furnace exhaust vent leaks or the gases are released in the vicinity of an air intake or window, hazardous levels of carbon monoxide gas may enter the air supply and be distributed to the occupied spaces within a building. There are "no warnings, only symptoms." Low-level carbon monoxide exposures may result in shortness of breath, mild nausea, and headaches. Higher level exposures progress to more severe nausea and headaches, dizziness, and light headedness. In extreme exposures, carbon monoxide may cause death. Awareness of HVAC furnace discrepancies goes a long way toward prevention!

Available air to a firebox is vital! If the HVAC furnace is in a tight building with little or no outdoor air to displace the air burned and exhausted—air taken from an air tight building and exhausted outdoor with limited outside air to replace that which has been exhausted. Subsequently, the firebox will be starved for air resulting in incomplete combustion. To avoid this, outdoor makeup air is piped directly into the burner. A design change can go a long way in avoiding the creation of toxic carbon monoxide backing up into the HVAC system—posing a deadly health hazard to the occupants.

Observable fuel-fired furnace vent/exhaust discrepancies include, but are not limited to, damaged vents (e.g., punctures and rust), poor connections (e.g., loose screws), poor exhaust locations (e.g., side of a building, around open windows), and poor slopping of vents (e.g., straight horizontal). Ninety degree vent elbows are often heat damaged at the transition point(s). Soot buildup can result in vent blockage, and debris/dirt buildup can block the air intake and exhaust vents.

Be forewarned! Damage to and improper installation of gas-fired vents and exhausts may occur during new construction.

Condensate Water Drains

Condensate water drainage is an often misunderstood AHU component whereby improper design and maintenance can significantly impact indoor air quality—noxious odors and irritating gases. How is this possible?

Flaws in the condensate water drain can lead to positive or negative air flow into and/or out of an air handler. A properly functioning condensate drain discharges water from the cooling coil drain pan into a drain line which has a U-trap that prevents water and gases from returning to the AHU. However, improperly functioning drains can have disastrous consequences.

- "No trap, no water in a trap," will result in negative pressure and air flow back into the AHU. The airstream has sufficient velocity to launch water droplets at the base of the cooling coil which, in turn, can propel moisture into other parts of the system. The resultant aerosol mist can be carried through the ducts and into the conditioned space, possibly causing bacterial growth and transmission. All too often overlooked problem with air inflow is the source of air. Drain lines that flow into wastewater lines may potentially draw rising sewer gases into the AHU. If they flow into an outdoor area, stored or sprayed toxins may potentially be drawn into the HVAC.
- A "shallow U-trap" can act as a "no water" trap as water may be depleted from the trap due to AHU start-up negative pressure, see Figure 13.1.
- A "deep U-trap" can prevent water drainage, causing the condensate to backup and overflow into the AHU.

In colder months when the heater is operating—there is no water in the U-trap. When the HVAC has been turned off for an extended period of time—there is no water in the U-trap. In these scenarios, the "no trap, no water in the trap" negative pressure can result in long-term collection and delivery of noxious gases into the occupied spaces.

Another discrepancy is that of the drain clean-out pipe. There is often confusion as to whether to leave the cap on or take it off. Well, this is a clean-out pipe, not a vent pipe. The cap needs to remain! If the cap is removed—depending on the clean-out pipes location prior to or after the U-trap—this can provide a means whereby the air within the vicinity of the drain outlet will be sucked into the AHU, delivered to the building occupants. In an attic, any gases, irritants, or other contaminants (e.g., rodent droppings) can be distributed from the attic to the building occupants. In a mechanical room that has a sewage drain, the sewer gases can be distributed to the building occupants. In outdoor roof units, exhaust vent gases and/or air contaminants from work being performed on the roof (e.g., asphalting) can be distributed to the building occupants.

Beyond the U-trap and drain clean-out pipe, improper gravity-feed drain sloping could be problematic as well. Each condensate drain component contributes to the whole, and each can contribute to poor indoor air quality. Although not a building material, condensate drainage systems are a likely conduit whereby emissions from building materials can be distributed from an unoccupied space into occupied spaces.

Summary

The plumbing, electrical, and mechanical building material contributions to poor indoor air quality are illusive—poorly understood and seldom acknowledged. Whereas product emissions are minimal, potable water contaminants, toxic gas leaks, water damage and mold growth, generation of toxic gases, and other sources of toxic emission are often overlooked, and deserve special comment.

Although unlikely sources of irritant/toxic emissions, plumbing lines may release irritant/toxic gases from deteriorating and/or leaking sewage pipes and vents. Damaged in-foundation sewage pipes—old and new—may leak sewer gases into the enclosed interior. The in-wall pipes may be inadvertently penetrated during construction and/or punctured after construction. Sewage pipes and vents are potential sources of irritant/toxic gas releases. The list of possibilities goes on! Then, the potable water pipes may contain lead. Punctured sewage and potable water pipes in wall cavities may result in water damage and mold growth—out-of-sight, detected only by odor. Older potable water pipes and service mains are potential sources of lead—a health hazard to the occupants drinking the water.

PVC coated electrical wiring and PVC conduit may emit irritant/toxic levels of hydrogen chloride when exposed to extreme temperatures and/or fires. Properly locating the conduit and wiring and proper sizing the wire go a long way to avoid thermal loading and fires.

Most HVAC systems have internal insulation which may emit and/or release irritants. Fiberglass insulation may emit formaldehyde and/or may be dislodged and released into the breathing air. New fiberglass emits formaldehyde, and old fiberglass when wet emits formaldehyde. Woven polymer mats may release irritating plasticizers as well.

Considerable attention has also been given to HVAC features—those that can contribute to poor air quality. Refrigerants in the cooling system can leak, releasing Freon, Freon-substitutes, and/or ammonia into the air stream. Incomplete combustion in a heating unit (e.g., natural gas) can introduce deadly levels of carbon monoxide into the distribution system.

Poor indoor air quality can also be associated with poorly designed and/or maintained condensate water drains. Poor drain design may result in (1) water backup into the HVAC unit (e.g., growth of mold and release of spores); (2) sewer gases; and (3) HVAC vicinity of hazardous contaminants (e.g., irritant/toxic gases).

In brief, poorly functioning, poorly maintained, leaking plumbing, and mechanical systems can contribute indirectly to illusive odors and extensive indoor air quality complaints in addition to building material emissions. Plumbing and mechanical systems are conveyances of the life's blood of a building. They are not to be taken lightly!

References

AIHA. 2013. *Odor Thresholds for Chemicals with Established Occupational Health Standards.* Akron, Ohio: American Industrial Hygiene Association.

Drinking Water Inspectorate. 2010. "Lead in Drinking Water." *Drinking Water Inspectorate.* January. Accessed March 24, 2016. http://dwi.defra.gov.uk/consumers/advice-leaflets/lead.pdf.

EPA. 2014. "What You Should Know about Refrigerants When Purchasing or Repairing a Residential A/C System or Heat Pump." *Ozone Protection System-Regulatory Program.* April 23. Accessed September 3, 2015. http://www.epa.gov/ozone/title6/phaseout/22phaseout.html.

Georg Fischer Harvel. 2015. "MSDS Rigid PVC and CPVC." *GF Technology Center.* Accessed August 28, 2015. http://www.harvel.com/technical-support-center/safety/msds-material-safety-data-sheet.

Georg Fischer Harvel LLC. 2015. "Temperature Limitations." *GF.* Accessed August 27, 2015. http://www.harvel.com/technical-support-center/engineering-design-data/temperature-limitations.

Gibson, Scott. 2016. "Electrical Issues in Old Houses." *Old House Web.* Accessed March 4, 2016. http://www.oldhouseweb.com/how-to-advice/electrical-issues-in-old-houses.shtml.

Goodway. 2009. "Ammonia as a Refrigerant: Pros and Cons." *Goodway (blog).* August 12. Accessed September 4, 2015. http://www.goodway.com/hvac-blog/2009/08/ammonia-as-a-refrigerant-pros-and-cons.

Green, Amanda. 2015. "A Brief History of Air Conditioning." *Popular Mechanics.* January 1. Accessed March 4, 2016. http://www.popularmechanics.com/home/how-to/a7951/a-brief-history-of-air-conditioning-10720229.

Kovacs, Betty. 2016. "Plastics." *Medicine Net.* Accessed March 15, 2016. http://www.medicinenet.com/plastic/page2.htm.

Plastics International. 2015. "Chemical Resistance Chart." *Plastics International.* Accessed August 28, 2015. http://www.plasticsintl.com/plastics_chemical_resistence_chart.html.

Steinmetz, Katy. 2010. "Brief History—Air Conditioning." *Time.* July 2. Accessed September 2015. http://content.time.com/time/nation/article/0,8599,2003081,00.html.

Walsh, Brian. 2014. "Lead Didn't Bring Down Ancient Rome—But It's Still a Modern Menace." *Time.* April 23. Accessed March 3, 2016. http://time.com/73555/ancient-rome-lead.

14

Thermal/Acoustical Insulation and Interior Wall/Ceiling Materials

Many Europeans think North Americans are spoiled as they demand stricter indoor "temperature tolerances." Short of living in a rock-insulated cave, man has been forced to technologically evolve and develop several approaches to artificially insulated manmade structures.

> Although cork was the first to be used to insulate ancient Roman foot wear, it was not until the industrial age that it was used to insulate ice houses. Later, cork was to be used to insulate water pipes.
>
> The predecessor manmade mineral wool, the first actual all-around building insulation, was Pele's hair—a molten lava glass fiber—which was encountered around volcanos, most notably the Hawaiian Islands. The fine, light fibers were crafted in the volcanoes and blown into the air, traveling miles until settling on remote areas as fluffy fibers which were used by the natives to insulate their huts. Later, manmade "glass fiber" made its debut.
>
> Then came corrugated asbestos paper and cording to insulate hot pipes and other hot building materials (e.g., gaskets on boiler doors). Also asbestos fibers were used in magnesia as a binder. Calcium silicate replaced it in the mid-fifties.

Interior wall and ceiling materials are often considered insulation material albeit to a less degree. Inside buildings, the demising walls and ceilings are acoustical barriers, and the exterior walls and ceilings sometimes serve as a minimal thermal insulation material. These materials have also undergone a transformation over time.

> Plaster is age old! It dates back to the Egyptian Pharaohs. Heated, dehydrated gypsum was mixed with water and applied as a plaster to strips of reed lath.
>
> In the 19th century, tin ceilings were introduced. They were originally rolled, later stamped in various forms. Its popularity was short lived, ending during World War I.
>
> Upon its fall from grace of tin ceilings, gypsum board found its way into the world of architecture. It has since been the dominant wall and ceiling material in construction.

Man has progressed from poorly functional, natural insulation and wall/ceiling materials to highly functional, manmade products. Many of today's products mimic the old. The old was less durable with limited resources. The new is more durable with greater resources. Herein, we discuss the components of most of these building materials from the twentieth century forward.

Thermal/Acoustical Building Insulation

About 40%–70% of the energy usage in homes is for heating and air conditioning. A conditioned building without insulation—to act as a barrier to heat gain and/or heat loss—will strain to maintain and use up to 70% of the energy usage in the building. No insulation and/or minimal insulation is a stark contrast to good insulation. For example, a small 1200 square foot residence with minimal insulation can have an energy usage bill as much as three times that of well-insulated 3400 square foot residence. Then, again, the same 3400 square foot residence without attic insulation can have an energy usage bill three times higher than the 1200 square foot residence. These examples are based on experience—hot summer conditions in Texas!

Well designed and installed building insulation can (1) save money on energy bills; (2) reduce reliance on heating and cooling systems; and (3) improve thermal comfort. Yet, there remains a potential for some insulation to negatively impact the indoor air quality. From natural fibers to cellulose to minerals-and-metals to fiberglass to the world of plastics, some advances in technology bring unintended consequences! See Table 14.1.

Natural Fibers and Cellulose Insulation

Natural fibers include, but are not limited to, cotton, sheep's wool, straw, and hemp. Cotton is generally a recycled material (e.g., old blue jeans) and generally treated with a fire retardant (e.g., boric acid and ammonium sulfate). Water-resistant sheep's wool is generally moth proofed and treated with a fire retardant (e.g., disodium octaborate). Straw and the woody component of hemp are also treated with a fire retardant (e.g., HempCrete®: lime is resistant to fire, insects, and mold).

Cellulose, the main component of tree and plant cell walls, is one of the most abundant organic compounds on Earth. The cellulose content of cotton is about 80%–90% (Hegde et al. 2004). The cellulose content of hemp bark is about 53%–75% (Van der Werf n.d.), and the cellulose content of wood is 40%–50% (Pettersen 1984). Foremost among its many uses, cellulose is processed to make paper which is used not only in the print media but also in thermal insulation—most of which is recycled paper. As a blown-in material

TABLE 14.1

Insulation R-Values, Irritant and Toxic Health Effects

Type of Insulation	R-Value per inch	Irritants	Toxins	Comments
Natural fiber insulation				
Cotton (e.g., batts and rolls)	3.4–3.7	–	–	Recycled cotton
Sheep's wool (e.g., rolls and loose fill)	3.25–3.5	–	–	–
Straw (e.g., panels and bales)	2.4–3.0	–	–	–
Hemp	1.4–2.0	–	–	–
Blanket insulation (e.g., batts and rolls)				
Fiberglass (high density)	3.6–5.0	X	Formaldehyde[a]	–
Fiberglass (medium density)	3.1–4.3	X	Formaldehyde[a]	–
Rock wool	3.0–3.85	–	–	Nonflammable
Cellulose	3.0–3.8	–	–	Fire retardants
Loose fill insulation				
Rock wool	3.0–3.83	–	–	Nonflammable
Cellulose	3.0–3.8	–	–	Fire retardants
Fiberglass	2.5–3.7	X	Formaldehyde[a]	–
Vermiculite	2.13–2.4	–	–	Possible asbestos
Rigid panel insulation				
Polyurethane (CFC/HCFC expanded)	6.25–8.0	X	Isocyanates[b]	Closed cell
Polyurethane (pentane expanded)	5.5–6.8	–	Isocyanates[b]	Closed cell
Polyisocyanurate (pentane expanded)	5.5–6.8	–	Isocyanates[b]	Closed cell
Phenolic	4.0–5.0	X	Formaldehyde[a]	95% closed cell
Polystyrene	3.85–5.0	–	–	Closed cell
Fiberglass	2.5		Formaldehyde[a]	–
Spray foam insulation				
Polyisocyanurate foam	4.3–8.3	–	Isocyanates[b]	Closed cell
Phenolic spray foam	4.8–7.0	–	Formaldehyde[a]	95% closed cell
Urea formaldehyde foam	4.0–4.6	X	Formaldehyde[a]	Open cell
Cementitious foam (inorganic)	3.9	–	–	Solid
Cutting edge (e.g., difficult to find insulation)				
Plastic fiber	3.8–4.3	–	–	Recycled polymers (e.g., plastic bottles)

[a] Possible formaldehyde emissions when new or when exposed to heat and moisture—UF and rigid panels are more likely to give up greater amounts of formaldehyde gas than others (e.g., loose fill fiberglass).

[b] Possible residual isocyanate emissions from new products.

or compressed rolls, cellulose is treated with a flame retardant and insect repellent (e.g., boric acid and borax).

In some instances, "cellulose insulation" is not always a reference to only cellulose. Recycled products are attaining greater importance—particularly within the Green Movement. For example, Cellulose Material Solutions' Ecocell Batt and Blanket components include greater than 40% newsprint, less than 15% jute, less than 15% cotton, and 15%–30% PET recycled plastic bottles.

Of the many fire retardants, the fine power-like borates pose a minimal health concern. Although toxic upon ingestion, boric acid, borate salts, and borax are mild eye, skin, and respiratory irritants. Boric acid has a low inhalation toxicity of greater than $500 \, \mu m^3$—equivalent to a thick dust cloud (National Pesticide Information Center 2015), and boron oxide is a dust-like, low toxicity regulated substance. Trace amounts of mineral oil and/or corn starch are also added to some of the natural fiber/cellulose insulation for "dust suppression."

Mineral Insulation

Inorganic, naturally occurring minerals—rock wool and vermiculite—are inherently fire and pesticide resistant. Another inorganic form of insulation is a cementitious foam—a recent addition to the marketplace.

Mineral wool, also referred to as rock wood and slag wool, insulation was first introduced in 1875 and has since remained popular until the 1950s. To this day, however, nonirritating mineral wool is a viable option to skin and respiratory irritating fiberglass insulation. Mineral wool is manufactured by heating a mixture of lava rock (e.g., basalt) and steel slag (e.g., silicon, aluminum, magnesium, and sulfur) to about 3000°F. The molten mix is spun and blown through a spinning screen that turns the mix to fine fiber. As a loose fill insulation, mineral wool is generally treated with an oil to keep the dust down. On the other hand, mineral wool rolls/batting require a binder (e.g., glue). In some instances, this binder may be, depending on the manufacturer, "cured urea extended phenolic formaldehyde." Wet batting could result in formaldehyde emissions.

Vermiculite is a naturally occurring mineral composed of shiny flakes, resembling mica. When heated to a high temperature, the shiny flakes expand eight to 30 times their original size. Expanded vermiculite is not only lightweight but fire-proof and inexpensive—ideal as a building insulation material. Thus, having been mined since the early 1900s, vermiculite became one of the first insulation materials to be used extensively in residential, office, and commercial buildings. Then, in the 1980s, the vermiculite that was born of Zonolite Mountain in Libby, Montana got bad press!

> Press releases estimated that at least 192 people, in the 2700 population town of Libby, had died, and another 375 people were diagnosed

of asbestos-related disease. Subsequently, a hailstorm of woeful tales and lawsuits spread from Libby miners to building owners who had Zonolite® insulation in their attics and/or wall cavities. With the insulation present in as many as 35 million American attics, legal repercussions for former Zonolite® insulation-owner W.R. Grace was once tremendous. In fact, the company was named in over 112,000 asbestos-related lawsuits in North America. In 2008, W.R. Grace settled a class action lawsuit filed against the company to the tune of $140 million. The payment for this claim is to be paid out over a 25-year period—going towards the cost of abatement, property-related damage, and general compensation.

Between 1919 and 1990, 70% of the world's supply of vermiculite came from the asbestos-contaminated Libby mine, according to the U.S. EPA, and Zonolite® insulation was first produced and sold in 1940 (The Mesothelioma Center 2015). However, not all vermiculite contains asbestos, and asbestos-containing vermiculite has not been banned in the United States. It is a fireproof, mold/pesticide resistant, low cost insulation, and there are no irritant/toxic VOC or formaldehyde emissions.

Be forewarned! Demolition and renovation of buildings built between 1940 and 1990 should be suspect of asbestos-containing vermiculite insulation in the attic and/or walls (e.g., loose fill in cement block cavities).

Cementitious foam, a cement-like metal insulation, is fireproof as well as pesticide and mold resistant. First introduced around 1990, cementitious foam (e.g., Air Krete®) has had a slow start—slightly more costly than polyurethane foam in terms of material and installation. The product is made of magnesium oxide, water, and air with some calcium and alumina in the absence of Portland cement. The manufacturers allege no irritant/toxic emissions both during and after application—no VOCs, no formaldehyde.

Fiberglass Insulation

The process for making the first glass fibers that we know today as fiberglass was discovered purely by accident. A young researcher for Corning Glass had been attempting to weld two glass blocks together to form an airtight seal. Unexpectedly, a jet of compressed air hit the molten glass and created a shower of glass fibers. Then, in 1935, Corning Glass and Owen-Illinois joined together and patented Fiberglas® which later became a generic household name—fiberglass.

Fiberglass, also referred to as glass fiber, is today produced by forcing molten glass through superfine holes, creating glass filaments that could be woven or unwoven. The unwoven, less structured filaments are the major components in "fiberglass" thermal/acoustical building insulation. Yet, as most purchasers buyers are unaware, fiberglass insulation requires a binder to keep the fibers loosely adhered to one another. Typically, the binder is and has been 1%–4% UF and/or PF. More recently, however, the binder is a urea-extended PF.*

According to a 2003 Formaldehyde Emissions Study, formaldehyde emissions from fiberglass batt insulation depend on Kraft paper facing and/or vapor barrier containment of the insulation (Lent 2009).

- Unfaced fiberglass, no vapor barrier, and no drywall—Formaldehyde emissions were high in the first 24 hours (51.1 µg/2 · hour), reduced by 50% after 58 days (25.4 µg/2 · hour) with a modeled office building air quality concentration of 16.6 ppb.

- Unfaced fiberglass, no vapor barrier, and unprepared drywall—Formaldehyde emissions were moderate in the first 24 hours (28.9 µg/2 · hour), increased by 14% after 58 days (33 µg/2 · hour) with a modeled office building air quality concentration of 20.3 ppb.

- Kraft faced fiberglass, no vapor barrier, and unprepared drywall—Formaldehyde emissions were moderate in the first 24 hours (25.8 µg/2 · hour), increased by 24% after 58 days (32.1 µg/2 · hour) with a modeled office building air quality concentration of 21.5 ppb.

- Unfaced fiberglass, poly vapor barrier, and unprepared drywall—Formaldehyde emissions were moderately low in the first 24 hours (17.8 µg/2 · hour), increased by 22% after 58 days (21.8 µg/2 · hour) with a modeled office building air quality concentration of 14.3 ppb.

- Unfaced fiberglass, poly vapor barrier, and prepared, painted drywall—Formaldehyde emissions were moderately low in the first 24 hours (18.9 µg/2 · hour), reduced by 21% after 58 days (14.9 µg/2 · hour) with a modeled office building air quality concentration of 9.8 ppb.

Unfaced fiberglass dissipates formaldehyde emissions fastest from an open wall cavity, whereas unfaced and Kraft faced with unprepared drywall surfaces tend to retain formaldehyde emissions—increasing formaldehyde levels in the wall cavity over time. The best approach appears to be "Unfaced fiberglass where the wall cavity is completely enclosed." However, it remains unclear as to formaldehyde gases that may be trapped within the wall cavity—even though the end result is a slight reduction in the modeled office building air quality over the other approaches. It should be noted that the

---

* A December 2008 survey of MSDSs from five major manufacturers of fiberglass batt material revealed that all were actually using similar CAS #25104-55-6 to reference the binder used in their fiberglass batt. The same CAS numbers were referred to as "Phenol formaldehyde resin cured," "Urea extended phenol formaldehyde resin." "Phenol formaldehyde urea polymer," and "Urea, polymer with formaldehyde and phenol" meaning the phenol formaldehyde binder is extended with urea (Lent, 2009). Another survey performed in 2015 by the author revealed similar findings—but some of the same manufacturer MSDSs identified by Tom Lent included free formaldehyde of less than 0.1%–0.01%, free phenol of less than 0.02%, and up to 30% binder (e.g., Guardian Fiberglass Inc.).

open wall cavity and closed wall cavity modeled formaldehyde levels are within the NIOSH recommended limits of 16 ppm, and all are within the ANSI/ASHRAE Standard 189.1 maximum concentration for formaldehyde (27 ppb). Although the levels of formaldehyde off-gassing are speculative, formaldehyde emissions from fiberglass insulation are undeniably a contributing source of formaldehyde to indoor air quality.

As public awareness deepens, there has been a scramble by the manufacturers to produce a formaldehyde-free product. Presently, Johns Manville stands out as having successfully developed a formaldehyde-free fiberglass building insulation. In their "Fire-Retardant Faced, Formaldehyde-free Fiber Glass Building Insulation," the binder is a thermoset resin (CAS not available), and the fire-retardant is antimony trioxide (may be facing or adhesive) (Johns Manville 2012). As formaldehyde based resins (e.g., phenol formaldehyde) are thermoset resins and the CAS not disclosed, the information provided on their SDS is not sufficient to confirm a formaldehyde-free product. However, in another SDS for "Fiber Glass Building and Flexible Duct Insulation, Formaldehyde-free, Antimony Trioxide-free," they disclose the resin as an acrylic thermoset resin—the CAS undisclosed (Johns Manville 2009).

Thermoset acrylic resins are cross-linked, nonrecyclable, two-part resins as opposed to thermoplastic acrylic resins which are straight chained, recyclable, polymers. The formaldehyde-free acrylic resins mentioned herein are thermoset two-part resins, and like all thermoset polymers, there is a potential for monomer excesses and off-gassing after the resin has cured—of trace amounts of monomer methacrylates (e.g., methyl methacrylate and butyl methacrylate) which have a relatively low toxicity and may be slightly irritating to the eyes, mucous membranes, and skin. Methyl methacrylate has a sharp, "plastic-like odor" at levels (0.049–0.34 ppm) well within their regulated exposure limits (50 ppm).

Spray and Rigid Panel Foam Building Insulation

In the 1970s, the world of plastics heralded building insulation that doubled, even tripled the R-value* of other forms of insulation. But it did not come without a price!

Whereas fiberglass insulation binders have a polymer component, the most sprayed-on and rigid foam building insulation is an organic polymer foam—with but one exception. See cementitious foam in the section herein on "Minerals-and-Metals" insulation.

All organic polymeric foams are potential new product regulated substance emitters. See different types of foam products in Table 14.1. Detectable isocyanate (e.g., methylene diphenyl diisocyanate) emissions are possible

---

* An R-value is the measure of a solid material's resistance to conductive heat transfer. In other words, the higher the R-value, the better the insulation.

from newly manufactured rigid panel polyurethane/polyisocyanurate insulation and recently sprayed polyisocyanurate foam insulation. On the other hand, significant levels of formaldehyde emissions are possible from rigid panel phenolic and sprayed-on phenolic and UFFI. Allegedly, urea formaldehyde foam is no longer manufactured and/or used. In buildings where it has long since been applied and dissipated, water damage and wet urea formaldehyde foam will likely result in the release of bound formaldehyde.

Interior Walls and Ceilings

Interior walls and ceilings run the gambit—materials of ill repute, emitters, dust hazards, and no hazards. Most wall materials are used likewise on ceiling, but very infrequently are dedicated ceiling materials also used for walls.

Gypsum Board

Gypsum board, also referred to as drywall, wallboard, sheetrock, and plasterboard, is a manufactured gypsum product—gypsum (e.g., calcium sulfate dihydrate), reinforcing fibers (e.g., asbestos, fiberglass, and/or cellulose fibers), binder (e.g., starch), special additives (e.g., boric acid and/or vermiculite), and water. All gypsum board has a front-and-back facing (e.g., paper, fiberglass, and/or vinyl) to maintain the integrity of the drywall which would otherwise crumble and fall apart. In the past, most gypsum board was made from mined gypsum, but there has been a recent shift from mined gypsum to synthetic gypsum.

Although the technology has been available for over 30 years, synthetic gypsum only became popular in the later part of the twentieth century. In 2009, about 30% of all drywall produced and 57% of that which was sold in the United States was "synthetic gypsum" (Kozictki 2015). Comprised of recycled industrial by-products, synthetic gypsum is calcium sulfate dehydrate—natural, mined gypsum. The most common synthetic gypsum is a product of power plant flue gas desulfurization (FGD). It is formed when coal-fired power plants remove sulfur dioxide (SO_2) from power plant emissions by passing the flue gases through a slurry of limestone ($CaCO_3$). The sulfur dioxide reacts with the limestone to produce calcium sulfite ($CaSO_3$) which is infused with water and converted to gypsum ($CaSO_4 \cdot 2H_2O$). A smaller portion of synthetic gypsum is created through various acid-neutralizing (e.g. phosphoric acid) industrial processes, and all gypsum board—natural and synthetic—can be recycled.

Be forewarned! Mined gypsum contains crystalline silica—a potential worker exposure hazard.

In its pure form, irritant/toxic emissions from gypsum board are unlikely. However, pure is not a reality—particularly as it applies to FGD synthetic gypsum. Some of the more noteworthy potential irritant/toxic emissions are mercury and sulfur compounds.

In 2005, U.S. Gypsum in cooperation with the U.S. Department of Energy performed a study to determine the "mercury content" of natural and FGD gypsum products using flux-chamber testing. The findings were used to estimate airborne concentrations of mercury in a room about 10 feet by 13 feet by 16 feet. They were 0.028–0.28 ng/m³ and 0.13–2.2 ng/m³, respectively. The U.S. EPA reference value is 300 ng/m³, and the Agency for Toxic Substances and Disease Registry minimal risk level (MRL) is 200 ng/m³. In summary, the study concluded that there were very low levels of mercury in the U.S. Gypsum wallboard (Marshall 2005).

Later, in 2009, the U.S. EPA performed a "Drywall Sampling Analysis" for elements, volatile organic compounds, metals, formaldehyde, and sulfides, to name a few. The drywall sample manufactures included four U.S. products (e.g., U.S. Gypsum/Hamilton) and two Chinese products (e.g., Knauf) (Singhvi 2009). Although the study was in response to the controversial tainted Chinese drywall, some of the news media attached themselves to the concept of mercury content in drywall which was inadvertently included in the metals screening. In other words, mercury was not the focus of the study. Yet, seek and you shall find, and so they did. Many came to similar conclusions, and an article entitled "Mercury in Gypsum Wallboard: Quietly Turning Toxic?" the author summarized other findings in the study that were not mentioned by the EPA.

> EPA is finding mercury in the synthetic gypsum itself, both Chinese and domestic. In fact, the mercury levels in one major source of the U.S. synthetic gypsum was the highest of six sources EPA tested—more than three times the highest Chinese sample (2.08 ppm versus 0.562 ppm) (Power 2010).

The U.S. EPA study supports the report. Additionally, the highest mercury concentration was found in U.S. Gypsum/Hamilton drywall, and the second highest overall, highest Chinese concentration was found in Knauf drywall (Singhvi 2009). These concentrations, however, neither confirm nor deny mercury emissions from synthetic drywall within indoor air environments! In 2009, "tainted Chinese drywall" became suspect of causing noxious odors (e.g., rotten egg odors), health complaints, and corrosion of copper building materials (e.g., electric wires and air conditioning cooling coils). The most common health complaints were eye irritation and respiratory problems (e.g., coughing, sneezing, difficulty breathing, bronchitis, and asthma). Other symptoms included nose bleeds, headaches, and allergy-like symptoms. Although not specifically identified, sulfur-containing contaminants in the tainted Chinese drywall were the alleged source of the complaints, and some researchers speculated that the Chinese drywall was FGD synthetic gypsum (Hess-Kosa 2011, pp. 305–319).

Toxic mercury vapor and irritating, corrosive sulfur-containing compound emissions are possible, especially from FGD synthetic gypsum. Whether emissions are likely or not has yet to be determined. As the use of recycled waste in the manufacture of synthetic drywall increases, further studies may be forthcoming.

Be forewarned! In 1946, asbestos was used for the reinforcing fiber component in drywall manufacturing. It was phase-out/banned in 1989.

Cement Board

Cement board, also referred by trade name Hardibacker®, Durarock®, Wonderboard®, and Versarock®, is waterproof wall board that replaces gypsum board in wet areas (e.g., around bathtubs and showers). It is made of a slurry of Portland cement, reinforcing fibers (e.g., cellulose and/or fiberglass), filler (e.g., sand), additives (e.g., colorants and extenders), and water. Furthermore, some manufacturers identify sand in the ingredients listing on their MSDS—at levels as high as 50% by weight (e.g., Hardiebacker® cement board). The thin slurry is then poured into a form and pressed into and covered with a fiberglass mesh.

Although the component of sand may have crystalline silica, there are no identified toxins/irritants emissions associated with cement board. Cement board is not combustible!

Lime Plaster

Lime plaster is a three part process (e.g., scratch coat, brown coat, and finish coat on a wood or metal mesh)—similar to stucco. Yet, whereas the main stucco ingredients are concrete, sand, and slake lime, the traditional plaster is hydrated slake lime (CaO) only. When water is added, it turns to an extremely caustic (pH 12) plaster (CaOH) until it properly cures, reverting back to limestone ($CaCO_3$) which considerably less caustic (pH 8.6). Without additives, the plaster curing time can be up to one month. Yet, special additives (e.g., pozzolan) the curing can be reduced considerably. Although plastering is a time consuming, expensive process, the final product is extremely hard and durable, lasting thousands of years.

$$CaCO_3 + heat \rightarrow CaO$$

$$CaO + H_2O \rightarrow CaOH$$

Beyond the basic mix, lime plaster additives may include, but are not limited, fibers (e.g., animal hair), linseed oil/tallow, sand/stone dust (e.g., silica), pozzolan (i.e., stone, brick dust and/or ash), stabilizers (e.g., plaster of Paris), and pigments.

Irritant/toxic emissions from lime plaster products are highly unlikely. However, the final uncured caustic plaster can cause chemical burns when mishandled.

Magnesia Wallboard

Magnesia wallboard, also referred to as magnesium oxide wallboard, magnesia cement board, and magnum board, promises to be the new twenty-first century be-all and end-all "drywall replacement." It is 20% heavier than gypsum board and 20% lighter than cement board. Magnesia wallboard is fire resistant, waterproof, mold/pesticide proof, and impact-resistant. In 2003, magnesia wallboards were approved for construction in the United States. As 70% of the world's magnesium oxide is in Asia, China is the world's leader in magnesia wallboard production.

Magnesium oxide (MgO) exists as a "rock" and is mined as such. It is ground up and pulverized to a fine powder, combined with additives, and water is added to make a cement-like slurry. The slurry is poured into forms and cured at ambient temperatures. All magnesia wallboards have reinforcing fiber additives within the core and/or glass mesh, fibers, or a scrim on the front and back surfaces.

Unidentified, undisclosed enhancers are alleged to increase flexibility and tensile strength. DragonBoard, however, has magnesium chloride and talcum powder in addition to magnesium oxide. The U.S. distributers of Magnumboard® list magnesium chloride, cellulose, perlite, fiberglass scrim, and proprietary additives. There we go again—proprietary additives. The Magnesium Oxide Board Corporation of Australia claims the addition of magnesium chloride, perlite, cellulose, and filler (glass fiber mesh and nonwoven fabric). Titanwall of Canada lists magnesium chloride, fiberglass nonwoven mesh, talc, and other nonhazardous ingredients.

Irritant/toxic emissions of magnesia wallboard are highly unlikely. Magnesia is noncombustible, unlikely to produce hazardous thermal decomposition products.

Solid Wood and Composite Wood Products

Wood wall coverings are used on interior and exterior walls as well. However, similar building materials to interior wood are rarely, if ever, treated with preservatives and are neither mold nor pesticide resistant. Natural wood may be stained and/or coated with a protective sealant. See Chapter 16, "Adhesives, Sealants, Surface Finishes, and More."

Laminated wallboard, plywood, and beadboard are composite wood products used on walls and ceilings. Laminated wallboard is surface finished at the factory whereas plywood and beadboard are surface finished after purchase. All composite woods are potential formaldehyde emitters, and composite woods that are intended for indoor applications are often comprised of urea formaldehyde which is likely not only to off-gas formaldehyde for up to a year after installation but years later whenever the UF composite wood gets wet due to water damage. Water damage to wall and ceiling composite wood may occur as a result of plumbing leaks and/or rainwater penetration through the exterior walls to the interior walls. That which generally gets attention in the case of water damaged composite wood is mold whereas formaldehyde off-gassing from the water damaged composite wood may be a contributor to poor indoor air quality—adding insult to injury.

Because of recent concerns regarding laminated flooring, some manufacturers are beginning to stray away from formaldehyde-based resins in all their laminated composite wood products. Whereas some manufacturers claim the adhesive resin component has no formaldehyde, others specify acrylates. Once acrylate adhesives have hardened, emissions are not likely.

Mined/Quarried Products

Many mined/quarried products are used both as exterior wall cladding and interior wall/ceiling coverings. The only difference is the need to take special moisture precautions—beyond that of interior walls—when constructing exterior walls. Interior applications may or may not have a backing such as porcelain tile on cement board as opposed to natural stone.

Natural stone, generally sedimentary stone, is usually comprised of varying amounts of quartz (e.g., crystalline silica)—a worker exposure health hazard when cutting and/or pulverizing natural stone—any means whereby fine, respirable dust may be generated. Although a potential crystalline silica hazard for workers, natural stone irritant/toxic emissions are unlikely.

Bricks are comprised of about 50%–60% sand which is comprised of silica—generally non-respirable sizes. Irritant/toxic emissions from bricks are highly unlikely.

Ceramic and porcelain tiles are comprised of about 5%–30% quartz (e.g., crystalline silica). The principal difference between ceramic and porcelain tile is that the porcelain tile has a highly refined and purified clay component. Other than posing as a potential worker exposure to crystalline silica, irritant/toxic emissions from ceramic and porcelain tiles are highly unlikely.

The use of metal as an interior design accent is becoming increasingly popular—particularly stainless steel. Other metals that may be popular are copper (e.g., designer wall panels) and tin (e.g., stamped ceiling panels). Torch cutting and/or welding stainless steel may result in the low toxicity chromium converting to carcinogenic hexavalent chromium. The higher the temperature, the greater the chance of generating hexavalent chromium. Unless subjected to extreme heat (e.g., during a fire), metals will not emit irritant/toxic substances.

Most quarried materials are bonded with mortar or grout—both potential sources of crystalline silica worker exposures. All mined/quarried products are nonflammable, and they are not likely to pose an emissions hazard.

Faux Stone and Stone Veneer Products

Back to the world of plastics—faux synthetic (e.g., plastic) stone is cheaper, lighter, and easier to install than natural stone. It is also claimed by manufacturers that synthetic stone looks like and is as durable as natural stone—an allegation that does not go unchallenged. There is a wide diversity of plastics that fall into the realm of synthetic stone, all of which are in the form of interlocking panels. Most of the synthetic stone is made of thermoset polyurethane foam, thermoplastic polypropylene, and thermoset polyester resin with fiberglass. Irritant isocyanate emissions are possible from rigid polyurethane panels, and most rigid polymers have additives (e.g., UV stabilizers). See Chapter 9, "Plastic Additives."

Beyond plastic faux stone, thin, interlocking, form-cast concrete, also referred to as precast stone veneer, are cost effective and "not plastic." They are installed with aluminum channel locks that are screwed into the wall, and some manufacturers recommend adhesive to bond inside corners. Mortar is not required. Other forms of precast stones are made in a similar fashion, but some are applied to a lath and plaster base, and the gaps are filled with mortar. Although worker exposures to crystalline silica are likely, emissions from concrete blocks and precast stones are highly unlikely.

Of the imitation stone products, the polymer "faux synthetic stone panels" are likely to result in irritant emissions. Concrete stone veneer is a silica-containing, nonflammable product while the combustion by-products of most synthetics are carbon dioxide and carbon monoxide with one exception. The combustion by-products of polyurethane include oxides of nitrogen and trace hydrogen cyanide.

Light Transmitting Concrete

Light transmitting concrete, also referred to as "see-through concrete" and Litracon™, exploded into the world of architectural marvels in 2001. Concrete blocks and panels are infused with optical glass fibers (i.e., fiber optics) and are, allegedly, able to transmit light through blocks up to 50 feet thick with minimal light loss (Gajitz 2015). They retain the load-bearing capacity of unaltered concrete blocks, and they can be used for interior walls as well as exterior walls. With that said, cut, sanded, and pulverized concrete can result in the release of silica—a worker exposure—and the very fine glass fibers are almost invisible to the naked eye and when broken off can penetrate and irritate human tissue. Yet, due to the expense of the material, it is unlikely that construction workers will cut, sand, or pulverize the material. Emissions are highly unlikely.

Suspended Ceiling Tiles and Panels

Between the fourteenth and sixteenth century, the Japanese originated the concept of decorative, aesthetically pleasing—coffered and bronze gilded wood-slat suspended ceilings. In 1596, the Blackfriars Theater in London, England, used dropped ceilings for acoustics. In 1919, "modern" suspended ceilings with interlocking tiles were created to hide a building's infrastructure (e.g., plumbing, electrical, and air ducts) and unsightly building faults (e.g., structural damage). Many of the ceiling tiles back in the early 1900s were mostly tin. Accessible, pop-up suspended ceilings grids and tiles did not emerge until 1958, and today the tile inserts have evolved from tin to mineral fibers to plastic. They serve as one , or a combination of, the following:

- To cover a plethora of pipes and wires
- To make an architectural statement
- To attenuate sound
- To manage lighting and thermal conditions

The most commonly encountered ceiling tiles today are mineral fiber, fiberglass, gypsum board, PVC, and melamine acoustical foam. Ceiling tiles, also referred to as ceiling panels, are structurally suspended ceilings, dropped ceilings, "false ceilings," and grid ceilings—to mention a few. They may be recessed, coffered, or concealed grid (an older type of suspended ceiling). See Table 14.2.

TABLE 14.2

Types of Suspended Ceiling Tiles

Material Type	Acoustical[a] NRC	Rust/ Corrosion Resistant	Moisture Resistant	Water Resistant	Antimicrobial	Irritant/ Toxic Emissions
Mineral fiber	0.55–0.7	X	X[b]	–	–	–
Fiberglass with vinyl cover	0.9–0.95	X	X	Partial	X	X[c]
Gypsum board	0.6–0.8[d]	X	–	–	–	–
Vinyl	–	X	X	X	X	–
Melamine acoustical foam	0.95–1.2[e]	X	X	X	X	–

[a] Noise reduction coefficients (NRC): 0–1.0; 0 = no reduction, 1.0 = complete noise reduction.
[b] Where surface coating on mineral fibers.
[c] Fiberglass tiles: Cured urea extended PF in the fiberglass and acrylic binder in fiberglass mat.
[d] Perforated gypsum: NRC is based on size of holes.
[e] 0.95 for 1 3/8 inch foam, 1.2 for 2 3/8 inch foam.

On the heels of tin tiles comes "mineral fiber tiles" which is typically comprised of perlite, cellulose, slag wool fibers, kaolin clay, limestone, and starch with possible crystalline quartz contaminants. Mineral fiber tiles are also referred to—in some of the home improvement centers—as fiberboard while referencing a name brand mineral fiber tile. By definition, fiberboard is a building material made of wood or other plant fibers compressed and glued (generally with a formaldehyde containing polymer) into rigid sheets. The only plant fiber component of mineral fiber tiles is cellulose, and the glue is starch. So, this is yet another example of a distributor misnaming a building product. Although the fibers are a physical irritant to the eyes, skin, and respiratory tract, mineral fibers are not likely to emit irritant/toxic chemicals.

Be forewarned! Asbestos was a component of cellulose ceiling tiles in older buildings until its ban in 1976.

Fiberglass ceiling tiles are usually covered on the visible side with a thin layer of vinyl. The fiberglass is typically impregnated with "cured" PF or urea extended PF, and the fiberglass mat is an acrylic based binder. Wet PF may potentially release formaldehyde.

Vinyl covered gypsum ceiling tiles are a newly added usage for gypsum which is used predominantly in tap-and-float wall/ceiling panels. As per the section on gypsum board, "toxic mercury vapor and irritating, corrosive sulfur-containing compound emissions are possible, especially from FGD synthetic gypsum." Whether mercury and sulfur-containing chemical emissions are likely has yet to be discovered.

Vinyl, also referred to as PVC, ceiling tiles can and do mascaraed as metal, wood, and special designer tiles—at a considerably reduced cost. Subjected to excessive heat, PVC will deteriorate, emit plasticizers and other plastic additives. Yet, high temperatures are not likely indoors—under normal conditions. Combustion of PVC will, however, result in the release of hydrogen chloride.

Most, if not all, acoustical panels are MF foam, a step up from UF sprayed-on insulation that was banished from the market due to formaldehyde emissions. Although less likely to emit formaldehyde, melamine acoustic panels may potentially emit formaldehyde—especially when wet.

Novelty drop-out ceiling tiles, also referred to as melt-out ceiling tiles, are made possible, once again, through the Wonderland of Plastics. They do that which others cannot! Drop-out ceiling tiles hide unsightly sprinkler systems. In the event of a fire, the heat sensitive tiles melt and drop from the suspension grid, allowing the fire suppression system to do its job. The most common materials are vinyl and expanded polystyrene (EPS). Vinyl softens at about 80°C (176°F), and unstabilized PVC starts to decompose at 100°C (212°F). EPS begins to soften at 70°C (158°F), and polystyrene decomposes at 220°C (428°F). Emissions are unlikely at indoor temperatures less than 38°C (100°F). PVC thermal decomposition products include hydrogen chloride.

Summary

Into the twenty-first century, building technology has expanded, reaching new highs in thermal/acoustical insulation and interior wall/ceiling building materials. Through these building materials, energy efficiency and climate comfort become increasingly compatible. Yet, efficiency and comfort comes at an expense. Sometimes it is in the indoor air quality. Sometimes it is in the cost of the latest-and-greatest. There are trade-offs!

The crown jewel to temperature and sound comfort, insulation materials have certainly seen a big transition. Starting with natural and ending in synthetic polymers, efficiency improves as we depart from natural products as does the expense.

Natural fibers and mined minerals are not without the influence of modern technology. Some cellulose insulation includes not-so-natural recycled plastic bottles. The binder for mineral wool may contain a phenolic resin that off-gases formaldehyde. Vermiculite may contain asbestos. Cementitious foam an extremely expensive metal insulation has no VOCs or formaldehyde emissions.

Fiberglass is the good old standby, used by most, due to low cost and efficiency. The binder for most fiberglass insulation is typically UF or PF that emit or will likely emit formaldehyde. A formaldehyde-free acrylic resin may emit irritating methacrylates, and some fiberglass has no binder, no emissions. The latter is difficult to find and expensive.

Oftentimes, interior walls and ceilings also provide insulation value to the interior of a building, albeit minimal, but they do contribute enough to count. Gypsum board is the most commonly used, potentially the most problematic. Tainted drywall may off-gas corrosive sulfur chemicals that may impact the indoor air quality and damage copper building materials. In some camps, drywall is also suspect of mercury emissions.

Whereas no emissions are likely from cement board, plaster, magnesia wallboard, quarried products, solid wood, and veneered stone, formaldehyde emissions from composite wood and faux stone are likely. Ceiling tiles not likely to emit irritant/toxic chemicals include mineral fiber ceiling tiles and vinyl ceiling tiles. Formaldehyde emissions are likely from vinyl covered fiberglass ceiling tiles and melamine acoustical foam ceiling panels. Irritant/toxic emissions are unlikely—short of a fire—from the very expensive vinyl and expanded polystyrene drop-out, thermal melting ceiling tiles.

As for the most commonly used thermal/acoustical insulation, formaldehyde emissions may contribute to other sources of formaldehyde in the interior of a building. And for the most commonly used interior wall/ceiling building material, irritating, corrosive gases are possible and potentially very damaging to other building materials. However, they are all too frequently ignored as potential contributors to poor indoor air quality. Buyers beware!

References

Gajitz. 2015. "Totally Translucent: High-Tech, Light-Transmitting Concrete." *Gajitz*. Accessed October 21, 2015. http://gajitz.com/totally-translucent-high-tech-light-transmitting-concrete/.

Hegde, Raghavendra R., Atul Dahiya, and M. G. Kamath. 2004. "Cotton Fibers." *Materials and Science Technology 554*. April. Accessed September 16, 2015. http://www.engr.utk.edu/mse/pages/Textiles/Cotton%20fibers.htm.

Hess-Kosa, Kathleen. 2011. *Indoor Air Quality: The Latest Sampling and Analytical Methods*. Boca Raton, Florida: CRC Press.

Johns Manville. 2009. "MSDS for Fiber Glass Building and Flexible Duct Insulation, ID: 1071." *Johns Manville SDS*. June 3. Accessed September 21, 2015. http://idi-insulation.com/PDFs/insulation/fiberglass/johnsmanville/MSDS.pdf.

Johns Manville. 2012. "MSDS for Fire-Retardant Faced, Formaldehyde-Free Fiber Glass, ID: 1021." *Johns Manville*. January 3. Accessed September 21, 2015. https://msds.jm.com/irj/go/km/docs/documents/Public/MSDS/200000000016_REG_NA_EN.pdf.

Kozictki, Chris. 2015. "6 Facts About Synthetic Gypsum." *Feeco*. Accessed September 23, 2015. http://feeco.com/6-facts-synthetic-gypsum/.

Lent, Tom. 2009. "Formaldehyde Emissions from Fiberglass Insulation with Phenol Formaldehyde Binder." *Healthy Building Network*. August 26. Accessed September 18, 2015. http://www.healthybuilding.net/uploads/files/formalde-hyde-emissions-from-fiberglass-insulation-with-phenol-formaldehyde-binder.pdf.

Marshall, Jessica. 2005. *Fate of Mercury in Synthetic Gypsum for Wallboard Production*. Topical Report, Task 1 Wallboard Plant Test Results, Chicago, Illinois: USG Corporation.

National Pesticide Information Center. 2015. "Boric Acid Technical Fact Sheet." *National Pesticide Information Center*. Accessed September 16, 2015. http://npic.orst.edu/factsheets/borictech.html.

Pettersen, Roger C. 1984. "The Chemical Composition of Wood." *Forest Products Laboratory, Forest Service*. Accessed September 16, 2015. http://www.fpl.fs.fed.us/documnts/pdf1984/pette84a.pdf.

Power, Matt. 2010. "Mercury in Gypsum Wallboard: Quietly Turning Toxic." *Green Builder*. April 14. Accessed September 24, 2015. http://www.greenbuilderme-dia.com/blog/mercury-in-gypsum-wallboard-quietly-turning-toxic.

Singhvi, Raj. 2009. "Drywall Sampling Analysis." *U.S. EPA*. May 7. Accessed September 24, 2015. http://nepis.epa.gov/Exe/ZyPDF.cgi/P1004UPQ.PDF?Dockey=P1004UPQ.PDF.

The Mesothelioma Center. 2015. "Zonolite Insulation." *asbestos.com*. January 30. Accessed September 17, 2015. http://www.asbestos.com/products/construc-tion/zonolite-insulation.php.

Van der Werf, Hayo M. G. 2015. "Hemp Facts and Fiction." *Hemp Food*. Accessed September 16, 2015. http://www.hempfood.com/IHA/iha01213.html.

15

Interior Finish-Out Components

Without the interior components, a building is but a shell! Whereas the exterior is the face, the interior is the character. And character comes at a cost. Today's interior components trend toward a dependence on the World of Plastics.

Once the interior walls and ceilings have been installed and finished (e.g., painted), the cabinets, countertops, flooring, and trim are installed. Many of these components are big contributors to poor indoor air quality of all building materials. Most are comprised of polymeric glues and plastics. Off-gassed chemicals are emitted directly into the building interior. Occupants have a tendency to be in close proximity to the components and their emissions.

All too often, purchasers and sometimes environmental professionals are deluded into thinking artfully crafted, veneered cabinets are solid wood. Synthetic countertops appear to be natural stone. Flooring is often misunderstood, and plastic trim has the appearance of wood. No problem here. Move on. Not so fast! Let's break it down and herein discuss reality as opposed to appearances.

Cabinets

From the Hoosier free-standing, solid wood cabinets of 1910 to the wall-mounted engineered wood and plastic cabinets of today, choices of cabinet materials have sprung forth with options to meet a broad spectrum of cost, design, functionality, and durability needs. Today, the most commonly encountered cabinets are wood, plastic laminate/other synthetics, stainless steel, and combinations thereof. However, out-of-sight, out-of-mind, the "core" to most modern cabinets (including some metal cabinets) is composite wood. Particleboard, MDF, and plywood support the cabinet facade.

Particleboard, encountered in most of today's cabinets, is used in the panels that makeup the boxes and the shelving. MDF, denser and heavier than particleboard, is used in cabinet doors, boxes, and shelving, and plywood, more rigid and stabile than MDF, is used in high-end products. That composite wood which is visible (i.e., box interior/shelving panel surface and edges as well as drawer fronts) is generally covered with a wood veneer, laminate, or metal. Some cabinets may be comprised of one or a combination of these composite wood—all of which potentially off-gas formaldehyde, particularly newly manufactured and/or wet composite wood.

Typically, the doors and visible surfaces may be comprised of one, or a combination of, solidwood, veneer, laminate, melamine, thermofoil, and metal. Other materials are limited only by the imagination—concrete, all glass tiles, cement board with porcelain tiles, and the list goes on. Here we shall discuss the most commonly used materials and their potential product emissions.

Solid Wood and Veneered Cabinets

Built-in kitchen cabinets are a relatively new concept—introduced around the 1920s. Solid wood cabinets emerged and were the initially created byhand by cabinet makers using dove tail joinery for the finer cabinets. With the passage of time, however, mass production resulted in nail and glue joinery, and post-World War II, wood veneered composite cabinets became popular.

Solid wood, non-emissions cabinets are expensive and subject to warping, and veneered composite cabinets are less expensive and not subject to warping. Today, more affordable wood cabinets are made of veneered composite wood.

A thin sheet of fine wood veneer is glued to the surface of the composite wood to impart a beautiful façade to its surface. The veneer masks that which is beneath. The disguise is complete—beauty and the beast.

Composite wood (e.g., plywood, pressboard, and MDF) used in manufacturing cabinets is generally comprised of UF. Newly manufactured UF-impregnated composite wood is likely to off-gas formaldehyde for an indeterminate period of time up to a year after installation. When it gets wet, the UF will deteriorate and, once again, off-gas formaldehyde. Melamine formaldehyde (MF) plywood is less likely to release formaldehyde than UF, and phenol formaldehyde is the least likely to off-gas formaldehyde. It should be noted that resins used in composite wood is not always known nor are the implications of the different resins and formaldehyde emissions completely understood—even by some manufacturers and/or cabinet makers. That said, most "solid wood cabinets" today are a combination of solid wood (e.g., door rail and stile and on the visible edges of plywood) and veneered composite (e.g., box and shelving). The more affordable cabinets are likely to be made entirely of veneered composite—impregnated with the less expensive UF resin—and pose an even greater potential for formaldehyde emissions.

Be forewarned! The term "solid wood" is often abused by cabinet manufacturers. Many manufacturers and fine cabinet makers represent veneered composite wood (e.g., thin layer of rosewood on pressboard) as solid wood or, more aptly, all wood. Indeed, veneered wood is solid wood. Yet, it may not be that which the purchaser envisioned which is a solid "single species of wood" (e.g., solid rosewood).

Now, to add insult to injury, UF glue may also be used, in some instances, to adhere the fine wood veneer to the composite wood. This is potentially another source of formaldehyde emissions. Alternative glues include hide glue, white polyvinyl acetate (PVA) woodworking glue, yellow aliphatic resin, epoxy glue, and contact cement. See Chapter 16: Adhesives. The thin veneer can lift and warp overtime, particularly when exposed to water. Aging veneer is not pretty!

Stainless Steel and Other Metal Cabinets

While they are strong, durable, and resist warping, modern, industrial appearing metal cabinets can be scratched and dented. Most are stainless steel but may be copper, brass, brushed nickel, or another metal yet to be divined by an interior designer. Although metal alone does not off-gas irritants and/or toxins, many metal cabinets actually have solid wood or composite wood (preferably plywood) at their core to help dampen the noise you can get when opening and closing the doors. Formaldehyde emissions are possible with newly manufactured core composite wood.

Synthetic Polymer Cabinets

The World of Plastics lives on in the polymer coatings—laminates, melamine, and thermofoil. Although they are durable and come in a wide variety of colors, styles, and designs, plastics will never match the beauty of wood. Yet, the versatility of plastics allows for cost containment and creativity!

Laminates are paper-based products that have been impregnated with resins. The overlay, also referred to as the wear layer, is comprised of a resin impregnated, printed "decorative" paper. The decorative paper is a refined, long fiber cellulose (similar to coffee filters). It is printed with solid colors, woodgrains, natural/abstract designs, and the printed decorative paper is then impregnated with MF. A Kraft paper layer, comprised of soft/hard wood and recycled paper (similar to paper grocery bags), is impregnated with phenol formaldehyde. The overlay is hard, scuff resistant, and the Kraft paper layer provides strength and flexibility. Both layers are then bonded through the application of high temperatures and pressure (Wilsonart 2015). High-pressure laminate (HPL) is subjected to 70–100 bars (1000–1500 psi) and 138–160°C (280–320°F). Low-pressure laminate, less durable and more likely to chip than HPL, is subjected to 20–30 bars (290–435 psi) and 168–191°C (335–375°F) (Nova 2012). The laminate is then glued to a composite wood which is another source of formaldehyde off-gassing. Thus, formaldehyde emissions are highly likely from newly manufactured or wet laminate cabinets.

Melamine, also referred to as thermally fused laminate, is similar to paper-based laminate—without the Kraft backing. The MF saturated decorative layer is, however, fused to form a smooth surface composite wood (e.g., particleboard and MDF) by heat and pressure. Formaldehyde emissions are possible from newly manufactured or wet laminates.

Thermofoil is a thin vinyl film that is used to cover cabinet boxes, doors, and drawer fronts. A rigid film vinyl is formed over the substrate material (e.g., composite wood) by heating a thin layer of PVC. Although it is extremely durable and is resistant to water, PVC will discolor with age and is heat sensitive. Thermofoil cabinets close to a heat source (e.g., oven) will break down and potentially off-gas hydrogen chloride. In addition, thermofoil PVC products contain plasticizers and heat stabilizers—potential nonregulated irritants.

Countertops

From the single species wood and stone counters of yesteryear to the marble and granite, fine woods and butcher block countertops, and occasional metals (i.e., zinc, tin, nickel, or galvanized iron) of the 1800s to tiles, stainless steel, and nickel in the late 1800s and early 1900s to the more extensive use of stainless steel and the introduction of laminate countertops in the 1940s to the laminate crazypost World War II period to the composite stone and plastics of today, countertops make a statement. Choice of countertop materials is fraught with options based upon cost, style, and durability (Lee 2014).

Whereas granite and marble are considered "elegant" and extremely costly, stainless steel is considered "contemporary, industrial" and is by far more costly than natural stone as are the "warm sensible" paper composites and "industrial chic" concrete. The materials used to manufacture countertops are herein classified as natural stone, metal, tiles, concrete, butcher block, and synthetic polymers.

Natural Stone Countertops

At the top of the food chain, granite, an igneous rock, is resistant to heat and scratches (e.g., knife nicks). However, if not routinely sealed, granite will stain. Irritant/toxic emissions from granite are highly unlikely.

Marble, a metamorphic rock, is resistant to heat, but not to scratches and chipping. It is also very susceptible to stains—even when routinely sealed. Irritant/toxic emissions from marble are highly unlikely.

Soapstone, a sedimentary rock, is more subdued than granite and marble. Although resistant to heat, soapstone is readily scratched, yet the scratches can be sanded out. It will darken with age, and it can crack over time. It fits best in "older, cottage-style" designs. Irritant/toxic emissions from soapstone are highly unlikely.

All natural rocks contain crystalline silica, a regulated worker exposure hazard. The more natural stone is manipulated (e.g., cut, sanded, and chiseled), the greater the exposure potential for the worker.

Metal Countertops

See "Stainless Steel and Other Metal Cabinets."

Ceramic and Porcelain Tile Countertops

Ceramic and porcelain tile countertops work well with a wide range of countertop designs from "country" to "majestic Old World." They resist heat, scratching, and stains—exclusive of unsealed grout.

Ceramic and porcelain tiles are comprised of about 5%–30% quartz (e.g., crystalline silica). The principal difference between ceramic and porcelain tile is that the porcelain tile has the highly refined and purified clay component. Other than posing as a worker exposure problem, irritant/toxic emissions from ceramic and porcelain tiles are highly unlikely.

Concrete Countertops

Concrete countertops can be cast into any shape, custom tinted (any color or combination of colors), and creatively inlayed (e.g., glass fragments, rocks, shells, and tiles) for a look and feel of "industrial chic" to "artistic elegance." Concrete is not as heat resistance as natural stone. If not sealed, concrete will stain, and with time and settling, concrete will crack. Yet, custom concrete can be very pricy. Irritant/toxic emissions from concrete tiles are highly unlikely. Worker exposures to crystalline silica are however possible.

Butcher Block Countertops

Butcher block countertops are single species or multiple species solid wood that has been joined and glued. Butcher block fits into "country" and "cottage-style" designs. Wood scratches and harbors bacteria, swells and contracts, and is readily stained. Water could, overtime, weaken the glue joins. The type of adhesive used in the joinery should be water resistant wood glue which once cured will not off-gas irritant/toxic gases. Thus, irritant/toxic emissions from butcher block are highly unlikely.

Synthetic Polymer Countertops

The synthetic polymers range from the poor man's "affordable" material which is low maintenance, easy to clean yet prone to scratching, heat burns, and peeling to a very pricy "warmly sensible, environmentally friendly" material which is heat resistant yet is not scratchproof and is susceptible to chemical damage—laminate to paper composite, respectively. Intermediate in price are "high tech" solid surface and synthetic granite/marble.

The first synthetic countertop was a thin hardwood look-alike surface coated with plastic the composite of which is referred to as a laminate. In the early 1900s, it was comprised of phenol formaldehyde impregnated decorator paper. Today, laminated countertops are manufactured with a slightly different twist. MF impregnated decorator paper is heat and pressure glued to a phenol formaldehyde Kraft backing. The laminate is then glued to a formaldehyde emitting composite wood. Formaldehyde emissions are possible from newly manufactured or wet laminated countertops. For details on the laminating process, see "Synthetic Polymer Cabinets."

Paper composite countertops are resin impregnated recycled paper. The resin is a "cured phenol formaldehyde," or "formaldehyde-free" thermoset resin. While most manufacturers of paper composites identify the resin as phenol formaldehyde, all claim the resin is an eco-friendly formaldehyde-free product. Yet, the combustion products include formaldehyde.

Solid surface countertops, also referred to as composite countertops, sometimes by tradename (e.g., Corianand Formica), are high tech, sculpted countertops that can take on different shapes and appearances (e.g., stone). These counters provide a nonporous surface, do not age and discolor or darken with time, and can be reshaped and sanded much as wood. They are a blend of acrylic or polyester resins, powdered fillers (e.g., alumina trihydrate), and additives (e.g., pigments and fire retardants). The resin and additives are mixed and poured into a cast to render any of a number of shapes, sizes, and designs. The most commonly used acrylic resin used for solid surface countertops is comprised of methyl methacrylate and butyl acrylate monomers that yield copolymers with glass-like clarity. Although monomer (e.g., methyl methacrylate) emissions are unlikely under most situations, thermal decomposition (such as hot cookware) can result in the release of the acrylates which are regulated substances and can at low levels cause eye, skin, and respiratory tract irritation. Acrylic resins are harder, more impact resistant, and less brittle than polyester resins. For this reason, polyester resins are less frequently utilized in the manufacture of solid surface counters which attains a higher polish and greater translucence than acrylic resins. Most polyester resins are manufactured by a condensation reaction between acids (e.g., phthalic anhydride) and glycols (e.g., polypropylene glycol). Unreacted glycols and water are vacuum extracted, and blended with styrene to reduce the thickness of the resin. The resin mix is cured by adding a peroxide catalyst (e.g., methyl ethyl ketone peroxide). Regulated/flammable styrene and flammable/explosive peroxides emissions from polyester are unlikely once the product has cured.

Synthetic granite/marble, also referred to as engineered stone, quartz surfacing, countertop is an acrylic solid surface composite with crushed quartz. The original quartz surfacing countertop was introduced in 1990 under the trade name of Silestone which has since become a generic term.

With pigments and quartz, acrylic solid surfacing can take on the appearance of granite and marble. Its cost is less than that of natural stone. It can be formed into any shape. It looks "similar" to real stone. Whereas granite has natural variegation, synthetic granite does not. It requires less maintenance than natural stone, and does not require sealant. As with acrylic solid surface countertops, monomer (e.g., methyl methacrylate) emissions are unlikely. However, thermal decomposition (such as with the use of hot cookware) can result in the release of the acrylates which are regulated substances and can at low levels cause eye, skin, and respiratory tract irritation.

Flooring Materials

From the Egyptian stone mosaics about 5000 years ago, to the Roman heated stone floors to rough planked wood floors in the Middle Ages, to the Chinese carpets from 960 AD to 1279 and Iranian Persian rugs from 1502 to 1736, to painted canvas floor coverings from the 1300s to linoleum originating in 1863, and the popularity of vinyl flooring post World War II, flooring building materials range from elegance to comfort. Natural stone to plastic flooring are a blend of the old and the new. Floor covering materials to be discussed herein are natural stone, wood, ceramic/porcelain tiles, carpeting, and resilient flooring.

Natural Stone

Most of the stone used for flooring has multiple uses. Natural stone is used in exterior walls, interior walls, countertops, backsplashes, base trim, fireplaces, and some specialty treatments (e.g., stone kitchen counter facing). Natural stone and types of rocks (e.g., igneous) have been previously discussed at length within the Chapter 12, "Exterior Enclosure Components."

From the granite flooring and sandstone pyramids of Egypt to the multiple-choice stone (e.g., local stone) and sometimes limestone castles of medieval times to the fifteenth century Venetian terrazzo, natural stone flooring is coveted for its durability and beauty. Granite, an igneous rock, is the most durable, expensive natural stone that is slightly stain resistant—followed by marble, a metamorphic rock that is not stain resistant. Slate, a metamorphic rock, is durable, less expensive than marble, and stain resistant. All contain crystalline silica, a regulated worker health hazard, and irritant/toxic emissions are unlikely.

Sandstone, a sedimentary rock, is comprised mostly of non-respirable, medium-grained sand (1/16–2 mm) with finer silt and clay, coarser gravel, and minerals (About Education 2015). Limestone, one of the softest

sedimentary rocks used for flooring, is composed of calcite (i.e., calcium carbonate) and mineral impurities—generally iron oxide (red and yellow) or
carbon (e.g., blue, black, and gray). As silica occurs naturally in most rocks,
limestone is likely to have some crystalline silica. Irritant/toxic emissions are
unlikely from sandstone and limestone.

Wood Flooring

During the Baroque era (1625–1714) in Europe, only the affluent and royal
could afford to upgrade from earth floors to elegant hand-rubbed, stained, and
polished wood floors in parquetry and marquetry patterns. Simultaneously,
due to the great abundance of old-growth forests in North America, the colo-
nialists brought into common use wide, thick plank flooring. The rough-sawn
unpolished planks were laid side-by-side and face-nailed into the floor joists.
These were later ship-lapped then tongue-and-groove joined. In the latter
eighteenth century, painted wood floors became popular, and the wealthy
gentry sported parquet and/or hand-planed, varnished/waxed floors. By
the nineteenth century, the average American began to use polished dimensional wood with orange shellac and wax. Today, wood floors are treated
with varnish or PU. Solid wood flooring remained popular until the mid-
twentieth century at which time old-growth lumber had become rare and
was being replaced by expensive, depleting supplies of domestic and exotic
imported hardwoods. In 1977, a solution to hardwood shortages by Pergo in
Sweden in the form of laminate tabletops, which technique found its way to
the shores of America in 1994, and laminated flooring was born. Solid wood
and laminated flooring are discussed herein.

Solid wood flooring is expensive and high maintenance, and it does not
off-gas formaldehyde. On the other hand, veneered wood flooring is less
expensive and low maintenance but not as durable, and it is a potential
source of formaldehyde. The exotic and finer domestic woods are hardness
rated which translates to durability, see Table 15.1.

Even more cost-effective, wood look-alike, laminated flooring is less costly
and low maintenance. Yet, it is even less durable than veneered wood, and
laminate flooring is notorious for formaldehyde emissions. It is created with
formaldehyde-based resins, and the laminate is glued to a composite wood
(generally MDF) which is impregnated with formaldehyde-based resins.
In addition, underlayment, or carpet-like padding, is glued to the bottom
of the composite wood in some laminate floors. The medley of formaldehyde-based and carpet pad components could be a potential disaster in the
making.

In early 2015, the new media started a hail storm of horrified consumers
stating, "Chinese-made laminate flooring is off-gassing toxic formaldehyde." Enough said! A stampede of people scrambling to learn if they
were victims of bad flooring and irritating/toxic levels of formaldehyde

TABLE 15.1

Janka Hardness Scale for Wood Flooring Species

Wood Flooring Species	Hardness[a]
Ipe/Brazilian Walnut/Lapacho	3684
Cumaru/Brazilian Teak	3540
Ebony	3220
Brazilian Redwood/Paraju	3190
Angelim Pedra	3040
Bloodwood	2900
Red Mahogany/Turpentine	2697
Spotted Gum	2473
Brazilian Cherry/Jatoba	2350
Santos Mahogany/Cabreuva/Bocote	2200
Pradoo	2170
Brushbox	2135
Karri	2030
Sydney Blue Gum	2023
Bubinga	1980
Cameron	1940
Tallowwood	1933
Merbau	1925
Amendoim	1912
Jarrah	1910
Purpleheart	1860
Goncalo Alves/Tigerwood	1850
Hickory/Pecan/Satinwood	1820
Afzelia/Doussie	1810
Bangkirai	1798
Rosewood	1780
African Padauk	1725
Blackwood	1720
Merbau	1712
Kempas	1710
Locust	1700
Highland Beech	1686
Wenge/Red Pine	1630
Tualang	1624
Zebrawood	1575
True Pine/Timborana	1570
Peroba	1557
Kambala	1540
Sapele/Sapelli	1510
Curupixa	1490
Sweet Birch	1470
Hard Maple/Sugar Maple	1450
Coffee Bean	1390

(Continued)

TABLE 15.1 (*Continued*)

Janka Hardness Scale for Wood Flooring Species

Wood Flooring Species	Hardness[a]
Natural Bamboo (represents one species)	1380
Australian Cypress	1375
White Oak	1360
Tasmanian Oak	1350
Ribbon Gum	1349
Ash (White)	1320
American Beech	1300
Red Oak (Northern) [benchmark]	1290
Carribean Heart Pine	1280
Yellow Birch	1260
Movingui	1230
Heart Pine	1225
Carbonized Bamboo (represents one species)	1180
Cocobolo	1136
Brazilian Eucalyptus/Rose Gum	1125
Makore	1100
Boreal	1023
Black Walnut	1010
Teak	1000
Sakura	995
Black Cherry/Imbuia	950
Boire	940
Paper Birch	910
Cedar	900
Southern Yellow Pine (Longleaf)	870
Lacewood/Leopardwood	840
Parana	780
Sycamore	770
Shedua	710
Southern Yellow Pine (Loblolly and Shortleaf)	690
Douglas fir	660
Larch	590
Chestnut	540
Hemlock	500
White Pine	420
Basswood	410
Eastern White Pine	380

[a] The Janka hardness of a particular wood is a good indicator of its resistance to wear and denting. The Janka hardness test is a measurement of the force necessary to embed a .444-inch steel ball to half its diameter in wood, expressed in pounds-force (*lbf*). It is the industry standard for gauging the ability of various species to tolerate denting and normal wear, as well as being a good indicator of the effort required either to nail or to saw a particular type of wood. The higher the number, the harder the wood.

sought the opinion of environmental professionals and lawyers. The big box home repair stores peddled formaldehyde test kits of questionable efficacy. And the AIHA professional organization scrambled to provide guidance to the warry public by publishing, "Is Formaldehyde from Laminate Flooring a Problem in My Home?"

Air monitoring and product emission testing had been and continues to be performed by distributors (e.g., Lumber Liquidators), environmental/industrial hygiene consultants, and others. Yet, clarity as to the proper sampling approach and interpretation of formaldehyde emissions from laminate flooring remains vague. According to the AIHA, formaldehyde emission rate tests that were conducted between 2014 and 2015 demonstrated that Chinese laminate flooring substantially exceeded the 2009 California Air Resources Board rule which regards formaldehyde emissions from composite wood products. Formaldehyde emissions from building materials are not U.S. federally regulated.

The AIHA publication went on to say, "The emissions of formaldehyde from [composite wood] products are highest after initial installation and decrease over time. The half-life, the time for the formaldehyde emissions to decrease by half, can range from a few months to a few years depending on the specific product."

Due to the complex nature of building materials, furnishings, and personal products, on-site air sampling is of questionable efficacy. According to the AIHA publication, sources of formaldehyde include, but are not limited to, cabinets, counters, particle-board shelving, particle-board subflooring, some insulation materials, permanent press fabrics, home furnishings, building finishes, and cigarette smoke.

Although the media focused on Chinese laminate flooring, formaldehyde emissions from laminate flooring in general is highly likely. Once again, the type of formaldehyde-based resins used in the manufacture of laminated flooring is relevant to the amount of formaldehyde emissions from the product.

Also, worthy of mention and not normally discussed, laminate floor padding is typically a synthetic rubber (i.e., styrene–butadiene (SB) rubber). SB rubber is the suspect source of 4-phenylcyclohexene which is an unregulated human irritant. For details, see subsection "Carpet Padding."

Ceramic/Porcelain Tiles

Ceramic and porcelain tiles are comprised of about 5%–30% quartz (e.g., crystalline silica). The principal difference between ceramic and porcelain tile is that the porcelain tile has the highly refined and purified clay component. Irritant/toxic emissions from ceramic and porcelain tiles are highly unlikely.

Carpeting

Carpeting is a composite of carpet face (also referred to as nap and/or pile), carpet backing, adhesive, and/or rubberized latex. Commercial carpet is manufactured by one of two processes: tufting and weaving. Although each process produces quality floor coverings, tufted carpet accounts for 95% of all carpet (The Carpet and Rug Institute 2015).

Tufting involves several hundred needles that stitch hundreds of rows of pile yarn "tufts" through a primary fabric backing. A secondary fabric backing is then bonded to the primary backing with an adhesive (e.g., rubberized latex, PU, and other glues).

Woven carpet is created on looms by simultaneously interlocking face yarns and backing yarns into a complete product, eliminating the need for a secondary backing. A small amount of rubberized latex is subsequently applied to the back of the carpet.

The subject of carpet emissions has come into play in recent years, and indoor environmental consultants have been attempting to get a handle on what the emissions are and what the actual source is. Product emission chamber testing has identified organic emissions from several types of carpeting (Hess-Kosa 2011, pp. 209–210), see Table 15.2. According to a 2002 NIOSH toxicological review, the irritant 4-phenylcyclohexene (4- PCH) frequently occurred in carpets backed by styrene–butadiene rubberized (SBR) latex (Haneke 2002).

While chamber testing is generally dedicated to organic compounds only, some rubberized backing materials contain nonorganic ammonia, see Figure 15.1. Ammonia is a regulated irritant/toxic chemical. Ammonia has

TABLE 15.2

Product Emissions from Carpet Components

Carpet	
Acetaldehyde	Nonanal
Benzene	Octanal
Caprolactam	**4-PCH (rubber backed carpeting)**
2-Ethylhexanoic acid	**Styrene**
Formaldehyde	**Toluene**
1-Methyl-2-pyrrolidinone	**Vinyl acetate**
Naphthalene	
Carpet adhesives	
Acetaldehyde	Phenol
Benzothiazole	**4-PCH (rubberized adhesives)**
2-Ethyl-l-hexanol	**Styrene**
Formaldehyde	**Toluene**
Isooctylacrylate	**Vinyl acetate**
Methylbiphenyl	Vinyl cyclohexene
I-Methyl-2-pyrrolidinone	Xylenes (*m-, o-, p-*)
Naphthalene	

Note: **Bold:** Common emissions encountered in carpet backing/face fibers and carpet adhesives.

FIGURE 15.1
After-market rubber rug backing—danger: contains ammonia.

a pungent/irritating odor at levels (mean odor threshold: 17 ppm) slightly below the ACGIH acceptable limit (25 ppm). Yet, ammonia may be detected by some individuals at considerably lower levels (low range: 0.043 ppm). To further confuse the issue, "pungent" is a term generally used to describe formaldehyde.

With the exception of wool, most carpet face fibers are comprised of polymers, and the most commonly encountered polymer fiber carpet materials are olefin polypropylene, acrylic, and polyester. Up until 1990, Dupont offered Orlon acrylic carpet fibers, but has largely replaced its manufacture with newer synthetic fibers such as polyesters (Dacron), polypropylenes, and polyimides. Olefin fibers makeup 30% of all carpeting made in the United States today. Nylon accounts for about 65% of U.S. carpet sales.

Acrylic Fiber Carpet

Acrylic fiber carpets were first developed by DuPont in the 1940s under the tradename of Orlon. They were mass produced in the 1940s and became popular in the 1960s. Other manufacturers followed, in suite, with their own acrylic fiber carpets under tradenames such as Acrilan, Creslan, and Courtell. In 1952, the first "wash and wear" consumer item was a blend of acrylic fibers and cotton.

Of all synthetic fibers, acrylic is closest to wool in appearance and feel. It is cheaper than wool and has some desirable characteristics such as it resists staining, is not susceptible to moth damage, resists fading in the sun, resists static electricity, and wicks moisture. Acrylic carpets do, however, become fuzzy and/or mat with age, and they are easily stained by oil and grease.

Acrylic fibers, spun thermoplastic polymers, are manufactured by the polymerization of about 80%–85% monomer acrylonitrile (Figure 15.2)

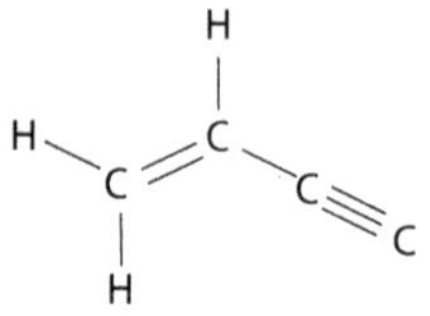

FIGURE 15.2
Structure of acrylonitrile monomer.

and about 2%–7% vinyl comonomer (e.g., vinyl acetate or methyl acrylate). Plasticizers and additives are included in the mixture. Once polymerized, the acrylic plastic is dissolved in special solvents (e.g., styrene) and spun into fibers which are woven to form carpet.

Another acrylic-like fiber, modacrylics are produced by replacing the vinyl comonomers with halogen-containing comonomers (e.g., vinyl chloride or vinylidene chloride). Modacrylics are flame resistant, soft, strong, resilient, chemically resistant, and nonallergenic. Although they burn when directly exposed to flame, they do not melt or drip and are self-extinguishing when the flame is removed. Encountered in scatter rugs and carpets, modacrylic fibers are also used in fur-like outerwear, high-performance fire resistant clothing (e.g., firefighter outerwear), and wigs (Kiron 2014).

Although the acrylic monomers are regulated toxic/flammable chemicals, the polymer has none of the hazardous properties of the monomer and comonomers. The thermal decomposition of polyacrylonitrile are carbon dioxide, carbon monoxide, hydrogen cyanide, and various nitrile compounds (Johnson et al. 1988). Unless subjected to extreme heat, acrylic fiber emissions are unlikely. At 322°C (611°F), the thermal decomposition product—hydrogen cyanide—may be produced.

Olefin Fiber Carpets

In 1957, an Italian discovered a catalytic method to make fibers out of olefin plastics. It is a fiber that is "light weight," tough, hard wearing, colorfast, and resistant to water, chemicals, and staining. Olefin fibers are used in indoor/outdoor carpeting, wallpaper, and house wrap (e.g., Tyvek®). In consumer products, some of the olefin fibers were used in cold weather gear (e.g., Thinsulate®), ropes, upholstery, cigarette filters, and diapers.

Olefin fiber carpets are predominately manufactured by the polymerization of propylene. A thermoplastic polymer, polypropylene is melted and mixed with additives (e.g., colorants and stabilizers). Then, it is extruded through a spinneret (i.e., a plate with small holes) into fibers. If, however, strength is desired, you get strength through an alternative method called gel spinning.

Gel spinning is an old technique, new use. The polymer is dissolved in a small amount of solvent (i.e., 1%–2%), resulting in a viscous solution. This solution is either dry- or wet-spun to fibers that, in the end product, retain most of the solvent, or a polymer and solvent gel. While in the gel state, the fiber can be stretched in order to pull the molecules of the polymer into an ultra-strong elongated fiber. These fibers are three to four times stronger than polyester, and some allege that gel spun olefin fibers are 40 times stronger than Kevlar (Bailey 2010). This method of fiber production is used mostly for ballistics protection, not in carpeting.

Regulated irritant/toxic emissions are unlikely from olefin carpets, but there may be unregulated irritant plasticizer and additive (e.g., colorants and stabilizers) emissions.

Nylon Carpets

Nylon carpeting, a silk-like thermoplastic, was introduced in 1947. Nylon is a generic designation for a family of synthetic polymers comprised of aliphatic or semi-aromatic polyamides. Two types of nylon are typically used in carpet fibers. They are Type 66, a premium carpet with strength and durability and Type 6, a lesser quality carpet. Type 6 nylon, also referred to as Nylon 6, is derived from a single monomer—caprolactam, see Figure 15.3.

Type 66 nylon, also referred to as Nylon 6-6, Nylon 6/6, or Nylon 6,6, is comprised of two monomers—hexamethylenediamine and adipic acid. Each of the monomers contain six carbon atoms which is the rationale for the assignment of the double six, see Figure 15.4.

Both Nylon 6 and Nylon 66 polymers are extruded through spinnerets. Although nylon carpets are luxurious, water resistant, durable, resistant to wear and heavy foot traffic, and easy to clean, they are easily stained and discolored, and tend to accumulate static electricity (as much as 12,000 volts) (How Stuff Works 2015). In response to the pitfalls, most

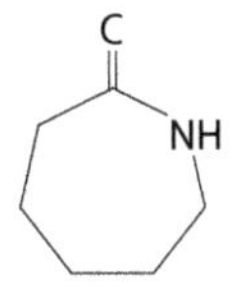

FIGURE 15.3
Nylon 2 monomer caprolactam.

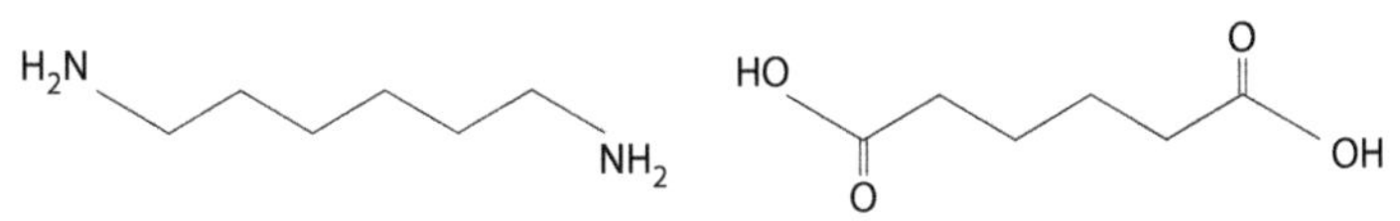

FIGURE 15.4
Nylon 66 monomers hexamethylenediamine (left) and adipic acid (right).

nylon fiber carpets today are treated with stain resistant chemicals and antistatic coatings.

Although regulated substance emissions are unlikely, many nylon carpets have additives (e.g., stabilizers, colorants, and flame retardants) that can potentially be irritating to the eyes and skin. See "Polymer Additives." Thermal decomposition of Nylon 6 and Nylon 66 may include ammonia and detectable levels of hydrogen cyanide and aldehydes.

Polyester Carpets

Polyester carpet, a soft thermoplastic, was first introduced in 1965. Beyond carpeting, polyester is used in textiles (e.g., clothing), plastic bottles and bottle caps, tire cord, food trays, and hoses. It is the other polyester products—mostly plastic bottle caps—that contribute to the making of polyester carpet. Thus, it has earned the heralded distinction as the one and only carpet made of "100% recycled products."

The polyester polymer is manufactured by mixing dimethyl terephthalate and ethylene glycol to form bis-terephthalate which is heated to 132°C (270°F) to form polyethylene terephthalate (PET). See Figure 15.5. Pellets of PET are reheated and extruded to form polyester fibers. See Figure 15.4. PET, in and of itself, is not likely to emit irritant/toxic chemicals. However, polyester carpet is not simply PET. There are also additives such as impact modifiers (e.g., rubber), compatibilizers, lubricants, colorants, and stabilizers. According to an Edinburg Plastics MSDS, the rubber modifier, identified by its CAS number, is poly(styrene-co-butadiene-co-methyl methacrylate). It is a nonregulated chemical which has been reported to cause eye irritation. Other additives (e.g., stabilizers), generally not identified by CAS number or name by the manufactures, may be irritants and/or sensitizers as well.

Polyester "fibers" are reputed to cause allergic contact dermatitis. When and where in direct contact with polyester fibers, sensitive individuals may experience mild to severe itching, hives, redness, and sometimes tenderness with an abnormal warm feeling at the site of contact. In extreme cases, severely sensitive people may also experience more generalized symptoms such as shortness of breath, tightness or pain in the chest, and respiratory difficulties (Allergy Symptoms 2016). Predominantly associated with polyester clothing, sensitivity to polyester carpeting may be

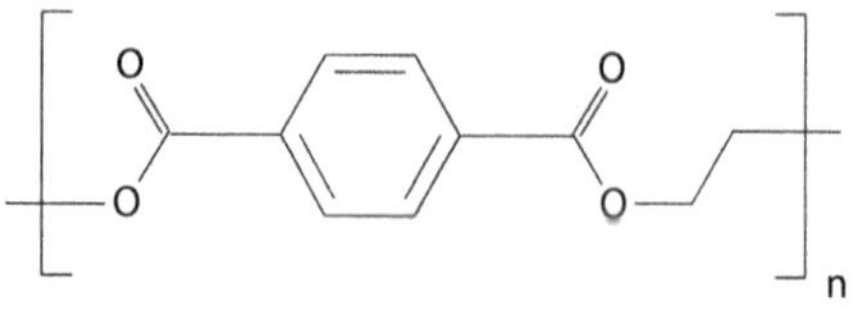

FIGURE 15.5
Structure of PET.

overlooked. Lying, walking, and sitting upon polyester carpeting may contribute to allergic contact dermatitis symptoms. It is not clear, however, as to whether the sensitivity is due to the polymerized fibers or to plasticizers and additives. Unbound plasticizers and additives are potential polyester carpet emissions. See Chapter 8, "Plasticizers" and Chapter 9, "Plastic Additives."

Polyester is not flammable and does not easily ignite. Once ignited, however, polyester melts as it burns at which time the melted polymer is likely to severely melt and damage exposed skin (e.g., burning polyester clothing to skin).

As presented by a MSDS for "polyester carpets," the thermal decomposition products from polyester carpets with their additives include ammonia and small amounts of hydrogen cyanide and aldehydes. This is likely to reflect all components—inclusive of the additives.

Wool Carpets

Wood is natural fiber, and wool carpets are luxuriously soft, durable, stain resistant, fire resistant, nonallergenic, and renewable. Yet, it is expensive; colors tend to fade in sunlight; and wool is susceptible to moth damage. Irritant/toxic emissions are not likely from untreated wool.

Carpet Padding

Carpet padding, also referred to as carpet cushion and carpet underlay, are predominantly polymers. They are urethane foam, rubber, and natural fiber.

Urethane foam padding is a thin, low-density pad whereas "bonded" urethane foam padding is thicker, higher density carpet padding. The bonded urethane pad is comprised of recycled high-density urethane foam used in furniture and automotive manufacturing. Potential emissions of regulated isocyanates and unregulated additives are possible from foam carpet padding. The combustion products include oxides of nitrogen and hydrogen cyanide. For more details, see "Polymeric Foams: Polyurethane."

Waffle rubber padding is a low density, "waffle-shaped" rubber pad of low density, and flat, rubber padding is high density "ultimate" rubber carpet padding. As discussed in the chapter regarding elastomers, nonregulated product emissions from most, if not all, rubber may cause eye, skin, and respiratory tract irritation. In the NIOSH review mentioned in the carpeting section, SBR latex adhesive for binding carpet secondary backing has typically been identified as the primary source of 4-PCH. In the same review, a German study reported headaches, eye irritation, and nausea associated with 4-PCH (Haneke 2002, p. 10). Yet, it still remains unconfirmed as to the source of eye, skin, and respiratory tract irritation. For details regarding rubber emissions, see Chapter 7, "Elastomeric Polymers." The thermal decomposition products of SBR include small amounts of sulfur dioxide (Janowska 2010).

Resilient Flooring

Resilient flooring is affordable and durable—able to spring back into shape after bending, stretching, or being compressed. Prior to the 1900s plastics and synthetic rubbers, linoleum, contrary to popular misconceptions, was developed from nature's bounty as were natural rubber tiles and cork flooring. In the 1900s, there came a transition period wherein synthetic vinyl flooring was introduced and began to outpace the others, and synthetic rubber tiles became a niche item.

Linoleum

Linoleum, the original sheet flooring, was introduced in the 1860s in Europe and sold under the name of Kampticon which was a popular cloth floor covering at the time. Yet, shortly thereafter, Kamticon was renamed linoleum after the Latin words "linum" (flax) and "olium" (oil), and manufacturing began in the United States in 1872.

As the term "linoleum" implies, its principal component is solidified linseed oil. Linseed oil is mixed with pine rosin, ground cork dust, wood dust, and mineral fillers (e.g., calcium carbonate) and poured over a pigmented or unpigmented burlap or canvas backing. Embossed inlaid linoleum was introduced in 1926. Due to its purported nonallergenic quality, linoleum sheeting and tiles are still in use. Although product emissions are unlikely, linoleum is flammable. The combustion products are non-impressive.

Cork Flooring

Cork is the outer bark of trees—most commonly the cork oak (*Quercussuber*). It is a nonallergenic, sound/temperature insulator, soft and resilient. Yet, it is high maintenance and susceptible to water/moisture damage. Solid cork is a natural fire retardant (Cork Link 2015). Irritant/toxic emissions are highly unlikely.

Rubber Flooring

Natural rubber floor tiles were developed in 1894 in the United States. Colors were limited. They were durable, sound-deadening, and easy to clean. They were, however, easily stained. They deteriorated overtime. Thus, natural rubber flooring was short lived!

Synthetic rubber flooring was reintroduced to the market after World War II, and synthetic rubber is recyclable. As discussed previously, synthetic rubbers are suspect of off-gassing nonregulated irritants and 4-PCH. For details, see *Carpet Padding* in this chapter.

Vinyl Flooring

Vinyl flooring is a synthetic polymer that is durable, versatile, affordable, and water/moisture resistant. Vinyl, PVC, is readily decomposed at elevated temperatures, liberating hydrogen chloride. Indoor elevated temperatures may occur where there is extreme solar loading, where in-floor heating is installed under vinyl floor tiles, and where radiator heating is in close contact with vinyl floor tiles. In elevated temperatures, vinyl flooring is likely to off-gas hydrogen chloride. Yet, the elevated temperatures required for vinyl tiles to emit hydrogen chloride are not normal. Under most conditions, irritant/toxic off-gassing is not likely.

Be forewarned! Vinyl floor tiles and black mastic floor tiles adhesives manufactured from origin until 1986 are likely to be composed of asbestos.

Terrazzo Flooring

Terrazzo is a composite material composed of chips and fine aggregates of granite, marble, quartz, glass, and/or other colorful chips and pieces (e.g., mother of pearl, onyx metal, and recycled porcelain/concrete). It is either sprinkled on or poured with a binder onto a substrate (e.g., cement). The binder may be cementitious, polymeric, or a combination thereof, and although the original fifteenth century Venetian terrazzo substrate was clay, most terrazzo substrate today is concrete. The binders were polyester and vinyl ester resins in the 1970s, and today the binder is typically an epoxy resin. It is generally made on-site but can be manufactured as tiles, and the final product is polished walls, floors, patios, and panels. Irritant/toxic emissions are unlikely from sandstone and limestone. For details regarding binders, see Chapter 16: Adhesives.

Trim, Molding, and More

Trim, molding, and more are the finishing touches, the final destination. From the traditional basics to splendor and elegance, the trim may be simply functional or intricately complex. Thus, the range of finishing touches to a structure is comprised of a cornucopia of building materials. For a description and listing of mold, trim, and more materials, see Table 15.3.

Wood trim/molding may be solid hardwood—exotic or domestic—or it may be composite wood. It may be unfinished (e.g., au natural cedar), stained, varnished, or painted. Composite wood used in trim is similar to that of

TABLE 15.3

Molding, Trim, and More Materials

Basic Description	Typical Component Material(s)
Baseboards (also referred to as base trim, base skirting, and base molding)	
Contrary to most lay persons' perception, its main purpose is to hide imperfections and gaps in wall materials and to cover gaps due to floor movement (e.g., expansion and contraction of hardwood flooring)	1,**2**,3,**4**,5,6,7,8
Its secondary purpose is aesthetics and protection from damage	
Cove base—typically a curved "vinyl baseboard" that is most frequently encountered installed in kitchens and/or commercial buildings	
Crown molding (also referred to as ceiling trim)	3,4
Amazing! The original purpose of crown molding was to provide a surface to hang pictures— wherein the early 1900s plaster walls were too hard to hang pictures on directly	1,**2**,3,4,5,6
Today, its purpose is aesthetics	
Window/door trim and window ledges	1,**2**,3,5,6,7
Trim (also referred to as molding)—covers joints around windows (and door openings) while making a design statement	
Window ledge (also referred to as an interior window sill)—a narrow horizontal surface resembling a shelf and projecting from the bottom of a window either on the inside or outside	
Stair rails	
Handrail—a rail that is fixed to posts or to a wall for people to hold onto for support	1,**2**,5,6
Baluster—(also referred to as a spindle or stair stick) a molded shaft or form, typically made of wood, wrought iron, or stone, that supports the handrail of a staircase	1,**2**,3,5,6,7
Newel—the central pillar or upright from which the steps of a winding stair radiate	1,**2**,3,5,6,7
Crash rails—generally flat rails mounted on a wall that prevent wall damage	1,**2**,3,4,5,6

Note: **Bold**: Likely irritant/toxic emissions from newly manufactured products.
1 Solid wood
2 Composite wood (e.g., laminate, veneer wood, and MDF)
3 Vinyl
4 Rubber
5 Polyurethane
6 Metal (e.g., stainless steel, aluminum, and wrought iron)
7 Natural and stone
8 Porcelain and ceramic tiles

wood flooring. It may be veneered wood or laminate, or it may be painted MDF. Formaldehyde emissions are likely from newly manufactured composite wood. For further information, see Wood Flooring within this chapter.

Vinyl, rubber, and PU trim/molding are a colored, extruded polymer. Where temperatures are less than 100°F, polymers are not likely to off-gas regulated irritant/toxic chemicals. Unregulated irritant emissions of unbound additives are, however, possible. For further information, see Section II, "Polymers."

Whereas metal is used in combination with other building materials, it is used mostly in stair rails. Disregarding occupational exposures to metal fumes when welding or cutting, metal building materials alone do not off-gas irritants and/or toxins.

The more pricey, the less frequently used is trim/molding stone, composite stone, and porcelain/ceramic tiles. Beyond occupational exposures to crystalline silica exposures, building materials emissions are not likely.

Summary

A witches' brew of formaldehyde and unregulated irritants arises from within the confines of the interior—finishing touches gone wild! As if we didn't have sufficient challenges with other building materials, the cabinets, countertops, flooring, trim, and other finish-out building materials just amped up contributions to poor indoor air quality.

Some of the contributors to detectable to high levels of formaldehyde emissions include

- Veneered cabinets
- Some metal steel cabinets
- Laminated cabinets
- Melamine cabinets
- Laminated countertops
- Veneered flooring
- Laminated flooring
- Paper-composite countertops
- Some carpet backing and adhesives

In reference to laminated flooring and air sampling for formaldehyde emissions after the flooring has been installed, the likelihood for testing for off-gassing from the laminate only is an absurd notion. This has been attempted and has likely ended up in court. The efficacy of any sample results, positive or negative to the case, is questionable. If all the sources of formaldehyde have not been identified, affirmation of emissions from a single source is an effort in futility.

Another component of the witches' brew is the irritant 4-phenylcyclohexane which has been targeted by green building programs like LEED and the ANSI/ASHRAE Standard 189.1. Some of the sources include

- Laminated floor rubber padding
- Rubber carpet backing
- Rubber carpet adhesives
- SBR latex carpet padding

Other product emissions likely include

- VOCs from carpet backing, face fibers, and adhesives
- Plasticizers and additives from synthetic carpets

The medley of possibilities is just the beginning! New product introduction will likely increase the cauldron of unknown irritant/toxic emissions, but you now have a good running start.

References

About Education. 2015. "All About Sandstone." *About*. Accessed November 17, 2015. http://geology.about.com/od/more_sedrocks/a/aboutsandstone.htm.

Allergy Symptoms. 2016. "Polyester Allergy." *Allergy SymptomsX*. Accessed November 14, 2016. http://allergysymptomsx.com/polyester-allergy.php.

Bailey, Jon. 2010. "UHMWPE: Gel Spinning." *Archimorph*. May 26. Accessed November 24, 2015. http://archimorph.com/2010/05/26/uhmwpe-gel-spinning/.

Cork Link. 2015. "The Amazing Natural Properties of Cork." *Cork Link*. Accessed December 2015. http://www.corklink.com/index.php/the-amazing-natural-properties-of-cork/.

Haneke, Karen. 2002. *4-Phenylcyclohexene Review of Toxicological Literature*. Literature Review, Research Triangle Park, North Carolina: NIOSH.

Hess-Kosa, Kathleen. 2011. *Indoor Air Quality: The Latest Sampling and Analytical Methods*. Boca Raton, Florida: CRC Press/Taylor & Francis Group.

How Stuff Works. 2015. "Choosing Carpet Fiber: Nylon." *How Stuff Works*. Accessed November 24, 2015. http://home.howstuffworks.com/home-improvement/carpet/choose-right-carpet-fiber1.htm.

Janowska, Grazyna. 2010. "Flammability of Diene Rubbers." *Springlink*. June 4. Accessed December 1, 2015. http://www.researchgate.net/publication/239115848_Flammability_of_diene_rubbers.

Johnson, P.K., E. Doyle, and R.A. Orzel. 1988. "Acrylics: A Literature Review of Thermal Decomposition Products and Toxicity." *Sage Journals*, 139–200. March/April. Accessed 24, 2015, November. http://ijt.sagepub.com/content/7/2/139.short.

Kiron, Mazharul I. 2014. "What is Modacrylic Fiber." *Textile Learner* (blog). Accessed November 2015. http://textilelearner.blogspot.com/2011/08/what-is-modacrylic-fiber-properties-of_9408.html.

Lee, Shannon. 2014. "Choosing the Countertop for Your Old House Remodel." *Old House Web* (blog). Accessed November 6, 2015. http://www.oldhouseweb.com/blog/choosing-the-countertop-for-your-old-house-remodel/.

Nova. 2012. "High-Pressure vs. Low-Pressure Laminate Furniture." *Nova Desk*. January 30. Accessed November 5, 2015. http://www.novadesk.com/blog/bid/50596/High-Pressure-vs-Low-Pressure-Laminate-Furniture.

The Carpet and Rug Institute. 2015. "Understanding Carpet Construction." *Carpet and Rug Institute*. Accessed December 3, 2015. http://www.carpet-rug.org/Carpet-for-Business/Specifying-the-Right-Carpet/Carpet-and-Rug-Construction.aspx.

Wilsonart. 2015. "Laminate Sustainability Information." *Wilsonart*. Accessed November 5, 2015. http://www.wilsonart.com/wilsonart-laminate-sustainability-information.

16

Adhesives, Sealants, Surface Finishes, and More

The glue that holds it all together, the sealant that closes the gaps, the finish that protects—topics not to be ignored. They are the cohesive components of a structure, assuring that the house of cards remains standing, air and water leaks are plugged, and surfaces are embellished and protected. All things considered, a medley of seemingly minor building materials is an important contributor to the whole.

Whereas the adhesives and sealants are generally out-of-sight, out-of-mind, they are used extensively throughout a structure. Their potential impact on the overall air quality certainly deserves attention. Surface finishes, inclusive of paints and varnishes, are already target building materials—part of the green movement.

Adhesives

Adhesives, also referred to as glues, are any organic material that forms a bond between two porous and/or nonporous objects. They are either natural or synthetic.

The natural glues date back to the Egyptian papyrus scrolls—a water plant fiber bonded with flour paste—and veneers/inlays. Bitumen, tree pitches, and beeswax were used as water proofing sealants and adhesives in ancient and medieval times. Gold leaf manuscripts were bound with egg white, and wood was bound with glues from fish, horn, and cheese. In the eighteenth century, advances were made in animal and fish glues, and rubber- and nitrocellulose-based cements were introduced in the nineteenth century.

Today, some of the natural adhesives that remain in use within building materials are blood albumen glues, sometimes referred to as nature's phenolic (used as a moisture resistant veneer glue), starch glues (used as a poor moisture resistant wallpaper glue), casein glue (used as a moderate moisture resistant veneer glue), and soy-protein glue (potential as a formaldehyde-free indoor plywood adhesive and veneer glue).

By the twentieth century, the stronger, more versatile, affordable synthetic adhesives began to replace the natural glues.

Today, almost all construction adhesives are manmade–synthetic polymers. Rarely are natural polymers used in construction. See Table 16.1. Synthetic adhesives are thermoplastic or thermoset glues. Thermoplastic glues melt when heated; they are soluble in organic solvents; they are creep resistant; and they are brittle unless plasticizer is added to the polymer. Thermoset adhesives do not melt when heated; they are insoluble in organic solvents; and they have a high resistance to creep. The synthetic glues are highly variable as far as their formulation and health hazards which are discussed herein, and most are suspended in a water-based or organic solvent. See Table 16.1.

Thermoplastic Adhesives

Thermoplastic polymer adhesives used in construction include acrylics, ethylene-vinyl acetate, and combinations thereof. These adhesives are polymers that have already completed polymerization—not the flammable component monomers. In other words, they do not require a catalyst and/or long curing times.

Thus, the glue is packaged in its polymeric form in such a manner that the polymer does not harden until it is used as a glue. All thermoplastic adhesives are suspended in a carrier agent. The polymer is either suspended in water or in an organic solvent. As the water dries or the solvent vaporizes, the adhesive hardens. VOC emissions are likely when the carrier agent is an organic solvent.

In order to attain flexibility, many thermoplastic polymers are comprised of unbound, unregulated plasticizers and plastic additives (e.g., stabilizers, flame retardants, and fillers). The unregulated plasticizers may cause eye, skin, and respiratory tract irritation. Yet, plasticizers and plastic additives are not generally disclosed by the manufacturers, because they are not regulated. A good rule of thumb may well be that if the thermoplastic glue is flexible, it is likely to contain plasticizer. Manufactures generally do, however, disclose the fillers—particularly where they include quartz and/or crystalline silica which usually has an occupational exposure warning regarding sanding and abrading the finished product. Fillers are generally comprised of limestone, kaolin, and/or quartz—potential crystalline silica exposures to workers. Once again, plasticizers and most plastic additives are typically undisclosed, unknown.

The most commonly used thermoplastic adhesives in construction are polyvinyl acetate and acrylic glues. They represent both the water suspended and organic solvent suspended glues.

Polyvinyl Acetate Adhesives

Polyvinyl acetate (PVAc) adhesives are a rubbery adhesive, a polymer comprised of the monomer vinyl acetate and an initiator (e.g., typically aluminum chloride) that is suspended in water. A milky white emulsion is instantly

TABLE 16.1

Construction Adhesives by Polymer

Construction Adhesives	Product Examples[a]
Natural glues	
Blood albumen glue (nature's phenolic)	Laminated wood and plywood—furniture
Starch and dextrin	Corrugated board and wallpaper adhesives
Casein glue	Veneer glue
Soy-protein glue	Formaldehye-free indoor plywood adhesive and veneer glue
Thermoplastic glues	
Acrylic adhesives (methacrylate)	Liquid Nails® LN-710 Paneling & Molding, Airtight Weather Ban Acrylic Sealant (siliconized)
	Liquid Nails® Ultra Quick Grip, Liquid Nails® Low VOC Heavy Duty Construction Adhesive
	Liquid Nails® Panel and Foam Adhesive, Liquid Nails® Subfloor and Deck Adhesive
PVAc	Elmer's Glue All, Titebond White Glue, LePage's Bondfast, GF Glue, Titebond II Wood Glue
	Locktite 375 Heavy Duty Construction Adhesive, Locktite Projects & Construction
	Adhesive, Grab Heavy Duty, Liquid Nails® Drywall Adhesive
	Liquid Nails® Fiberglass RPP Adhesive
PVAc-ethylene and proprietary	Scotch-Weld Hot Melt Adhesice
Acrylic/PVAc mixture	Liquid Nails® Extreme Heavy Duty
Thermoset glues	
Epoxy glue (2 part resin)	Elmer's Epoxy Glue, Scotch-Weld, Cold Cure Epoxy, Super Glue Plastic Fusion Epoxy
	Scotch-Weld Epoxy Adhesive, Perma-Loc, Pronto Instant Bond and Loctite
PU adhesives	Gorilla Glue, Loctite PL Premium Polyurethane Adhesive, Liquid Nails® Marble and Granite Adhesive
Synthetic rubber adhesives	Liquid Nails® Heavy Duty Construction Adhesive, Liquid Nails® Projects & Construction
	Titebond Heavy Duty Construction Adhesive

[a] Manufacturer product inclusions and/or omissions in Product Example listings are based on availability of information. Some claim "no hazardous ingredients." Some manufacturers only include regulated hazardous ingredients without identifying the type of adhesive (e.g., product type: liquid). Some declare "proprietary" information (e.g., no information). And very few actually disclose a comprehensive list of "components."

processed upon moisture extraction and/or water evaporation. Some PVAc adhesives are also comprised of an ethylene and/or alcohol (e.g., ethylene glycol) copolymers. Vinyl acetate and some of the copolymers are regulated. Yet, in its polymer form, PVAc monomers emissions are unlikely.

Acrylic Adhesives

Acrylic adhesives are more brittle adhesives, a polymer comprised of methyl methacrylate and butyl acrylate monomers. Without additives, acrylic polymers are transparent having a glass-like clarity. Many acrylic adhesives, however, contain plasticizers for a softer, easier-to-manipulate adhesive, and all contain fillers (e.g., limestone). The polymer is typically suspended in organic solvents. They are likely to emit regulated and unregulated organic solvents when applied.

Thermoset Adhesives

Most thermoset adhesives used in construction are two-part epoxies, two-part resorcinol, and two-part polyesters. They are each comprised of a complex multicomponent polymer with additives and a hardener which has a catalyst to decrease curing times.

Epoxy Resins

Epoxy resins, also referred to as epoxides and polyepoxides, are two part resins comprised of a resin and a hardener which is the curing agent. Whereas resins are generally unregulated and nonirritating, many hardeners are likely regulated toxins.

The hardeners may include a one, or a combination, of the following— amines, acids, phenols, alcohols, and thiols. And rarely, not to be overlooked a hardener worthy of mention is a mercaptan (e.g., methyl mercaptan) which is a regulated toxin, highly odiferous at low levels (smells like a skunk), and highly flammable.

> One of the most common construction epoxy resins is Bisphenol A. Whereas Bisphenol A is the resin, epichlorohydrin is the hardener. See Figure 16.1. The mixing ratio of resin to hardener is typically, but not always, 1:1. Now, if the resin and hardener are poorly mixed, the resin will remain tacky, fail to cure. On the other hand, if there is too much resin or hardener in the mix, the resin/hardener excess will off-gas after the finished product has cured. Although the Bisphenol A resin is described as a known endocrine disrupter, it is not an airborne regulated toxin—an excess of resin is unlikely to affect human health. However, an excess of uncured epichlorohydrin will emit a irritating, chloroform-like (some describe it as pungent, garlic-like) odor. It is an OSHA regulated toxin—an eye, skin, and respiratory irritant, a corrosive to the skin and eyes, and a "probable human carcinogen."

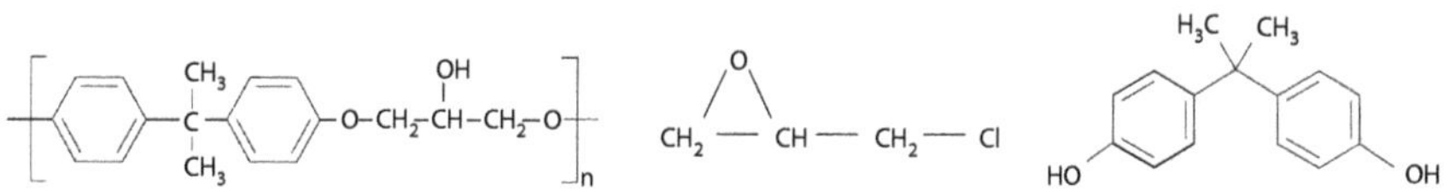

FIGURE 16.1
Epoxide structure (left); epichlorohydrin (center); and bisphenol A (right).

As the drum beats on, the complexity of the thermoset resin chemistry is truly complex, outside the scope of this book. Let it suffice to say that epoxies have a long curing time during which various irritant/toxic emissions are possible—particularly during the mixing and curing process. Epoxy resin emissions are reported to cause mild eye irritation, redness and skin irritation as well as possible sensitization. The resin may be hard enough to sand after a couple of hours, but it may not be completely cured for up to 2 weeks (Sentry Air 2010). Dust generated prior to the completion of the curing process can result in respiratory inflammation, irritation, and with continued exposure can result in respiratory sensitization and asthma (Sentry Air Systems 2010).

The chemical reaction between resin and hardener is "exothermic." In other words, when the epoxy is mixed, it "will" generate heat. If left unattended to cure in a large mass, it can generate enough heat to burn your skin and/or ignite surrounding combustible materials. The larger and/or thicker the uncured epoxy mass—the greater the heat generated. A 100-g mass of mixed epoxy can reach 400°F and cause a fire (West System 2016).

PU Foam Adhesives

PU construction adhesives are monomer isocyanates and polyol (e.g., glycerol)—suspended in organic solvents. They typically require a 24-hour minimum solvent evaporation cure time. During the hardening process, organic solvent (e.g., mineral spirits) emissions are likely. Some of the suspended organic solvents are regulated toxins and generally have a "petroleum" odor. Then comes the isocyanate monomer.

Free isocyanates are typically residual in the end products. It is similar to UF resins whereby the formaldehyde is typically residual in the end product—off-gassing for the product(s) after the curing process has been completed. PU foam adhesives are very similar. Yet, the isocyanates are residual and emissions are likely. Isocyanates have no odor and the exposure symptoms are eye, skin, and respiratory tract irritation, progressing to skin and/or respiratory sensitization.

Synthetic Rubber Construction Adhesives

Elastomeric construction adhesives are synthetic rubber generally suspended in organic solvents. The synthetic rubber generally encountered is

SB—the result of the copolymerization of styrene (i.e., ethenylbenzene) with 1,3-butadiene. The adhesives are typically suspended in multiple organic solvents—including, but not limited to, acetone, hexane, toluene, cyclohexane, and heptane. During the application and curing processes, organic solvent emissions are likely in standard SB adhesives.

> Whereas the suspended organic solvent, or volatile organic compound (VOC), content in most SB construction adhesives may be as high as 400 grams per liter, manufacturers claim VOC-compliant "green glue" wherein the SB construction adhesive has up to 70 grams per liter VOCs.
> Some manufacturers claim the solvents are "de-aromatized hydrocarbons." These are hydrocarbons without the more toxic aromatic benzene and benzene derivatives. This includes the more commonly known toxic organics such as toluene, xylene, benzene, and ethyl benzene. The list of de-aromatized hydrocarbons does not, however, include all toxic organic such as cyclohexane and n-butanol.

In construction, SB adhesives are used in heavy duty construction glues and in rubberized carpet backing. SB construction glues are strong, water/weather resistant, and flexible. It is recommended for use on countertops, cabinets, brick veneer, plywood and wafer board, foil insulation board, tile board, treated lumber, drywall, and most common building materials (concrete, lumber, etc.).

Be forewarned! Two different manufacturers may have the same assigned tag line (e.g., Heavy Duty Construction Adhesive) but not the same components. For instance, Liquid Nails® Heavy Duty Construction Adhesive is an SB synthetic rubber adhesive; Liquid Nails Low VOC Heavy Duty Construction Adhesive an acrylic adhesive; and Loctite® 375 Heavy Duty Construction a PVAc adhesive. All are referred to as "general purpose" construction adhesives.

Caulks and Sealants

In the not too distant past, caulk was a term used in boat construction, and sealant a term used in building construction. Recently, however, caulk describes an all-purpose waterproofing material, and sealant describes high-performance products. More often than not, however, the terms are used interchangeably. Their purpose is to fill gaps between building materials, prevent water and/or air movement—fix cracks, fill less than ½ inch gaps in structures, and seal around windows. They can be classified as water-based, synthetic-rubber, silicones, elastomeric polyurethane, and butyl rubber. Each is specific for different building materials (e.g., wood, glass, concrete, metal, rubber, vinyl, window, and door frames). See Table 16.2. Most caulks

TABLE 16.2

Types of Sealants

Sealant Type	Examples
All contain: Oxydipropyl dibenzoate	
Vinyl acetate–ethylene copolymer	Liquid Nails® Acoustical Sealant, Liquid Nails Window Glazing, All Purpose Window Caulk
	Liquid Nails Energy Saving Multipurpose Caulk
SB rubber	Liquid Nails Clear Seal All Purpose Caulk
Acrylic	Liquid Nails Concrete Repair, Liquid Nails Window and Door Supercaulk
Silicone	Liquid Nails Silicone Sealant

are synthetic organic polymers, and some are inorganic polymers. Some are suspended in water some are suspended in organic solvents. Most are similar to adhesives with the differences being the amount and type of plasticizers and additives which are generally not identified in manufacturer MSDSs. They are not considered hazardous ingredients, and proprietary information guards against full disclosure of all components—with but a few exceptions. Very few manufacturers disclose all component information—regulated and nonregulated.

Water-Based Caulks

Water-based caulks, also referred to as latex caulks, are smooth, easy to apply, clean up with water, have little or no odor, and are generally paintable. They are best applied in warm, dry weather. Humid conditions slow the curing process.

They are generally comprised of thermoplastic polymers, similar to polyvinyl acetate and acrylic latex adhesives with, up to 10% plasticizer (e.g., oxydipropyl dibenzoate). After curing, monomer emissions are unlikely whereas plasticizer emissions are possible when caulked building materials are exposed to extreme temperatures (e.g., radiant heat in close proximity to caulked building materials). In brief, water-based caulks emit little or no VOCs and are nonflammable.

Silicone Caulk

Silicones have a long history, evolving over 50 years. Although silicone caulks cannot be painted, silicone stands up to extreme weather. It cures soft and remains flexible and is sometimes "inappropriately referred to as a rubber caulk." Silicones are best used on glass, metal, and tiles.

Silicone is an inorganic polymer. Irritant/toxic emissions are highly unlikely. In brief, there are no VOC emissions from silicone caulks, and the silicone caulks are non-flammable. In brief, silicone caulks emit no VOCs and are nonflammable.

Synthetic Rubber Caulk

Synthetic rubber caulks are the most elastic of all the caulks. They stretch and recover up to 500% without cracking. For this reason, rubber caulks are excellent for expansion and contraction adhesion—filling large areas. They are paintable. They adhere to most substrates. They can be applied in wet and cold weather, and they are mildew resistant.

During the application and curing processes, organic solvent emissions are likely—to a greater extent in standard synthetic rubber caulk and to a lesser extent in low VOC caulks. Also, due to their organic solvent content, synthetic rubber caulk is highly flammable particularly during application and curing. The end product may, also, be flammable. In brief, synthetic rubber caulks emit flammable VOCs.

Elastomeric Polyurethane Caulk

Elastomeric polyurethane (PU) caulk is the good, the bad, and the ugly. It is the toughest, most abrasion-resistant caulk in the lineup of choice adhesives. It is flexible, water resistant, paintable, and dries rigid. Yet, it does have its Achilles' heel.

The regulated irritant/toxic isocyanates are likely to off-gas for an undisclosed, unclear period of time—during the 4 to 7 day curing time and potentially thereafter. Organic solvent emissions are also likely. In brief, elastomeric polyurethane caulks may emit irritating isocyanates, and they are combustible.

Butyl Rubber Sealant

Butyl rubber sealants stretch like chewing gum but "do not recover." It has an unsightly tar-like in-appearance but butyl rubber sealant is the most water-resistant product available. It is used mostly for water proofing exterior surfaces (e.g., gutters). Butyl rubber monomers are nontoxic polyisobutylene and isoprene. The polymer is suspended in organic solvents that are generally flammable and toxic. Organic vapor emissions are highly likely. For more in-depth information regarding emissions, see Chapter 7. In brief, butyl rubber caulks may emit VOCs, and they are flammable.

Be forewarned! Caulk manufacturers often used asbestos before the 1970s as a fire retardant. Common places in where it was used are around ovens, fireplaces, boilers, pipe joints, ducts, brickwork, and exteriors. Caulk can be damaged, release asbestos due to aging, excessive water, impact, drilling, sanding, scraping, or attempts to remove it.

Paints

From cave drawings and petroglyphs to art to architectural coatings, the most durable paints have been comprised of "lead." The Egyptians painted murals and hieroglyphics in their tombs and temples with paint comprised of lead as well as ground glass, semiprecious stones, colored earth, animal blood mixed with oils, resins, and fats. To this day, over 5000 years later, the Egyptian murals retain their brilliance. Some attribute this to the use of lead. The Greeks and Romans used similar formulations to paint murals.

Later, artists mixed white-lead paste with linseed oil, turpentine, and other color pigments. Prior to the 1600s, paint was almost exclusively used by artists for illustrating images on walls, ceiling, and canvases. During the nineteenth century, paint gained popularity as a means to both beautify and to protect buildings. Linseed oil (i.e., oil-based) paint protected, and lead gave the paint long-lasting durability. Lead-based paints stuck where the paint was applied, and they weathered well.

During World War II, linseed oil and solvents became scarce. Necessity is the mother of invention! So, in the middle of 1950s, polymer paints were formulated, and they have been found to outperform linseed oil and natural ingredients. Although banned by the U.S. EPA in 1978 from artists' and house paints, lead is still widely used in industrial applications and for painting road markings.

All paints, both past and present, have four basic components. They are pigments, binders, solvent(s), and additives. The following discussion predominantly regards the paint formulations of today.

Pigments

All pigments provide color while hiding the underlying building material. Many also provide corrosion (e.g., lead, chromium, zinc, and magnesium) and/or mold resistance (e.g., mercury-containing pigments). In a U.S. EPA nanomaterials study, completed in 2000, researchers made an unconfirmed, startling comment regarding inorganic pigments.

> Inorganic color pigments, based on hazardous metals such as chromium, mercury, cobalt, lead, etc., account for over 95% of the United States and worldwide pigments consumption. Color pigments are commonly used by numerous industries and in various consumer products. It is estimated that they are one of the largest vehicles of hazardous metal-based chemicals commerce. A technology that can provide cost-effective, nonpolluting alternative to pigments with nonhazardous formulation while maintaining color performance would make a major impact in preventing pollution. The feasibility study successfully demonstrated that precision nanolaminated materials with tailored optical properties have a strong potential for application as specialty high performance pigments and colored coatings.

Many paint manufacturers today disclose "trade secret" nontoxic pigments in their SDSs for paints. It is not clear as to whether nanolamination of toxic pigments has been universally accepted and implemented. Furthermore, the manufacturers do not generally disclose the potential release of toxic metals during a fire.

Paint pigments include primary pigments and extenders. Primary pigments—inorganic and organic—contribute color. Inorganic pigments yield durable, long-lasting color, and organic pigments yield brighter colors which are, however, neither durable or long lasting inorganic pigments. Extenders add thickness. Binders hold the paint together. Carriers are the solvents. And additives modify. All are discussed herein.

Primary Inorganic Pigments

Inorganic pigments may be made of finely ground powder which does not dissolve. It is merely suspended in the solvent (e.g., water and/or oil). The primary pigments are not bound within a matrix along with the extenders. And beyond their function as a primary colorant, inorganic pigments sometimes serve other purposes, such as zinc oxide which is used to prevent corrosion and mold growth.

The primary inorganic pigments may be comprised of nontoxic minerals (e.g., titanium oxide) and/or regulated toxic minerals (e.g., lead chromate). While lead is generally considered the number one toxic pigment of concern, there are other inorganic pigments that are as toxic as or more toxic than lead. Other toxic pigments encountered in paints include hexavalent chromium and arsenic. In addition, many toxic inorganic pigments are either used exclusively in artists' paints or are infrequently encountered in commercial paints. See Table 16.3.

Lead-Containing Pigments

Lead-containing pigments in paint are on the U.S. EPA's "Most Dangerous Hazardous Building Materials" list. Yet, contrary to common perception, the ban on lead in paint is largely focused on structures and environments where young children are likely to be exposed—not on all structures and environments.

> Lead-containing paint is the most common source of lead exposure to preschool children. Containing up to 50 percent lead, (lead-containing) paint was in widespread use until the 1950s when it became common knowledge that exposures to lead posed a health hazard. Thereafter, exterior lead-based paint" and decreased usage of interior lead-based paint continued until 1978.
>
> About 74 percent of privately owned, occupied housing units in the United States built prior to 1980 were coated with lead-based paints. Exposures to preschool children result when they ingest paint chips or paint-contaminated dust and soil. Many exposures also result when

TABLE 16.3

Toxic Components in Inorganic Paint Pigments

	Color	Pigment Component	Toxic Component
Antimony	White	Sb_2O_3	
Arsenic	Bright yellow	As_2S_3	Inorganic arsenic
Barium	White	Barium sulfate ($BaSO_4$)	
Copper	Purple		
	Blue	Synthetic	
	Paris green	Cupric acetoarsenite ($C_4H_6As_6Cu_4$)	Cobalt
	Scheel's green	$CuHAsO_3$	Inorganic arsenic
Cobalt	Violet	Cobaltous organophosphate $[Co_3(PO_4)]$	Cobalt
	Blue	Cobalt(II) stannate (CoO_3Sn)	Cobalt
	Cobalt yellow	Potassium cobaltinitrite $[Na_3Co(NO_2)_6]$	Cobalt
Manganese	Violet	Manganese ammonium pyrophosphate ($NH_4MnP_2O_7$)	Manganese
	Blue	Manganese oxide	Manganese
Cadmium	Cadmium yellow	Cadmium sulfide (CdS)	Cadmium
	Cadmium orange	Cadmium sulfoselenide (Cd_2SSe)	Cadmium
	Red	Cadmium selenide (CdSe)	Cadmium
Chromium III	Viridian (dark green)	Chromic oxide [Cr_2O_3 (Cr+3)]	–
	Green	Hydrated chromic oxide [Cr_2O_3 (Cr+3).H_2O]	–
Chromium VI	Chrome yellow	$PbCrO_4$	Lead chromate
	Chrome orange	$PbCrO_4$ + PbO	Lead chromate
	Strontium yellow (deep lemon yellow)		Strontium chromate
	Zinc yellow	Zinc chromate ($ZnCrO_4$)	
Iron	Yellow ochre	$Fe_2O_3.H_2O$	–
	Red	PR103	–
	Red ochre	$Fe_2O_3.H_2O$	–
	Raw umber (brown)	Fe_2O_3 + MnO_2 + nH_2O + Si + AlO_3 Clay	–
	Burnt umber (brown)	Heated raw umber	–
	Raw sienna (red-brown)		–
	Black	Fe_3O_4	–
Lead	Naples yellow	Lead antimonate [$Pb(SbO_3)_2$/$Pb_3(SbO_4)_2$]	Lead

(Continued)

TABLE 16.3 (*Continued*)

Toxic Components in Inorganic Paint Pigments

	Color	Pigment Component	Toxic Component
	Red lead (orange red)	Lead tetroxide (Pb_3O_4)	Lead
	White lead	Lead carbonate [$PbCO_3.Pb(OH)_2$]	Lead
Mercury	Vermilion (orange-red)	Mercuric sulfide (HgS)	Mercury
Titanium	Titanium yellow		
	Titanium white	Titanium oxide (TiO_2)	–
Tin	Mosaic gold	Stannic sulfide	–
Zinc	White	Zinc oxide (ZnO)	–

homes are remodeled or renovated without the proper precautions being taken. Lead-containing paint is typically found on kitchen and bathroom walls, doors, and wood trim of houses. It was used on children's playground equipment where small children may chew the surfaces and in house paints for interior/exterior windows. The latter poses a particular problem, because the surfaces are abraded and worn by opening and closing the windows.

In 1978, the Consumer Product Safety Commission banned the manufacture and use of paints containing more than 0.05 percent lead by weight on the interiors and exteriors of residential surfaces, toys, and furniture. With proper labeling (i.e., "Warning: Contains Lead. Dried Film of This Paint May be Harmful if Eaten or Chewed."), the following products are exempt from the ban:

1. Agriculture and industrial equipment paints
2. Industrial/commercial buildings, maintenance equipment, and traffic/safety line paints
3. Graphic art paints (i.e., paints marketed solely for billboards, road signs, identification markings in industrial buildings, and other similar uses)
4. Touch-up paints for agricultural equipment, lawn/garden equipment, and appliances
5. Catalyzed coatings marketed solely for use on radio-controlled model powered aircraft

The following products are exempt without proper labeling:

1. Mirrors with lead-containing paint in the backing
2. Artists' paints and related materials
3. Metal furniture articles, not for use as children's attire and bearing factory-applied lead-containing paint

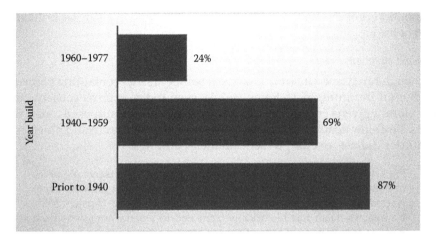

FIGURE 16.2
Older structures likely to contain lead-based paint.

Lead exposure hazards are most likely during renovation projects where lead-based paint is disturbed—sand blasting, chipping, and scraping as well as welding and torch cutting structures that have been painted with lead-based paint. The exposures may drift and not only affect the workers on a construction site but also others in the vicinity and/or on an open HVAC system. See Figure 16.2 for the U.S. EPA published information as to older buildings that may require special attention—especially during renovation and/or demolition projects.

During construction, however, regulated toxic lead exposures may occur while spray painting structures (e.g., structural steel beams). Whereas worker exposures are more likely, overspray and fine paint particles may pose an exposure to others in the vicinity of spray painting and/or to future building occupants. Overspray may collect on thermal insulation and/or HVAC units—a source of lead exposure to the occupants.

Trivia! Lead has a sweet taste. Is it no wonder young children may wish to gnaw on lead-based-paint painted walls and relish in delectable paint chips?

Hexavalent Chromium Pigments

Some lead-based paints also contain hexavalent chromium (e.g., lead chromate). Hexavalent chromium is a highly toxic, carcinogenic regulated substance. Some non-lead-based paints contain strontium chromate which has been identified by the ACGIH as even more toxic than hexavalent chromate. These paints are highly durable and are used primarily on steel structures (e.g., beams and bridges), and yellow lead chromate is typically used on roads to mark traffic no pass zones. Exposures may occur during renovation/

demolition projects, new construction spray painting in a similar fashion to that of lead.

Arsenic Pigments

Although rarely encountered in modern paints, arsenic-containing pigments harbor an intriguing past. Some people, even today, express concerns that the vibrant greens encountered in arsenic pigments are still used today. Old tales die a slow death! Yet, famous artists from the eighteenth and nineteenth centuries did use arsenic pigments in their paintings.

> Experiments at the end of the 19th century proved that arsenic pigments in damp or rotting wallpaper were lethal. The mold that grew on damp wallpaper emitted a toxic odor that smelled of garlic (Austen 2010). Napoleon's death has historically been tied to arsenic emissions from moldy wallpaper.
>
> Scheele's Green was a colouring pigment that had been used in fabrics and wallpapers from about 1770. It was named after the Swedish chemist Scheele who invented it. The pigment was easy to make and was a bright green colour but under certain circumstances the copper arsenite could be deadly. Gosio discovered that if wallpaper containing Scheele's Green became damp and then became mouldy, the mould could carry out a chemical process to get rid of the copper arsenite. It converted it to a vapor form of arsenic, normally a mixture of arsine, dimethyl and trimethyl arsine which was very poisonous. If Napoleon's wallpaper had been green, it could possibly have contained arsenic, and this could have been the source of the arsenic in the hair sample. Napoleon might have been an early victim of Gosio's disease (Ball 2002).
>
> The green pigment did contain arsenic and it began to look as if Napoleon might have been a victim of Gosio's disease, poisoned not by the British authorities, but inadvertently by the British wallpaper makers. Many of the other people who were with Napoleon on St Helena also became ill and complained of the "bad air". Arsenic poisoning causes stomach pains, diarrhea, shivering and swollen limbs; Napoleon's butler did actually die. Dr Jones' conclusion was that the amount of arsenic in Napoleon's wallpaper was not particularly great and consequently the amount of arsenic vapor in the air would not have been large, otherwise more people would have become sick or died. Although the arsenic was not enough to have killed Napoleon, once he was already ill with a stomach ulcer, the arsenic would have exacerbated his condition. Certainly some of the symptoms he complained about do correspond to those of arsenic poisoning (Ball 2002).

Generally, pre-twentieth century homes had arsenic in the wallpaper pigments. However, copper acetoarsenite has been previously used in paint and still is used in paints on ships and submarines (New Jersey Department of Health 1999). Arsenic used presently to paint interior and/or exterior building materials is highly unlikely.

All toxic inorganic pigments in paint may pose an exposure problem to firemen during a fire and to others during the cleanup. Older structures are more likely to pose a greater risk than those built recently.

Organic Pigments

Organic pigments, also referred as dyes, have molecules that are made of carbon atoms along with hydrogen, nitrogen, or oxygen atoms. Natural organic pigments, derived from plant and animal components, have been around since the beginning of time. Beyond cave paintings, natural pigments were used to make body ornaments and to dye textiles. Yet, they are short lived. The colors fade and degrade with time. Then, when paint coatings came into fashion, natural pigments were ultimately replaced by synthetic pigments.

Synthetic organic pigments, also referred to as dyes and colorants, are the product of modern chemistry. They are carbon based and are made from petroleum compounds, and they are mostly solid powders—azo and polycyclic pigments.

Azo pigments are characterized by the presence of one (i.e., monoazo) or two (i.e., disazo) azo bonds (i.e., $-N=N-$) in the molecule. Generally, yellow, orange, red violet, and brown in color, azo pigments represent about 70% of all organic pigments used worldwide, not only in paint but also in inks, food, cosmetics, and plastics. As an unmixed powder, azo pigments can cause eye irritation and are reputed to be highly flammable (e.g., diazomethane may spontaneously explode at 100°C).

Polycyclic pigments include a wide variety of chemical structures, but in general consist of mostly aromatic six- and/or five membered condensed carbon ring systems, and in part aromatic heterocyclic systems containing nitrogen, oxygen, and/or sulfur. By far the most important group of the polycyclic pigments is represented by the copper phthalocyanine structures upon which all blue and green shades of organic pigments are based. Another similar structure is cobalt phthalocyanine which is a brilliant cobalt blue pigment—cobalt is a regulated toxin.

Other polycyclic pigment types of commercial importance are quinacridones and perylene pigments covering orange and red shades. Toxic metals may be associated with any of the various organic pigments. Quinacridone pigments generally produce brilliant, intense colors ranging from deep yellow to even vibrant violet. 2,9 Dimethyl quinacridone is the ever familiar magenta. Yet, the toxicology and flammability of the polycyclic pigments is presently unknown!

Extenders

Extenders, also referred to as fillers, are designed to add bulk. Not well-suited to hide surface flaws, extenders do, however, influence the paint sheen, color retention, and texture. Containing 1%–10% by weight of primary pigments, many paints may contain up to 50% extender. The more expensive the paint, the higher the percentage of primary pigment. Typical extenders are talc, finely ground quartz (e.g., silica), clay, and/or crushed limestone. Silica and silicates are increasing the paint's durability. With the exception of quartz, all extenders are nontoxic. Although the finely ground quartz would pose worker exposure hazards, powdered crystalline silica emissions are highly unlikely unless the paint extender is mixed on site. Job site mixing of quartz extenders could likely result in occupational exposures to crystalline silica.

Binders

The binder, also referred to as the resin, is the main paint component that holds the paint together and carries the pigment. When the paint dries or cures, the binder provides film integrity and adhesion. Binders are either water-based latex or oil-based alkyds. As a water-based latex binder dries, the water evaporates, and the binder forms a solid film on top of the pigments. On the other hand, oil-based alkyd paints cure. They don't evaporate as do the latex paints but remain in suspension within the binder. Glossy paints tend to have more binder and less extender than matt paint which has less binder and more extender. There is a trade-off. In the end, the binder differentiates water-based latex from oil-based alkyd paints. They are not compatible!

Water-Based Latex

The term "latex" originally referred to the use of rubber in one form or another as a binder in paint formulations. The solvent was water. Today, however, many paints made with a water carrier are not rubber. Thus, by extension, non-rubber water soluble paints are referred to as latex, or "acrylic latex."

Most, if not all, modern day paints are synthetic polymers. The most commonly encountered water-based binder is "acrylic" latex (i.e., polymethyl methacrylate). Many are polyvinyl resins—vinyl-acrylic and vinyl acetate/ethylene. Some of the more durable, long-lasting water-based paints are two-part PU and epoxy resins. The latter are considerably more costly and generally used in wet environments (e.g., exterior wood decks). Binder emissions are highly unlikely from water-based latex after the paint has dried.

Oil-Based Alkyd

Developed in the 1920s, alkyd-based enamel paints were one of the most important types of surface coatings. Yet, due to the incorporation of toxic organic solvents and to their low durability on exterior surfaces, alkyd paints gave rise to the low VOC, more durable polymer paints as discussed above.

Alkyd resins are a group of resins derived from reacting dicarboxylic acids (e.g., phthalic and maleic acid) or their anhydrides (e.g., phthalic anhydride and maleic anhydride) with polyvalent alcohols (e.g., glycol and glycerol) to form an ester. With the addition of an unsaturated oil (e.g., tung oil, linseed oil, or dehydrated castor oil) to the ester compound(s), a branched polyester-containing fatty acid is formed. When applied to a surface, the oil portion of the polyester undergoes a cross-linking in the presence of oxygen. As the oil-modified polyester dries in the presence of a catalyst, a tack-free film, or coating, is formed.

Although the tack-free oil-modified polyester emissions are highly unlikely. Alkyd paints are generally suspended in organic solvents. Consequently, organic solvent emissions are highly likely.

Solvents

Solvents, often referred to as the carriers, regulate the paint "viscosity" (i.e., thickness). Latex paint solvent is water, and alkyd paint carriers are organic solvents. Organic solvents include, but are not limited to, turpentine, mineral spirits, and petroleum distillates—aliphatics (e.g., hexane), aromatics (e.g., benzene, toluene, xylene, and ethylbenzene), alcohols (e.g., glycerol), ketones (e.g., methyl isobutyl ketone), esters (e.g., n-butyl acetate), and glycol esters (e.g., ethylene glycol). Most of the organic solvents in paint are regulated VOCs. They have an odor, and organic solvents can cause various health problems from typical indoor air quality health complaints to systemic organ damage. Builders are typically well aware of the need to reduce and/or eliminate entire VOC emissions from paint, but sometimes emissions are unavoidable either due to a requirement for an oil-based paint or lack of awareness as to VOC contributions by color mixing systems.

Color mixing systems which are used in latex as well as alkyl paints contain organic solvents. Pigments that are suspended in organic solvents are often added to a base paint. Whereas the content of a base paint may be "low VOCs," the addition of organic mixing system colors is likely to increase the actual VOC content, potentially altering the claim that a paint has "low VOCs." The significance of the increase is largely dependent upon the amount of organic colorant added.

Additives

Additives are special liquid paint components used to modify the properties of the paint. Some are thickeners (e.g., additional pigments and less solvent) that provide greater coverage of surface flaws. Some are surfactants (e.g., sodium stearate) that evenly disperse the paint components to ensure even coating of all components. Some are biocides (e.g., phenyl mercury chloride, copper fungicides, and methyl chloroisothiazolinone) that resist mold growth in damp environments. Some are defoamers (e.g., silica and mineral oil) that burst tiny bubbles created in the mixing process. Some are light stabilizers to prevent ultraviolet light discoloration of pigments. Some are antifreeze (e.g., ethylene glycol) that resists freeze damage. Some are plasticizers (e.g., phthalate esters) that increase flexibility and adhesion of the surface film (e.g., binder). Some are anti-marring agents (e.g., silicone-based compounds) that resist scratching and marring of the paint film. Some oil-based paints may also have fire retardant additives. And the list goes on!

For details regarding some of the possible plasticizer and additive emissions, see the Chapter 8, "Plasticizers" and Chapter 9, "Plastic Additives." Otherwise, the sheer numbers and complexity of all additives is a daunting topic unto itself. Innovation and proprietary secrets further confuse the issue. And so it is with the possible irritant/toxic emissions.

Stains and Varnishes

Stains and varnishes ride on the coattails of the wide world of paints—with but a different emphasis—and wood is the beneficiary. Stains impart color. Varnishes protect and waterproof.

Stains are pigments. As with paint, pigments may be inorganic pigments and/or organic dyes. Most commercial stains contain both. Pigments tend to be more opaque, and dyes tend to be more translucent. Pigments tend to not penetrate hard woods, and dyes tend to penetrate better. All stains are comprised of pigment, solvent (e.g., oil or water), and small amounts of binder. The greater the amount of binder, the greater the protection from weathering. Wood stain sealants protect wood but provide very poor penetration. The greater the amount of binder, the greater the protection. Product emissions are highly unlikely.

Varnishes are binders (e.g., resins) and solvents. With the exception of natural shellac, most varnishes contain resin, oil, and a solvent. The most common components are as follows:

- Oils—linseed oil and tung oil
- Resin (e.g., polymer)—alkyd, phenolic, and PU
- Solvents—mineral spirits, naptha, or paint thinner

When a varnish is made, the ratio of oil to resin can have a dramatic effect on the way the varnish will behave. For instance, using a small amount of oil and a large amount of resin will produce a very hard but somewhat brittle finish which is not necessarily suitable for outdoor applications. Outdoor finishes are subject to weather extremes. More oil imparts a softer, more flexible finish that will not crack when wood expands and contracts.

Oil type is another consideration. The most common oil used to make varnish is linseed oil. It is less expensive therefore the oil of choice. But many believe that the more pricy tung oil is a higher quality oil. Subsequently, many of the high-end marine varnishes are made with tung oil instead of linseed oil.

Generally speaking, phenolic resins are best-suited for outdoor use. But don't be deceived. Spar varnish and phenolic resins are not always one and the same. Some outdoor formulations use alkyd and urethane resins. A popular finish like Helmsman Spar Urethane contains PU-modified alkyd resins. However, phenolic-modified alkyd resins provide more durable, long-lasting outdoor coverage.

The major resins used in various varnishes used in construction are as follows:

- Shellac—a natural resin suspended in alcohol; imparts a fine finish to furniture. It is harvested bark where the female lac insect secretes a sticky substance.
- Alkyd varnish—a polyester resin mixed with oil and organic solvent. It penetrates well and often contains UV-blockers.
- Spar varnish—typically, not always, a phenolic resin mixed with a modified tung oil [obtained by pressing the seed from the nut of the tung tree (i.e., *Vernicia fordii*)] and organic solvent. Phenolic resins are phenol formaldehyde.
- PU varnish—PU mixed with a drying oil and a solvent. It is known for its hardness, abrasion-resistance, durability, and glossy surface.
- Acrylic varnish—an acrylic resin suspended in a water solution. It has poor penetrating and good UV-resistance qualities (e.g., does not yellow).

Formaldehyde emissions are possible from phenolic resin varnishes (e.g., spar varnishes). During solvent evaporation, VOC emissions are also likely. Organic solvents also pose a flammability hazard as well.

Summary

Adhesives, sealants, and surface finishes are everywhere, within and without. Most are comprised of plastics, solvents, and often undisclosed additives. Although they comprise only a small volume of the building materials used in construction, they are, however, potential contributors to poor indoor air quality.

All are similar to each other in basic content—polymer (e.g., thermoplastic, rubber, and silicone), carrier (e.g., water or organic solvent), and additives (e.g., plasticizer and UV block). Considerations are as follows:

- All polymers are potential sources of plasticizers and/or polymer additives that may cause eye, skin, and respiratory irritation. Many of the adhesive, caulk, paint, and varnish polymers are already polymerized prior to formulation, but the plasticizers and additives remain unbound.

- Volatile organic solvents are contained within some of the adhesives, many of the caulks, and some of the paints and varnishes. Many manufacturers allege low VOC emissions whereas a few had third party testing performed and so labeled their products. Others have had third party testing and chosen not to participate in a costly green program.

- Toxic metals (e.g., lead) and metalloids (e.g., arsenic) are pigments used in some paints. Paints with toxic metal pigments do not pose a toxic emissions problem—either before or after application. Paint chips from dry paints or welding on metal that has previously been painted with toxic metals and/or metalloids can, however, pose a problem.

- Phenol formaldehyde is a component of the better marine/outdoor varnishes—and a possible, not likely, source of formaldehyde.

All building material both large and small in volume are part and parcel of the whole. A little here. A little there. Many building material components are known, and many are unknowns. The health effects of multiple building component emissions—even at low levels—add up. Subsequently, we are compelled to confront, "A cauldron of toxins that lurk within—all modern buildings!"

References

Austen, Jane. 2010. "Emerald Green or Paris Green, the Deadly Regency Pigment." *Jane Austen's World*. March 5. Accessed January 2, 2016. https://janeaustensworld.wordpress.com/2010/03/05/emerald-green-or-paris-green-the-deadly-regency-paint/.

Ball, Hendrik. 2002. "Arsenic Poisoning and Napoleon's Death." *The Victorian Web*. January 14. Accessed January 2, 2016. http://www.victorianweb.org/history/arsenic.html.

New Jersey Department of Health. 1999. "Hazardous Substance Fact Sheet: Copper Acetoarsenite." *New Jersey Government*. January. Accessed January 4, 2016. http://nj.gov/health/eoh/rtkweb/documents/fs/0529.pdf.

Sentry Air Systems. 2010. "Hazards of Epoxy Fumes." *Sentry Air*. May 19. Accessed December 18, 2015. http://sentryair.com/blog/health/hazards-of-epoxy-fumes/.

West System. 2016. "Other Epoxy Related Hazards." *West System*. Accessed December 15, 2015. http://www.westsystem.com/ss/other-epoxy-related-hazards/.

Glossary

Acute exposure: A single exposure to a toxic substance which may result in severe biological harm or death. Chronic exposure.

Acute toxicity: As per OSHA (1910.1200 App A), acute toxicity is an adverse effect that occur following oral or dermal administration of a single dose of a substance, or multiple doses given within 24 hours, or an inhalation exposure of 4 hours.

Anneal: A heat treatment that alters the physical and sometimes chemical properties of a material to increase its ductility and reduce its hardness, making it more workable.

Bitumen (asphalt): A black viscous mixture of hydrocarbons obtained naturally or as a residue from petroleum distillation.

Borates: Boric acid, salts of borates (e.g., boron sulfate), and boron oxide.

Bronchospasm: A sudden constriction of the muscles in the walls of the bronchioles that results in difficulty in breathing which can be very mild to severe.

Cellulose: A complex polysaccharide, the skeleton of most plant structures and plant cells.

Chronic exposures: Multiple exposures occurring over an extended period of time or over a significant fraction of an animal's or human's lifetime.

Cladding: The exterior component of a building that is intended to control the infiltration of weather elements and/or for aesthetics.

Combustible: Any substance that can catch fire and burn.

Compatibilizer: Any substance used to stabilize blends of immiscible polymers.

Composite wood (also referred to as engineered wood): Manmade wood, or manufactured board; includes a range of derivative wood products which are manufactured by binding or fixing the strands, particles, fibers, or veneers or boards of wood together with adhesives, or other methods of fixation to form composite materials.

Conduit, electrical: A pipe or tube through which electrical wire passes.

Corrosion: Destroy, damage, and/or weaken a metal by chemical action (e.g., acid rain and salt water).

Cushion, carpet (padding)": The layer of material that lies between the carpet and floor.

Cutback: The reduction process such as dissolving a semisolid/solid asphalt in a solvent.

Demising wall: Demising wall is a wall that separates spaces belonging to different tenants, as well as separating private tenant areas from common areas. Such walls may be treated separately in the building code to address concerns about energy efficiency, noise pollution, and tenant safety. A simple example of a demising wall can be found in an apartment complex, where the shared wall between two apartments represents a separation between areas that belong to separate tenants.

Diene: An unsaturated hydrocarbon containing two double bonds between carbon atoms.

Diffusion: The process whereby particles of liquids, gases, or solids intermingle as the result of their spontaneous movement caused by thermal agitation and in dissolved substances move from a region of higher to one of lower concentration.

Diffusion rate: The rate of passage of a liquid, gas, or vapor passes through a solid (e.g., housewrap).

Diffusivity: A measure of the capability of a substance or energy to be diffused or to allow something to pass by diffusion.

Ductility: Ability to deform under stress without fracturing.

Edema: The accumulation of fluid.

Elastomer: A synthetic, rubber-like material capable of rapid, reversible extension (e.g., flexible and elastic). The term "elastomer" (from "elastic polymer") refers to any member of a class of polymeric substances that possess the quality of elasticity, that is, the ability to regain shape after deformation. Elastomers are the base material for all rubber products, both natural and synthetic, and for many adhesives.

Emission: A substance or substances discharged into the air.

Emulsion: A fine dispersion of minute droplets of one liquid in another in which it is not soluble or miscible.

Engineered wood: See composite wood.

Extrusion (plastic extrusion): Raw plastic is melted and formed into a continuous profile. Extrusion produces items such as pipe/tubing, weather stripping, fencing, deck railings, window frames, plastic films and sheeting, thermoplastic coatings, and wire insulation.

Extrusion process: A material is pushed or pulled through a die of the desired cross-section; pushed or pulled through a die.

Faux: An imitation.

Fenestration: Openings in the walls of a structure.

Flammable: Any liquid which has a flashpoint less than 140°F (60.5°C) [GHS SDS]; any liquid which has a flashpoint less than 100°F (37.8°C) [OSHA General Industry].

Formica: Trademark for laminated sheets produced from melamine/phenolic plastics.

Friable: Description of a material that, when dry, can be crumbled, pulverized, or reduced to powder by hand pressure.

Galvanneal: To zinc coat iron or steal and heat treat the protected iron or steel in order to increase its ductility, reduce its hardness; products are corrosion resistant and paintable without a primer.

Galvanize: To coat iron or steel with a protective layer of zinc.

Glass transition temperature (T_g): Temperature at which there is a reversible change in an amorphous polymer between a viscous, rubbery condition and a hard, brittle one.

Glazing: Panes or sheets of glass set or made to be set in frames, as in windows, doors, or mirrors.

Globally harmonized system (GHS): Globally harmonized system of classification and labeling of chemicals.

Glulam: Glue laminated lumber is structural timber that has been fabricated by joining two or more layers of wood with an adhesive so that the grains of all the layers are approximately parallel.

Gypsum: A mineral consisting of hydrated calcium sulfate.

Hemp: Woody component from a low tetrahydrocannabinol-producing industrial *Cannabis sativa*.

Hydrolysis: Cleavage of chemical bonds by the addition of water.

Injection molding: Injecting material into a mold; material is fed into a heated barrel, mixed, and forced into a mold cavity, where it cools and hardens to the configuration of the cavity.

Inorganic compound: A compound that does not have carbon atoms and is not derived from living matter.

Lacquer: A protective coating comprised of a resin, cellulose ester, or both, dissolved in a volatile solvent, sometimes with pigment added.

Lath: A latticework backing for plaster or stucco.

Lower respiratory tract: The trachea and all surfaces of the lungs.

MERV: A measure of filter efficiency.

Metal: An element that is typically hard, opaque, shiny, and has good electrical and thermal conductivity.

Metalloid: An element that has properties of both metals and nonmetals such as arsenic and silicon.

Micron: A unit of length equal to 0.001 mm.

Mil: A unit of length equivalent to one thousandth (10^{-3}) of an inch (0.0254 mm).

Mineral: A solid inorganic substance of natural occurrence.

Mold: (1) Microscopic plants, belonging to the fungi kingdom, which are composed of vegetative thread-like structures (called hyphae) that absorb water and digest organic food, and reproduce by creating spores. (2) A shaped cavity used to give a defined form such as molded plastic cabinet hardware.

Mould: British term for mold.

Off-gas: A gas that is given off, emitted as a by-product of an industrial process or that is given off by a manufactured object or material.

Organic compound: A compound that contains carbon atoms exclusive of carbon-containing alloys (carbon steel), metal carbonates (e.g., calcium carbonate; limestone), simple oxides of carbon (e.g., carbon monoxide), and cyanides (hydrogen cyanide), and allotropes of carbon (e.g., graphite).

Padstone: A simple type of building foundation consisting of a stone which both spreads the weight of a wooden building out on the ground and keeps the wood off the ground.

Perlit: An amorphous volcanic glass that has a relatively high water content, typically formed by the hydration of obsidian.

Permeance: Property of a material that prevents fluids (such as water or water vapor) to diffuse through it to another medium.

Petrochemical: A petroleum distillate, chemical products derived from petroleum.

Plasticizer: A chemical that is added to synthetic polymers to give them flexibility and resilience.

Polyolefin: Any of a class of polymers that are produced the polymerization of a single olfin (i.e., alkene) monomer.

Pozzolan: A type of silicon/alumina material that occurs naturally and is produced as a by-product of coal combustion.

Quarry: An excavation or pit from which stone or other minerals are extracted.

Raceway, electrical: A channel for loosely holding electrical wires in buildings.

Resin: A natural or synthetic compound that begins in a highly viscous state and hardens upon application.

Resinate: To treat or impregnate with a resin.

Rust: Any film or coating on metal caused by oxidation such as the red or orange coating that forms on the surface of iron when exposed to air and moisture, consisting chiefly of ferric hydroxide and ferric oxide formed by oxidation.

R-value: A unit thermal resistance for a particular material or assembly of materials (such as an insulation panel). The R-value depends on a solid material's resistance to conductive heat transfer.

Semi-volatile organic compounds (SVOC): Organic chemical compounds whose composition makes it possible for them to volatilize under normal indoor atmospheric conditions of temperature and pressure. However, semi-volatile compounds are the least likely compounds to volatilize. SVOCs will have boiling points above 240°C.

Shake: A shake is a basic wooden shingle that is made from split logs. Shakes have traditionally been used for roofing and siding applications around the world. Higher grade shakes are typically used for roofing purposes, while the lower grades are used for siding purposes. In either situation, properly installed shakes provide long lasting weather protection and a rustic aesthetic, though they require more maintenance than some other more modern weatherproofing systems. The term shake is sometimes used as a colloquialism for all wood shingles, though shingles are sawn rather than split. In traditional usage, "shake" refers to the board to which the shingle is nailed, not the shingle. Split wooden shingles are referred to as shag shingles.

Shingles: Shingles are a roof covering consisting of individual overlapping elements. These elements are typically flat, rectangular shapes laid in courses from the bottom edge of the roof up, with each successive courses overlapping the joints below. Shingles are made of various materials such as wood, slate, flagstone, fiber cement (e.g., asbestos-containing Transite), metal, plastic, and composite material (e.g., asphalt shingles).

Slag: The vitreous, glass-like mass left as a residue by the smelting of metallic ore.

Soffit: The underside of an architectural structure such as an arch, a balcony, or overhanging eaves.

Stabilizer: A chemical used to inhibit the reactions of plastics, intended to prevent degradation during processing and use.

Thermal decomposition: Chemical decomposition caused by heat. The decomposition temperature of a substance is the temperature at which the substance chemically decomposes.

Thermoplastics (thermoplastic polymers): A polymer shaped by heating and cooling, or re-heating; based on linear (crystalline) and branched (amorphous) polymers, become rigid when cooled and soften at varying elevated temperatures—depending on the polymer resin type and additives; polymer chains are only weakly bonded, free to slide past one another when sufficient thermal energy is supplied, making the plastic formable and recyclable.

Thermoplastics, amorphous: Polymer chains acquire a bundled structure, like a ball of thread disordered, an amorphous structure that lends elastic properties to thermoplastic materials.

Thermoplastics, crystal: Polymer chains acquire an ordered and compacted structure that lends mechanical properties of resistance to stresses or loads and the temperature resistance of thermoplastic materials.

Thermoset (thermosetting polymers): A polymer which is preshaped, or molded, and cannot be softened or melted by heat; high mechanical strength (e.g., high stress and heavy loads) and chemical/physical resistance (solvents and high temperatures); polymer chains form strong cross links; thermosets cannot be reflowed once they are *cured* (i.e., once the cross-links form). Instead, thermosets can suffer chemical degradation (altered properties) if reheated excessively.

Transformer: A transformer is a device that transfers electrical energy from one circuit to another through inductively coupled conductors—the transformer's coils.

Unplasticized: A plastic that does not have plasticizer.

Upper respiratory tract: The nose, mouth, sinuses, and throat.

Vapor: A gaseous liquid or solid.

Varnish: A resin dissolved in a liquid for applying on wood, metal, or other materials to form a hard, clear, shiny surface when dry.

Veneer: A thin decorative covering of fine wood applied to a coarser wood or other material such as a composite wood.

Vinyl: (1) Short for polyvinyl chloride. (2) An unsaturated hydrocarbon radical $-CH=CH_2$, derived from ethylene by removal of a hydrogen atom.

Volatile organic compounds (VOC): Organic chemicals that have a high vapor pressure at ordinary room temperature.

Units of Measurement

Volume

1 liter (L) = 1.06 quarts
1 milliliter (mL) = 10^{-3} liter
1 microliter (μL) = 10^{-6} liter

Length

1 meter (m) = 3.281 feet = 39.37 inches
1 centimeter (cm) = 10^{-2} meter = 0.039 inch
1 millimeter (mm) = 10^{-3} meter
1 micrometer (μm) = 1 micron (μ) = 10^{-6} meter

Weight

1 gram (g) = 0.035 ounce
1 milligram (mg) = 10^{-3} gram
1 microgram (μg) = 10^{-6} gram

Temperature

$1°$ Fahrenheit (F) = $[(1.8)\,(X°C)] + 32$
$1°$ Centigrade (C) = $[X°F - 32]/1.8$

Airborne Contaminants

ppm = parts of contaminant per million parts of air

ppb = parts of contaminant per billion parts of air

ppt = parts of contaminant per trillion parts of air

mg/m^3 = milligrams of contaminant per cubic meter of air

Chamber Testing

mg/m^2-hour = milligrams of contaminant per square meter of material in 1 hour

mg/X-hour = milligrams of contaminant per item, or composite unit, in X hour(s) of emissions testing

$mg/hour$-m^3 = milligrams of contaminant per item in 1 hour within a cubic meter space

Atmospheric Pressure

760 torr

14.695 pounds per square inch

1013.25 millibars

101. 325 kilopascal

29.921 inches mercury (0°C)

Abbreviations and Acronyms

4-PCH	4-Phenylcyclohexane
ABS	Acrylonitrile–Butadiene–Styrene
ACD	Alkaline Copper DCOI
ACGIH	American Conference of Government Industrial Hygienists
ACQ	Alkaline Copper Quaternary
ACZA	Ammoniacal Copper Zinc Arsenate
AHU	Air Handling Unit
ANSI	American National Standards Institute
APP	Atactic Polypropylene
ASHE	American Society of Healthcare Engineers
ASTM	American Society for Testing and Materials
AWPA	American Wood Protection Association
BFR	Brominated Fire Retardants
BHT	Butylated Hydroxytoluene
BUF	Built Up Roofing
CA	Copper Azole
CAS	Chemical Abstracts Service
CCA	Chromated Copper Arsenate
CFC	Chlorofluorocarbon
CLT	Cross Laminated Timber
CO	Carbon Monoxide
CO_2	Carbon Dioxide
COEHHA	California Office of Environmental Health Hazard Assessment
CPSC	Consumer Product Safety Commission
CREL	Chronic Reference Exposure Limits (California)
CRI	Carpet and Rug Institute
CSPE	Chlorosulfonated Polyethylene
Cu-HDO	Bis-(N-Cyclohexyldiazeniumdioxy)copper
CuN-W	Copper Naphthenate-Water Carrier
CX-A	See Cu-HDO
DCOI	4,5-Dichloro-2-N-Coty-4-Isothiazolin-3-One
DEHA	Di(2-Ethylhexyl) Adipate
DOT	Disodium Octaborate Tetrahydrate
EFIS	Exterior Finish Insulating System
EIFS	Exterior Insulation and Finish System
EPA	Environmental Protection Agency
EPDM	Ethylene Polypropylene Diene Monomer
EPM	Ethylene–Propylene
EPS	Expanded Polystyrene
FEMA	Federal Emergency Management Agency

FGD	Flue Gas Desulfurization
FRP	Fiber-Reinforced Plastic
GFRP	Glass Fiber-Reinforced Polymer
GHS	Globally Harmonized System
GH SDS	Globally Harmonized System Safety Data Sheet*
HALS	Hindered Amine Light Stabilizers
HBCD	Hexabromocyclododecane
HBFC	Hydrobromofluorocarbon
HCFC	Hydrochlorofluorocarbon
HCN	Hydrogen Cyanide
HDI	Hexamethylene Diisocyanate
HDPE	High-Density Polyethylene
HFC	Hydrofluorocarbons
HPV	High Production Volume
HVAC	Heating, Ventilation, and Air Conditioning
IAQ	Indoor Air Quality
IARC	International Agency for Research on Cancer
ICF	Insulated Concrete Foam
IDLH	Immediately Dangerous to Life and Health
IR	Infrared
KDS	Impralit Trade Name for Wood Preservative
kPa	KiloPascal
LD_{50}	Lethal Dose 50%
LDPE	Low-Density Polyethylene
LEED	Leadership In Energy and Environmental Design
LLDPE	Linear Low-Density Polyethylene
LRT	Lower Respiratory Tract
LVL	Laminated Veneer Lumber
MB	Modified Bitumen
MCQ	Micronized Copper Quaternary
MDF	Medium-Density Fiberboard
MDI	Methylene Diphenyl Diisocyanate
MDPE	Medium-Density Polyethylene
MERV	Maximum Efficiency Reporting Value
MF	Melamine Formaldehyde
$\mu g/m^3$	micrograms per cubic meter (10^{-6} g/m^3)
mg/m^3	milligrams per cubic meter (10^{-3} g/m^3)
MMA	Methyl Methacrylate
MSDS	Material Safety Data Sheet*
NIOSH	National Institute for Occupational Safety and Health

---

* Within this book, MSDS references apply to MSDSs, SDSs, and GH SDSs. The compliance transition date for chemical manufacturers, importers, distributors, and employers to change their format from MSDSs to SDSs in the United States and was June 1, 2015. As of 2016, some MSDSs were still available of manufacturer web sites.

OSB	Oriented Strand Board
PAH	Polycyclic Aromatic Hydrocarbons
PC	Polycarbonate
PCB	Polychlorinated Biphenyls
PCF	Pounds per Cubic Foot
PDMI	Polymeric Diphenylmethane Diisocyanate (i.e., Polyurethane)
PE	Polyethylene
PEL	Permissible Exposure Limit
perm	permeance
PET	Polyethylene Terephthalate
PEX	Cross-linked Polyethylene
PF	Phenol Formaldehyde
PIR	Polyisocyanurate
PM10	Particle Matter (with an aerodynamic diameter less than or equal to 10 µm [upper respiratory track])
PM2.5	Particle Matter (with an aerodynamic diameter less than or equal to 2.5 µm [lower respiratory track])
PMMA	Polymethyl Methacrylate
PNA	Polynuclear Aromatic Hydrocarbons
PP	Polypropylene
ppb	parts per billion
PPE	Personal Protective Equipment
ppm	parts per million
PS	Polystyrene
psia	Pounds per Square Inch Absolute
PSL	Parallel Strand Lumber
PTI	Propiconazole–Tebuconazole–Imidicloprid
PU	Polyurethane
PVAc	Polyvinyl Acetate
PVC	Polyvinyl Chloride
PVC-U	Polyvinyl Chloride-Unplasticized
REL	Reference Exposure Limits (California)
REL	Recommended Exposure Limit (NIOSH)
RF	Resorcinol Formaldehyde
RH	Relative Humidity
SBR	Styrene–Butadiene Rubberized-latex
SBS	Styrene–Butadiene–Styrene
SBX	Sodium Borate Preservative
SDS	Safety Data Sheet*
SMACNA	Sheet Metal and Air Conditioning National Association

---

* Within this book, MSDS references apply to MSDSs, SDSs, and GH SDSs. The compliance transition date for chemical manufacturers, importers, distributors, and employers to change their format from MSDSs to SDSs in the United States and was June 1, 2015. As of 2016, some MSDSs were still available of manufacturer web sites.

SVOC	Semi-Volatile Organic Compounds
TCDD	2,3,7,8-Tetrachlorodibenzo Para Dioxin
TDI	Toluene Diisocyanate
T_g	Glass Transition Temperature
TLV	Threshold Limit Value
T_m	crystalline Melting Temperature
TSCA	Toxic Substances Control Act
TVOC	Total VOC
U.S.	United States of America
UCS	Use Category System
UF	Urea Formaldehyde
UFFI	Urea Formaldehyde Foam Insulation
UL	Underwriting Laboratory
ULE	Underwriting Laboratory Environmental
UPVC	Un-Plasticized PVC
URT	Upper Respiratory Tract
UV	Ultraviolet
VLDPE	Very Low Density Polyethylene
VOC	Volatile Organic Compound
WHO	World Health Organization

Index

Note: Page numbers followed by *"fn"* indicate footnotes.

.